CAFFEINE and BEHAVIOR
Current Views and Research Trends

CAFFEINE and BEHAVIOR
Current Views and Research Trends

edited by
B.S. Gupta, Ph.D.
Department of Psychology
Banaras Hindu University
Varanasi, India

Uma Gupta, Ph.D.
Department of Basic Principles
Institute of Medical Sciences
Banaras Hindu University
Varanasi, India

CRC Press
Boca Raton   London   New York   Washington, D.C.

Library of Congress Cataloging-in-Publication Data

Caffeine and behavior: current views and research trends / edited by
    B.S. Gupta and Uma Gupta
        p. cm.
    Includes bibliographical references and index.
    ISBN 0-8493-1166-7 (alk. paper)
    1. Caffeine—Physiological effects. 2. Caffeine—Psychological aspects. I. Gupta, B.S.
II. Gupta, Uma.
QP801.C24C33 1999
615'.785—dc21
                                                            99-12161
                                                                   CIP

© 1999 by CRC Press LLC

No claim to original U.S. Government works
International Standard Book Number 0-8493-1166-7
Library of Congress Card Number 99–12161
Printed in the United States of America 1 2 3 4 5 6 7 8 9 0
Printed on acid-free paper

About the Editors

B.S. Gupta is a professor of psychology at Banaras Hindu University, Varanasi, India. He received his M.A. and Ph.D. degrees from the Department of Psychology at Panjab University, Chandigarh, India. Before coming to Banaras Hindu University, he taught at Meerut University in Meerut and Guru Nanak Dev University in Amritsar, India. He has published numerous research articles and book chapters on behavioral effects of drugs and Ayurvedic herbs, drug abuse, verbal conditioning, personality, perceptual judgment, cognition and environment, talent search and development, and adjustment and coping.

He has served as consulting editor on a number of journals, including the *International Journal of Psychology*. He is now coeditor of *Pharmacopsychoecologia* and is president of the Pharmacopsychoecological Association. He is a member of the Advisory Council of the International Biographical Centre, Cambridge, as well as the Research Board of Advisors of the American Biographical Institute, North Carolina. His research currently focuses on validation of tribal and Ayurvedic herbs, relaxation techniques, and drug effects on behavior.

Uma Gupta is a research scientist in the Department of Basic Principles, Faculty of Ayurveda, Institute of Medical Sciences, Banaras Hindu University. She received her M.A., M.Phil., and Ph.D. degrees in psychology from Meerut University, Meerut, India. She has completed a number of research projects sponsored by the Indian Council of Social Science Research and the University Grants Commission, New Delhi. She has contributed numerous research articles to journals, including the *British Journal of Psychology, Psychopharmacology, Neuropsychobiology,* and *Pharmacology, Biochemistry and Behavior* on pharmacopsychology, the biochemistry of behavior, personality, memory, drug abuse and addiction, relaxation techniques, organizational behavior, and environmental schematization.

She has recently coauthored a chapter on drugs in personality research in *Personality Across Cultures*. She has also coedited two books: *Advances in Psychopharmacology, Neuropsychology & Psychiatry* and *Readings in Environmental Toxicology and Social Ecology.* She is now coeditor of *Pharmacopsychoecologia* and is an invited honorary member of the Research Board of Advisors of the American Biographical Institute, North Carolina.

Contributors

Robert J. Carey, Ph.D.
Department of Psychiatry
College of Medicine
SUNY Health Science Center and
  VA Medical Center
Syracuse, New York

John W. Daly, Ph.D.
Laboratory of Bioorganic Chemistry
NIDDK
National Institutes of Health
Bethesda, Maryland

Ernest N. Damianopoulos, Ph.D.
Department of Psychiatry
College of Medicine
SUNY Health Science Center and
  VA Medical Center
Syracuse, New York

Mark T. Fillmore, Ph.D.
Department of Psychology
University of Waterloo
Waterloo, Ontario, Canada

David V. Gauvin, Ph.D.
Drug Enforcement Administration
Office of Diversion Control
Washington, DC

B.S. Gupta, Ph.D.
Department of Psychology
Banaras Hindu University
Varanasi, India

Uma Gupta, Ph.D.
Department of Basic Principles
Institute of Medical Sciences
Banaras Hindu University
Varanasi, India

Stephen J. Heishman, Ph.D.
Clinical Pharmacology Branch
NIDA Addiction Research Center
National Institutes of Health
Department of Psychiatry &
  Behavioral Sciences
The Johns Hopkins University
  School of Medicine
Baltimore, Maryland

Jack E. Henningfield, Ph.D.
Department of Psychiatry &
  Behavioral Sciences
The Johns Hopkins University
  School of Medicine
Baltimore
and Pinney Associates
Bethesda, Maryland

Frank A. Holloway, Ph.D.
Psychobiology Laboratories
Department of Psychiatry and
  Behavioral Sciences
University of Oklahoma Health
  Sciences Center
Oklahoma City, Oklahoma

Kenneth A. Jacobson, Ph.D.
Laboratory of Bioorganic Chemistry
NIDDK
National Institutes of Health
Bethesda, Maryland

David A. Johnson, Ph.D.
Division of Pharmacology and
  Toxicology
Graduate School of Pharmaceutical
  Sciences
Duquesne University
Pittsburgh, Pennsylvania

Malcolm H. Lader, M.D., Ph.D.,
D.Sc.
Department of Psychiatry
Section of Clinical
  Psychopharmacology
Institute of Psychiatry
University of London
London, United Kingdom

Monicque M. Lorist, Ph.D.
Faculty of Psychology
Rijks Universiteit Groningen
Groningen, The Netherlands

Mark Mann, M.A.
Department of Psychology
University of Maryland
College Park, Maryland

Devon E. McVey, M.S.
Health America, Inc.
Pittsburgh, Pennsylvania

J. Patrick Myers, M.S.
Mount Aloysius College
Cresson, Pennsylvania

Astrid Nehlig, Ph.D.
INSERM U 398
Faculté de Médécine
Strasbourg, France

Olga Nikodijević, M.D.
University Medical School
Skopje, Macedonia

Jennifer Rusted, Ph.D.
Department of Experimental
  Psychology
University of Sussex
Brighton, United Kingdom

Dan Shi, M.D.
Laboratory of Bioorganic Chemistry
NIKKD
National Institutes of Health
Bethesda, Maryland

Andrew P. Smith, Ph.D.
Department of Experimental
  Psychology
Health Psychology Research Unit
University of Bristol
Bristol, United Kingdom

Barry D. Smith, Ph.D.
Department of Psychology
University of Maryland
College Park, Maryland

Jan Snel, Ph.D.
Department of Psychonomics
Faculty of Psychology
University of Amsterdam
Amsterdam, The Netherlands

Odin van der Stelt, Ph.D.
Department of Child and Adolescent
  Psychiatry
Academic Medical Center
University of Amsterdam
Amsterdam, The Netherlands

Kenneth Tola, M.A.
Department of Psychology
University of Maryland
College Park, Maryland

David M. Warburton, Ph.D.
Department of Psychology
University of Reading
Reading, United Kingdom

Jason M. White, Ph.D.
Department of Clinical and
   Experimental Pharmacology
University of Adelaide
Adelaide, South Australia

Contents

Preface.. xiii

Chapter 1   The role of adenosine receptors in the central action
of caffeine 1
John W. Daly, Dan Shi, Olga Nikodijević, and
Kenneth A. Jacobson

Chapter 2   Caffeine in the modulation of brain function 17
J. Patrick Myers, David A. Johnson, and Devon E. McVey

Chapter 3   Cerebral energy metabolism and blood flow: Useful tools
for the understanding of the behavioral effects of caffeine 31
Astrid Nehlig

Chapter 4   Caffeine effects on locomotor and reward behavior................ 49
Ernest N. Damianopoulos and Robert J. Carey

Chapter 5   Behavioral effects of caffeine coadministered
with nicotine, benzodiazepines, and alcohol.. 75
Jason M. White

Chapter 6   Caffeine and arousal: A biobehavioral theory
of physiological, behavioral, and emotional effects.................................... 87
Barry D. Smith, Kenneth Tola, and Mark Mann

Chapter 7   Is caffeine a drug of dependence?
Criteria and comparisons 137
Stephen J. Heishman and Jack E. Henningfield

Chapter 8   Caffeine withdrawal 151
Malcolm H. Lader

Chapter 9   Caffeine, caffeine withdrawal and
performance efficiency 161
Andrew P. Smith

Chapter 10   The association of anxiety, depression and
headache with caffeine use 179
David M. Warburton

Chapter 11   Caffeine, impulsivity and performance....................................191
Uma Gupta and B.S. Gupta

Chapter 12   Behavioral effects of caffeine: The role
of drug-related expectancies 207
Mark T. Fillmore

Chapter 13   Caffeine and cognitive performance:
Effects on mood or mental processing? 221
Jennifer Rusted

Chapter 14   Caffeine and attention 231
Odin van der Stelt

Chapter 15   Caffeine and fatigue..241
Jan Snel and Monicque M. Lorist

Chapter 16   The subjective effects of caffeine: Bridging the
gap between animal and human research...257
David V. Gauvin and Frank A. Holloway

Overview 283
B.S. Gupta

Index 289

Preface

Caffeine-containing beverages are consumed widely throughout the world. It is, therefore, quite natural not only for the specialist and the physician but also for the curious layman to obtain objective information about the substance which is consumed extensively in both the developed and the developing countries.

The approach followed in this book is mainly experimental, and the major focus is on highlighting the current activities in research related specifically to the psychobehavioral effects of caffeine. The first chapter deals with the role of adenosine receptors in the central action of caffeine, while the second is devoted to evaluating the role of caffeine in the modulation of neurotransmission. The third chapter represents a powerful approach to the understanding of the behavioral effects of methylxanthines specifying caffeine effects on cerebral energy metabolism and blood flow. Of the remaining 13 chapters, 9 chapters are devoted to experimental research in regard to caffeine's effects on important aspects of psychobehavior, including Chapter 11, which is specifically devoted to highlighting the role of personality variables, especially the trait of impulsivity, in research related to caffeine's effects on human performance. In recent times, research has also focused on ascertaining whether caffeine can be placed in the category of addicting drugs. Chapter 7 presents the current state of the art by focusing on the crucial issues related to the possible addictive potential of caffeine. Chapter 8 highlights valuable information concerning caffeine withdrawal. Chapter 10 is based on an extensive survey of the general population regarding caffeine consumption in Great Britain. The final chapter proposes a model that relates synthesized experimental findings in animals to the state changes operating in human caffeine consumers and thus attempts to bridge the gap between animal and human research.

The chapters in this book have been authored by specialists in their fields. The book provides a wide range of coverage and will be of interest to professionals and students researching caffeine.

B.S. Gupta/Uma Gupta

chapter one

The role of adenosine receptors in the central action of caffeine

John W. Daly, Dan Shi, Olga Nikodijević,
and Kenneth A. Jacobson

Contents

I. Central sites of action of caffeine...1
II. Chronic effects of caffeine 4
III. Adenosine analogs and xanthines 9
IV. Summary 11
Acknowledgments..12
References...12

I. Central sites of action of caffeine

The widespread use of caffeine-containing beverages has focused research on the mechanisms underlying the central effects of caffeine.[1-3] While the effects of moderate doses of caffeine on behavior are complex, it appears likely that blockades at A_1- and A_{2A}-adenosine receptors are the primary molecular sites of action for caffeine. There are at least four types of adenosine receptors in brain.[4,5] The A_1-class can be inhibitory to adenylate cyclase, stimulatory to potassium channels, inhibitory to calcium channels, and stimulatory to phosphoinositide breakdown. Selective agonists and selective xanthine and nonxanthine antagonists are available for A_1-receptors. The A_{2A}- and A_{2B}-subclasses are stimulatory to adenylate cyclase. The A_{2A}- and A_{2B}-receptors differ in affinity and in agonist selectivity. Selective agonists and antagonists for A_{2A}-receptors are available. Selective agents for A_{2B}-receptors are not available. Caffeine is nearly equipotent as an antagonist at A_1-, A_{2A}-, and A_{2B}-receptors. The A_3-adenosine receptor also occurs in brain, is inhibitory to adenylate cyclase, and is remarkable in being insensitive to blockade by caffeine and other xanthines, at least in rodents.

Direct effects of caffeine on receptors, other than adenosine receptors, have not been reported. Indirect effects of caffeine on systems served by receptors other than the adenosine receptor will occur due to the blockade by caffeine of the tonic inhibitory input by adenosine through A_1-receptors on release of norepinephrine, dopamine, serotonin, acetylcholine, GABA, glutamate, and perhaps even neuropeptides. Since adenosine receptors can modulate responses of brain second messenger systems to norepinephrine, serotonin, and histamine,[6] both the cyclic AMP-generation and the phospho-inositide-breakdown mediated by receptors for those neurotransmitters would be expected to be altered by caffeine. Thus, caffeine through blockade of adenosine receptors, would be expected to indirectly influence the function of many other neuronal pathways in the brain. There is evidence for central effects of caffeine *in vivo* and *in vitro* on noradrenergic, dopaminergic, serotoninergic, cholinergic, GABAergic, and glutamatergic systems.[2]

Current research is focused on the interrelated adenosine–dopamine–acetylcholine systems of the striatum.[3] In striatum, activation of A_1 receptors opposes the effects of dopamine on D_1-dopamine receptors.[7,8] Similarly, activation of A_{2A}-adenosine receptors reduces affinity of dopaminergic agonists for D_2-dopamine receptors.[9,11] Caffeine, thus, could affect dopamine function through blockade of tonic adenosine input to both A_1- and A_{2A}-adenosine receptors of the striatum. The mechanisms involved require further research.

At the present, it appears that glutamatergic excitatory input to GABAergic neurons in the striatum is modulated by both A_1-adenosine and D_1-dopamine receptors.[7,8] Activation of D_1-receptors causes release of glutamate, while activation of A_1-receptors inhibits release of glutamate.[12] Blockade by xanthines of tonic adenosine-mediated activation of A_1-receptors, thus, will augment the excitatory effects of dopaminergic input to D_1 receptors.

The A_{2A}-receptors of striatum are localized in GABAergic neurons that express D_2-dopamine receptors.[13] Such A_{2A}-receptors stimulate adenylyl cyclase. Activation of the A_{2A} receptors actually decreases the potency of D_2-receptor agonists.[9,10] Thus, blockade of adenosine-mediated tonic activation of striatal A_{2A}-receptors will augment the effects of dopaminergic input to D_2-receptors. Recently it was reported that mice lacking the A_{2A}-receptor were not behaviorally stimulated by caffeine.[14]

Further evidence for effects of caffeine on activity of striatal neurons is based on effects of caffeine on expression of immediate early genes, a measure of neuronal activity. Caffeine in low doses reduced expression of such genes in certain striatal neurons.[15] In contrast, high depressant doses of caffeine enhanced expression of early genes in GABAergic neurons of the striatum.[16,17]

Molecular sites of action other than adenosine receptors are known for caffeine. Historically, the first site of action of caffeine to be identified was stimulation of release of calcium from intracellular storage sites. Caffeine binds to a site on a calcium-channel, which is associated with the intracellular, so-called *calcium-sensitive* pool of calcium, and thereby enhances calcium-dependent activation of the channel.[18] This calcium-channel is the one blocked by ryanodine. Caffeine is now a widely-used tool for studies of the role of this pool in nerve and muscle function, particularly with regard to oscillations in membrane potentials and calcium levels. Caffeine, however, has a very low affinity for such sites with thresholds for effects on release of intracellular calcium at about 250 μM, while 5 to 20 mM concentrations are required for robust effects. This is in contrast to the higher affinities of caffeine as an antagonist for adenosine receptors, where thresholds are less than 10 μM, and K_i values are 40 to 50 μM, well within plasma and brain levels attained by humans and animals with behaviorally effective doses of caffeine. Certain xanthines that are more potent than caffeine as calcium-releasing agents[19] may prove to be valuable in probing the role of the intracellular calcium-sensitive calcium channels in the behavioral pharmacology of caffeine.

Historically, the second site of action of caffeine to be identified was inhibition of phosphodiesterases. Other xanthines that are much more potent than caffeine as phosphodiesterase inhibitors have been developed, and most have proved rather nonspecific as inhibitors of various phosphodiesterase isozymes. Caffeine itself has IC_{50} values for phosphodiesterase isozymes ranging from 500 μM to 1 mM, again well above the range at which caffeine blocks adenosine receptors. Xanthines that are potent phosphodiesterase inhibitors, in particular toward a brain calcium-independent cyclic AMP phosphodiesterase (rolipram-sensitive type IV isozyme), are behavioral "depressants," in contrast to the behavioral stimulant activity of caffeine and other xanthines that are weak inhibitors of that isozyme.[20] The "depressant" part of the bell-shaped dose–response curve of caffeine with respect to open-field locomotor activity (Figure 1.1) may be due to inhibition of the calcium-independent phosphodiesterase, which would become significant only at the highest doses of caffeine.

In the late 1970s, caffeine was found to inhibit binding of benzodiazepines to sites on the GABA$_A$-receptor channel.[21] Although exciting from the standpoint of the anxiogenic properties of caffeine, the affinity of caffeine (K_i 280 μM) was several-fold higher than *in vivo* concentrations of caffeine that would be reached at nontoxic doses of caffeine. Interactions at GABA$_A$-receptors may be relevant to the convulsant activity of caffeine. Xanthines more potent than caffeine at benzodiazepine sites have not been developed.

Thus, in spite of extensive studies on possible biochemical sites of action for caffeine *in vivo*, only adenosine receptors have the requisite 10 to 50 μM affinities for caffeine. Other sites, such as intracellular calcium-sensitive calcium release channels, phosphodiesterases, and GABA$_A$ receptors require >200 μM concentrations of caffeine. At such concentrations, caffeine is a convulsant *in vivo*.

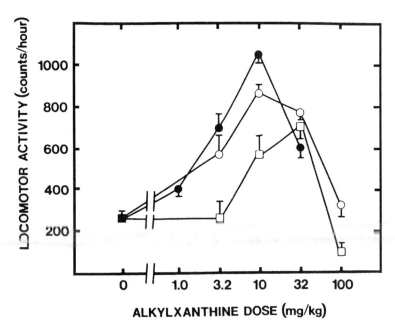

Figure 1.1 Typical bell-shaped dose-response curves for effects of xanthines on open-field locomotor activity in mice. Caffeine (●), theophylline (□), 3,7-dimethyl-1-propargylxanthine (○). Activity measured for 60 min in a circular arena after intraperitoneal injection of xanthine to male NIH Swiss strain mice. (From Daly, J.W. Mechanism of action of caffeine, in *Caffeine, Coffee and Health*, S. Garattini, Ed., Raven Press, New York, Chapter 4, pp 97-150, 1993. With permission.)

II. Chronic effects of caffeine

Chronic treatment of animals with caffeine, not surprisingly, results in an up-regulation of A_1-adenosine receptors, as first reported in 1982 for caffeine[22] and for theophylline.[23] Most subsequent studies on chronic caffeine or theophylline have documented an increase in cortical A_1-adenosine receptors. The A_{2A}-adenosine receptors do not appear to be up-regulated,[24,25] although there is one report of an increase in levels of A_{2A}-adenosine receptors in striatum after chronic ingestion of caffeine by mice.[26] The A_{2B}-receptor-mediated stimulation of cyclic AMP in rat brain slices did not appear altered after chronic caffeine.[22,27]

Most chronic studies related to alterations in levels or function of adenosine receptors have been with oral administration of caffeine or theophylline. In the 1980s, chronic administration of N^6-R-phenylisopropyladenosine was shown to reduce the analgetic and locomotor depressant effects of caffeine.[28] Levels of A_1-adenosine receptors were unaltered. Recently, chronic injections of the A_1-selective agents 8-cyclopentyl-1,3-dipropylxanthine and N^6-cyclopentyladenosine were shown to have effects on NMDA-induced seizures.[29] Chronic xanthine treatment greatly reduced the effects of NMDA, while chronic treatment with the adenosine analog enhanced the NMDA-elicited seizures. Levels of A_1-adenosine receptors were unaltered.

Adenosine receptors are not the only central receptors whose levels are altered after chronic caffeine ingestion. This is not surprising, since removal of tonic adenosine inhibition of neurotransmitter release might be expected to increase neurotransmitter release and lead to a down-regulation of the relevant neurotransmitter receptor. However, in most cases an up-regulation rather than a down-regulation of receptors occurs. There has been only one extended study of effects of chronic caffeine on levels of central receptors.[25,30,31] Chronic caffeine ingestion in male NIH Swiss strain mice was found to affect the density of receptors subserving noradrenergic, serotoninergic, cholinergic, and GABAergic pathways (Table 1.1). Remarkably, since a variety of evidence indicates that caffeine affects dopaminergic function,[9] the levels of dopaminergic receptors appeared unaffected. The levels of cortical and striatal A_1 adenosine receptors were increased by 15 to 20% by chronic caffeine, while the level of striatal A_{2A}-adenosine receptors was unaltered. The levels of cortical β_1- and cerebellar β_2-adrenergic receptors were reduced by about 25%, while the levels of cortical α_1- and α_2-adrenergic receptors were not significantly altered. The levels of striatal D_1- and D_2-dopaminergic receptors were not altered. Levels of cortical muscarinic and nicotinic receptors were increased by 40 to 50%. The apparent up-regulation of nicotinic receptors may actually represent conversion of nicotinic receptors to a desensitized state. The level of cortical benzodiazepine-binding sites associated with $GABA_A$-receptors was increased by 65%, and in this case the affinity for diazepam appeared slightly decreased. The level of cortical MK-801 binding sites associated with NMDA-glutamatergic receptors appeared unaltered. The level of cortical delta-opioid receptors was increased by 25%, while the levels of cortical mu- and kappa-opioid receptors were unchanged. The level of cortical sigma receptors was unchanged. The density of cortical nitrendipine-binding sites associated with L-type calcium channels was increased by 18%. Thus, there is an incredible array of alterations in levels of central receptors elicited by chronic caffeine ingestion in NIH Swiss strain mice. In addition, basal levels of striatal adenylate cyclase were decreased after chronic caffeine, while stimulations *via* D_2-dopamine receptors or A_{2A}-adenosine receptors were unaltered.[30] In rats, the up-regulation of adenosine receptors (see Reference 2), the down-regulation of β-adrenergic receptors,[32-34] and an up-regulation of benzodiazepine sites, associated with $GABA_A$-receptors,[35,36] have been reported after chronic caffeine or theophylline. Effects on receptors other than adenosine receptors do not appear to have been examined systematically in rats. The levels of forskolin-binding sites, associated with adenylate cyclase, have been reported to be increased in rat cerebral cortex after chronic caffeine.[37]

Clearly, the plethora of biochemical alterations in mice after chronic caffeine will make it difficult to interpret behavioral alterations in the chronically caffeine-treated animal. The most studied behavioral alteration has been tolerance to caffeine. An "insurmountable tolerance" has been reported

Table 1.1 Effect of Chronic Caffeine Ingestion on Receptors and Ion Channels in Brain Membrane from Male NIH Swiss Strain Mice[a]

Receptor (ligand)	B_{max} (fmol/mg protein)	
	Control	Chronic caffeine
A_1-Adenosine		
(CHA)	911 ± 23	1089 ± 39**
(CHA, striatum)	668 ± 14	767 ± 28**
A_{2A}-Adenosine		
(CGS 21680, striatum)	872 ± 57	884 ± 44
α_1-Adrenergic		
(Prazosin)	175 ± 7	189 ± 12
α_2-Adrenergic		
(Clonidine)	200 ± 3	193 ± 7
β₁-Adrenergic		
(DHA)	224 ± 9	167 ± 5*
β_2-Adrenergic		
(DHA, cerebellum)	158 ± 12	115 ± 11*
D_1-Dopamine		
(SCH 23390, striatum)	3097 ± 81	3165 ± 66
D_2-Dopamine		
(Spiperone, striatum)[b]	729 ± 21	725 ± 55
5-HT$_1$		
(Serotonin)	361 ± 14	474 ± 48**
5-HT$_2$		
(Ketanserin)	275 ± 11	347 ± 11*
Nicotinic		
(Nicotine)	34 ± 2	50 ± 3*
(Nicotine, striatum)	37 ± 1	39 ± 2
Muscarinic		
(Quinuclidinyl benzilate)	1153 ± 56	1509 ± 47**
NMDA		
(MK-801)	2653 ± 97	2588 ± 46
GABA$_A$		
(Diazepam)	1061 ± 69	1741 ± 100*
Opioid		
mu (DAMGO)	119 ± 6	134 ± 5
delta (DPDPE)	83 ± 2	104 ± 7**
kappa (U69593)	339 ± 65	230 ± 48
Sigma		
(DTG)	2580 ± 90	2560 ± 170
L-Type Calcium Channel		
(Nitrendipine)	314 ± 6	369 ± 12**

[a] The mice ingested caffeine (approximately 100 mg/kg/day) for 4 days, followed by 2 to 4 hours withdrawal for caffeine clearance.[25,30] Binding was of radioligands to cortical membranes or as noted to cerebellar or striatal membranes from control mice and chronic caffeine mice. Values for B_{max} are means ± S.E.M.[25,30]
[b] No change when assayed with [³H]raclopride.
* $P < 0.01$
** $P < 0.05$

in rats.[38-40] An explanation of how up-regulation of A_1-adenosine receptors could lead to an "insurmountable" tolerance to an agent, caffeine, that acts as an antagonist has not been forthcoming. The answer may lie in the biphasic dose–response curve to caffeine (see Figure 1.1), where low doses of caffeine cause stimulation, while higher doses cause depression of locomotor activity. Thus, after chronic caffeine the depressant effects may predominate, leading to the appearance of an "insurmountable tolerance" with respect to behavioral stimulation. The effects of chronic caffeine on open-field locomotor activity have been evaluated thoroughly, not only in Sprague-Dawley rats, but also in NIH Swiss strain mice. Tolerance does not occur in these mice, and indeed the threshold for stimulatory effects of caffeine is significantly lowered (Figure 1.2). Sensitization to behavioral effects of caffeine after chronic caffeine has also been reported in rats.[42] Behavioral depression by high doses of caffeine was perhaps slightly enhanced after chronic caffeine ingestion by mice.[41] The choreiform (dance-like) movements elicited in mice by high doses of caffeine were significantly reduced after chronic caffeine ingestion.[43]

Consonant with the up-regulation of A_1-receptors is the observation that the behavioral depressant effects of an A_1-selective adenosine analog, N^6-cyclohexyladenosine (CHA), were slightly enhanced after chronic caffeine ingestion in mice.[41,44] However, the behavioral depressant effects of an A_{2A}-selective adenosine analog, APEC, were also slightly enhanced, as were those of a potent mixed A_1/A_2-adenosine analog, NECA. A simple interpretation of these results is complicated by the fact that there appears to be a synergism between the behavioral depressant effects of activation of A_1-receptors and A_{2A}-receptors by selective adenosine analogs in mice.[45]

Since dopaminergic systems are strongly linked to caffeine pharmacology, it was reasonable to examine alterations in behavior subserved by dopaminergic pathways after chronic caffeine. Behaviorally, the stimulation of open-field locomotor activity by amphetamine, which releases dopamine, and cocaine, which blocks reuptake of dopamine, were little affected by chronic caffeine ingestion in NIH Swiss strain mice.[41] This is at least consonant with the lack of change in density of dopamine receptors. However, it should be noted that a 1 mg/kg dose of amphetamine had significantly less effect after chronic caffeine ingestion.[41] Further studies are needed, since a large body of evidence suggests that chronic caffeine ought to affect, *via* blockade of striatal adenosine receptors, the function of dopaminergic systems, leading to changes in dopaminergic receptors and/or dopaminergic sensitivity.[3,11,40] Dopamine systems do undergo homeostatic changes as a result of denervation, chronic receptor activation, or chronic receptor-blockade. In rats, chronic treatment with caffeine has been reported to elicit tolerance to the locomotor stimulatory effects of selective dopamine receptor agonists.[47] However, the rats were not tolerant to a combination of D_1- and D_2-receptor agonists. Antagonists for D_1 and D_2-dopamine receptors blocked caffeine-induced stimulation of locomotor activity in rats,[48,49] providing further evidence of the importance of dopamine systems to the behavioral

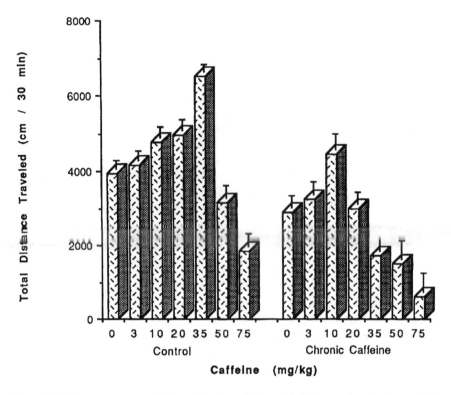

Figure 1.2 Dose-response relationships for effects of caffeine on locomotor activity in mice. Activity was measured for 30 min in a circular arena after intraperitoneal injection of caffeine in control and chronic caffeine-treated male NIH Swiss strain mice. Caffeine ingestion (100 mg/kg/day) was for 7 days, followed by 2 to 4 h withdrawal for caffeine clearance. (From Nikodijević, O., Jacobson, K.A., and Daly, J.W., Locomotor activity in mice during chronic treatment with caffeine and withdrawal, *Pharmacol. Biochem. Behav.* 44, 199-216, 1993. With permission.)

pharmacology of caffeine. Chronic treatment with adenosine analogs has been shown to result in an attenuation of both A_{2A}-adenosine and D_1-dopamine receptor-mediated stimulation of striatal adenylyl cyclase.[50] Dopamine denervation was reported to enhance A_{2A}-adenosine and D_2-dopamine receptor interactions in rat striatum.[46] Thus, it is clear that dopamine and A_{2A}-adenosine systems are subject to interrelated homeostatic regulation.

Cholinergic systems also appear intimately linked to function of striatal dopaminergic systems and to caffeine-sensitive adenosine systems. Chronic caffeine ingestion in NIH Swiss strain mice did cause alterations in levels of both muscarinic and nicotinic receptors.[25] Behaviorally, the stimulation of open-field locomotor activity by the muscarinic agonist scopolamine in mice was significantly changed after chronic caffeine ingestion.[41] Higher doses of scopolamine were required for the same degree of locomotor stimulation, suggesting an increased tonic input of acetylcholine to muscarinic receptors.

However, the depressant effects of a muscarinic agonist, oxotremorine, appeared only somewhat diminished after chronic caffeine ingestion. Nicotine/caffeine behavioral interactions have been extensively studied.[51,52] Behavioral depressant effects of the nicotinic agonist nicotine were absent in NIH Swiss strain mice after chronic caffeine ingestion,[30,41] in spite of an apparent up-regulation of nicotine receptors. Probably, as is the case for tolerance to nicotine elicited by chronic nicotine,[53] these "up-regulated" receptors are actually desensitized and nonfunctional. Nicotine, in combination with caffeine, had little effect on locomotor activity in control mice, but did cause behavioral stimulation after chronic caffeine.[30] It should be noted that the open-field locomotor activity of the NIH Swiss strain mice had been reduced by nearly 40% as a result of chronic ingestion of caffeine, and it was postulated that this might be due in part to enhanced cholinergic function.[41] Mecamylamine, a nicotinic antagonist, caused a somewhat greater depression in mice after chronic caffeine ingestion.[30]

Chronic caffeine ingestion reduced the depressant effects of ethanol, but chronic ethanol ingestion had no effect on the locomotor stimulation evoked by caffeine.[54]

Behavioral studies on noradrenergic, serotoninergic, $GABA_A$, and calcium channel function after chronic caffeine ingestion are also needed, since levels of receptors subserving such systems are altered after chronic caffeine ingestion by male NIH Swiss strain mice.[25]

III. Adenosine analogs and xanthines

The synergy with respect to behavioral depression observed with combinations of A_1- and A_{2A}- selective agonists[45] may explain the high potency of the mixed A_1/A_2 agonist NECA as a behavioral depressant, and might explain the enhanced depressant effects of all adenosine analogs after the chronic caffeine-elicited up-regulation of A_1-adenosine receptors. The converse to synergisms for agonists appears to apply with respect to the behavioral stimulation elicited by xanthines. Thus, 8-cyclopentyltheophylline (CPT), an A_1-selective antagonist, is a weak behavioral stimulant, while 8-cyclopentyl-1,3-dipropylxanthine (CPX), an even more A_1-selective antagonist, is not a behavioral stimulant.[44,45] 8-(3-Chlorostyryl)caffeine (CSC), an A_{2A}-selective antagonist, is a very weak behavioral stimulant.[55] However, a combination of CPX and CSC resulted in a synergistic stimulation of open-field locomotor activity (Figure 1.3). Thus, caffeine and other xanthines that are relatively nonselective as adenosine receptor antagonists may owe their effectiveness as behavioral stimulants to concomitant blockade of both A_1- and A_{2A}-adenosine receptors.

Germinal studies in the early 1980s proposed a correlation of A_1-receptor affinity and behavioral stimulation for xanthines,[56,57] but A_{2A}-receptors have also been proposed to have a major role in regulation of behavioral activity.[58] Subsequent studies provided further evidence that stimulation of locomotor

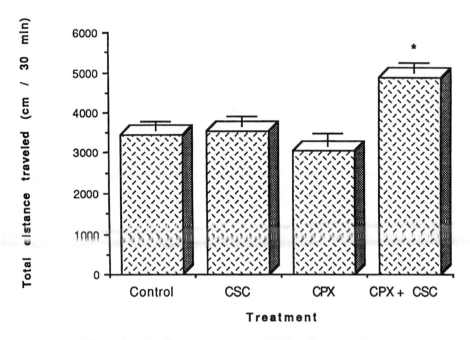

Figure 1.3 Effects of an A_1-selective antagonist (CPX) and an A_{2A}-selective antagonist (CSC) on locomotor activity in mice. Activity was measured for 30 min after intraperitoneal injection of 8-cyclopentyl-1,3-dipropylxanthine (CPX, 0.25 mg/kg), and/or 8-(3-chlorostyryl)xanthine (CSC, 1.0 mg/kg) in male NIH Swiss strain mice. (From Jacobson, K.A., Nikodijević, O., Padgett, W.L., Gallo-Rodriguez, C., Maillard, M., and Daly, J.W., 8-(3-Chlorostyryl)caffeine (CSC) is a selective A_2-adenosine antagonist *in vitro* and *in vivo*, *FEBS Lett.* 323, 141-144, 1993. With permission.)

activity by caffeine and development of tolerance to caffeine are closely related to blockade of A_1-adenosine receptors.[59,60] However, the demonstration of synergistic interactions of A_1- and A_{2A}-adenosine receptors in control of locomotor activity,[45,55] suggests that questions as to the relative importance of blockade of A_1- vs. A_{2A}-adenosine receptors to the behavioral effects of caffeine will be difficult to answer. It is worth noting that the effectiveness with which caffeine and other xanthines reverse the depressant effects of adenosine analogs is usually greater than their ability to cause behavioral stimulation alone.[57,61] But any conclusions as to the relationship between tonic effects of endogenous adenosine and the effects of administration of adenosine analogs are tenuous. The analogs will undoubtedly activate adenosine receptors that are not tonically activated by endogenous adenosine.

One other aspect of behavioral effects of adenosine analogs and xanthines further illustrates the complexity of interactions of adenosine analogs and xanthines. In the early 1980s it was noted that the combination of caffeine and N^6-R-phenylisopropyladenosine not only reversed the behavioral depressant effects of the adenosine analog, but actually caused a behavioral stimulation greater than that elicited by the xanthines alone.[56,57,62] This occurs

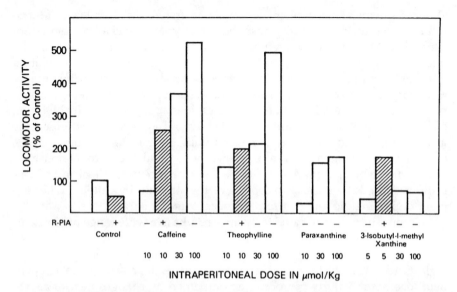

Figure 1.4   Effects of xanthines and the adenosine analog N^6-R-phenylisopropylad-enosine (R-PIA) on locomotor activity in mice. Open-field activity is for the second 30-min period after intraperitoneal injections of male ICR mice with xanthines alone or in combination with R-PIA (0.2 μmol/kg). Note the depressant effect of the lowest dose of caffeine and the stimulatory effect of R-PIA in combination with that dose of caffeine. (From Daly, J.W. Mechanism of action of caffeine, in *Caffeine, Coffee and Health*, S. Garattini, Ed., Raven Press, New York, Chapter 4, pp 97-150, 1993. With permission.)

with caffeine and theophylline and even with a nonstimulatory xanthine, isobutylmethylxanthine (Figure 1.4). An explanation was not apparent in 1980, nor has one been forthcoming. However, on examining dose–response effects on open-field locomotor activity for the adenosine analogs CHA, NECA, and APEC in combination with caffeine, it appears that the syner-gistic stimulatory effect of xanthine–adenosine analog combinations mani-fests itself as a stimulatory "bump."[30,44] It is noteworthy that the stimulatory "bump" is diminished after chronic caffeine ingestion, as are the synergistic depressant effects of A$_1$- and A$_{2A}$-agonists.[44]

IV. Summary

A large body of evidence suggests that A$_1$- and A$_{2A}$-adenosine receptors are the most likely targets for the behavior stimulant effects of caffeine, partic-ularly those adenosine receptors associated with the dopaminergic reward system of the striatum. Only the A$_1$- and A$_{2A}$-receptors of the central nervous system would be expected to be activated by endogenous adenosine under physiological conditions, while A$_{2B}$- and A$_3$-receptors would require the higher levels of adenosine that pertain during ischemia and convulsions.

Other biochemical mechanisms of action of caffeine, such as release of intra-cellular calcium, inhibition of phosphodiesterases and blockade of regulatory sites on GABA$_A$-receptors, require much higher concentrations than the micromolar concentrations of caffeine associated with behavioral stimulation in rodents and humans. Selective adenosine agonists and xanthine antagonists have provided insights into central roles for adenosine receptor subtypes. Thus, behavioral stimulation by xanthines appears to require blockade of both A$_1$- and A$_{2A}$-receptors, while behavioral depression by adenosine analogs appears to require activation of both A$_1$- and A$_{2A}$-receptors. Thus, caffeine may owe many central-mediated behavioral effects to its nonselective ability to block both A$_1$- and A$_{2A}$-adenosine receptors. Blockade of tonically activated A$_1$- and A$_{2A}$-receptors of the striatum by caffeine will augment dopaminergic function for both receptors.

Chronic ingestion of caffeine by mice results in a wide range of biochemical and behavioral alterations in the central nervous system.[25,30] The densities of A$_1$-adenosine receptors are increased, as are noradrenergic, cholinergic, GABAergic, and serotoninergic receptors. A depression in basal open-field locomotor activity can occur, accompanied by changes in responsiveness to caffeine and other xanthines, to adenosine analogs, and to cholinergic agents. Changes in behavioral responsiveness to dopaminergic agents in mice are minimal.

At present, further research is needed on the sites and mechanisms involved in behavioral effects of acute and chronic caffeine. Specific localization of the sites involved in alterations in expression of early genes,[15-17,63] changes in glucose utilization,[64] and up- and down-regulation of receptors[25,30] provide foci for such further research.

Acknowledgments

The support of the International Life Science Institute for our program on "Mechanism of Action of Caffeine and Theophylline" is gratefully acknowledged.

References

1. Nehlig, A., Daval, J.L., and Debry, G., Caffeine and the central nervous system: Mechanisms of action, biochemical, metabolic and psychostimulant effects, *Brain Res. Rev.*, 17, 139, 1992.
2. Daly, J.W., Mechanism of action of caffeine, in *Caffeine, Coffee and Health*, Garattini, S., Ed., Raven Press, New York, 1993, 97.
3. Fredholm, B.B., Arslan, G., Johansson, B., Kull, B., and Svenningsson, P., Adenosine A$_{2A}$ receptors and the actions of caffeine, in *The Role of Adenosine in the Nervous System*, Okada, Y., Ed., Elsevier, Amsterdam, 1997, 51.
4. Jacobson, K.A., Van Galen, P.J.M., and Williams, M., Adenosine receptors: Pharmacology, structure-activity relationships, and therapeutic potential, *J. Med. Chem.*, 35, 407, 1992.

5. Olah, M.E. and Stiles, G.L., Adenosine receptor subtypes: Characterization and therapeutic regulation, *Annu. Rev. Pharmacol. Toxicol.*, 35, 581, 1995.

6. Daly, J.W., *Cyclic Nucleotides in the Nervous System*, Plenum Press, New York, 1997.

7. Ferré, S., O'Conner, W. T., Svenningsson, P., Bjürklund, L., Lindberg, J., Tinner, B., Stomberg, I., Goldstein, M., Ogren, S. O., Ungerstedt, U., Fredholm, B. B., and Fuxe, K., Dopamine D_1 receptor-mediated facilitation of GABAergic neurotransmission in the rat strioentopenduncular pathway and its modulation by adenosine A_1 receptor-mediated mechanisms, *Eur. J. Neurosci.*, 8, 1545, 1996.

8. Ferré, S., Popoli, P., Tinner-Staines, B., and Fuxe, K., Adenosine A1 receptor-dopamine interaction in the rat limbic system: modulation of dopamine D1 receptor antagonist binding sites, *Neurosci. Lett.*, 208, 109, 1996.

9. Ferré, S., Fuxe, K., Von Euler, G., Johansson, B., and Fredholm, B.B., Adenosine-dopamine interactions in the brain, *Neuroscience*, 51, 501, 1992.

10. Ferré, S., Von Euler, G., Johansson, B., Fredholm, B., and Fuxe, K., Stimulation of high affinity A-2 receptors decreases the affinity of dopamine D-2 receptors in rat striatal membranes, *Proc. Natl. Acad. Sci. U.S.A.*, 88, 7237, 1991.

11. Ferré, S., Fredholm, B. B., Morelli, M., Popoli, P., and Fuxe, K., Adenosine dopamine receptor-receptor interactions in the basal ganglia, *Trends Neurosci.*, 20, 482, 1997.

12. Harvey, J. and Lacey, M.G., A postsynaptic interaction between dopamine D1 and NMDA receptors promotes presynaptic inhibition in the rat nucleus accumbens via adenosine release, *J. Neurosci.*, 17, 5271, 1997.

13. Svenningsson, P., Le Moine, C., Kull, B., Sunagara, R., Bloch, B., and Fredholm, B. B., Cellular expression of adenosine A_{2A} receptor messenger RNA in the rat central nervous system with special reference to dopamine innervated areas, *Neuroscience*, 80, 1171, 1997.

14. Ledent, C., Vaugeois, J.M., Schiffmann, S.N., Pedrazzini, T., El yacoubi, M., Vanderhaeghen, J.J., Costentin, J., Health, J.K., Vassart, G., and Parmentier, M., Aggressiveness, hypoalgesia, and high blood pressure in mice lacking the adenosine A_{2A} receptor, *Nature*, 388, 674, 1997.

15. Svenningsson, P., Nomikos, G.G., Ongini, E., and Fredholm, B.B., Antagonism of adenosine A_{2A} receptors underlies the behavioral activating effect of caffeine and is associated with reduced expression of messenger RNA for NGFI-A and NGFI-B in caudate-putamen and nucleus accumbens, *Neuroscience*, 79, 753, 1997.

16. Svenningsson, P., Nomikos, G.G., and Fredholm, B.B., Biphasic changes in locomotor behavior and in expression of mRNA for NGFI-A and NGFI-B in rat striatum following acute caffeine administration, *J. Neurosci.*, 15, 1712, 1995.

17. Johansson, B., Lindstrom, K., and Fredholm, B.B., Differences in the regional and cellular localization of c-fos messenger RNA induced by amphetamine, cocaine and caffeine in the rat, *Neuroscience*, 59, 837, 1994.

18. McPherson, P.S., Kim, Y.K., Valdivia, H., Knudson, C.M., Takekura, H., Franzini-Armstrong, C., Coronado, R., and Campbell, K.P., The brain ryanodine receptor: a caffeine-sensitive calcium release channel, *Neuron*, 7, 17, 1991.

19. Müller, C.E. and Daly, J.W., Stimulation of calcium release by caffeine analogs in pheochromocytoma cells, *Biochem. Pharmacol.*, 46, 1825, 1993.

20. Choi, O.H., Shamim, M.T., Padgett, W.L., and Daly, J.W., Caffeine and theophylline analogues: Correlation of behavioral effects with activity as adenosine receptor antagonists and as phosphodiesterase inhibitors, *Life Sci.*, 43, 387, 1988.
21. Marangos, P.J., Paul, S.M., Parma, A.M., Goodwin, F.K., Synapin, P., and Skolnick, P., Purinergic inhibition of diazepam binding to rat brain (*in vitro*), *Life Sci.*, 24, 851, 1979.
22. Fredholm, B.B., Adenosine actions and adenosine receptors after 1 week treatment with caffeine, *Acta Physiol. Scand.*, 115, 283, 1982.
23. Murray, T.F., Up-regulation of rat cortical adenosine receptors following chronic administration of theophylline, *Eur. J. Pharmacol.*, 82, 113, 1982.
24. Johansson, B., Ahlberg, S., Van der Ploeg, I., Brene, S., Lindefors, N., Persson, H., and Fredholm, B.B., Effect on long term caffeine treatment on A_1 and A_2 adenosine receptor binding and on mRNA levels in rat brain, *Naunyn-Schmideberg's Arch. Pharmacol.*, 347, 407, 1993.
25. Shi, D., Nikodijević, O., Jacobson, K.A., and Daly, J.W., Chronic caffeine alters the density of A_1-adenosine, β-adrenergic, serotonin, cholinergic, and $GABA_A$ receptors and calcium channels in mouse brain, *Cell. Mol. Neurobiol.*, 13, 247, 1993.
26. Hawkins, M., Dugich, M.M., Porter, N.M., Urbancic, M., and Radulovacki, M., Effects of chronic administration of caffeine on adenosine A_1 and A_2 receptors in rat brain, *Brain Res. Bull.*, 21, 479, 1998.
27. Zielke, C.L. and Zielke, H.R., Chronic exposure to subcutaneously implanted methylxanthines. Differential elevation of A_1-adenosine receptors in mouse cerebellar and cerebral cortical membranes, *Biochem. Pharmacol.*, 36, 2533, 1987.
28. Ahlijanian, M.K. and Takemori, A.E., Cross-tolerance studies between caffeine and (–)-N^6-(phenylisopropyl)-adenosine (PIA) in mice, *Life Sci.*, 38, 577, 1986.
29. Von Lubitz, D.K.J.E., Paul, I.A., Ji, X.D., Carter, M., and Jacobson, K.A., Chronic adenosine A_1 receptor agonist and antagonist: Effect on receptor density and N-methyl-D-aspartate-induced seizures in mice, *Eur. J. Pharmacol.*, 253, 95, 1994.
30. Shi, D., Nikodijević, O., Jacobson, K.A., and Daly, J.W., Effects of chronic caffeine on adenosine, dopamine and acetylcholine systems in mice, *Arch. Internat. Pharmacodyn. Therap.*, 328, 261, 1994.
31. Daly, J.W., Shi, D., Nikodijević, O., and Jacobson, K.A., The role of adenosine receptors in the central action of caffeine, *Pharmacopsychoecologia*, 7, 201, 1994.
32. Goldberg, M.R., Curatolo, P.W., Tung, C.S., and Robertson, D., Caffeine down-regulates β adrenoreceptors in rat forebrain, *Neuroscience Lett.*, 31, 47, 1982.
33. Fredholm, B.B., Jonzon, B., and Lindgren, E., Changes in noradrenaline release and in beta receptor number in rat hippocampus following long-term treatment with theophylline or L-phenylisopropyladenosine, *Acta Physiol. Scand.*, 122, 55, 1984.
34. Green, R.M. and Stiles, G.L., Chronic caffeine ingestion sensitizes the A_1 adenosine receptor-adenylate cyclase system in rat cerebral cortex, *J. Clin. Invest.*, 77, 222, 1986.
35. Wu, P.H. and Coffin, V.L., Up-regulation of brain [³H]diazepam binding sites in chronic caffeine-treated rats, *Brain Res.*, 294, 186, 1984.
36. Wu, P.H. and Phillis, J.W., Up-regulation of brain [³H]diazepam binding sites in chronic caffeine-treated rats, *Gen. Pharmacol.*, 17, 501, 1986.

37. Daval, J.L., Deckert, J., Weiss, S.R.B., Post, R.M., and Marangos, P.J., Up-regulation of adenosine A_1 receptors and forskolin binding sites following chronic treatment with caffeine or carbamazepine: A quantitative autoradiographic study, *Epilepsia*, 30, 26, 1989.
38. Holtzman, S.G., Complete, reversible drug-specific tolerance to stimulation of locomotor activity by caffeine, *Life Sci.*, 33, 779, 1983.
39. Holtzman, S.G. and Finn, I.B., Tolerance to behavioral effects of caffeine in rats, *Pharmacol. Biochem. Behav.*, 29, 411, 1988.
40. Holtzman, S.G., Mante, S., and Minneman, K.P., Role of adenosine receptors in caffeine tolerance, *J. Pharmacol. Exp. Therap.*, 256, 62, 1991.
41. Nikodijević, O., Jacobson, K.A., and Daly, J.W., Locomotor activity in mice during chronic treatment with caffeine and withdrawal, *Pharmacol. Biochem. Behav.*, 44, 199, 1993.
42. Meliska, C.J., Landrum, R.E., and Landrum, T.A., Tolerance and sensitization to chronic and subchronic oral caffeine: Effects on wheelrunning in rats, *Pharmacol. Biochem. Behav.*, 35, 477, 1990.
43. Nikodijević, O., Jacobson, K.A., and Daly, J.W., Acute treatment of mice with high doses of caffeine: An animal model for choreiform movement, *Drug Develop. Res.*, 30, 121, 1993.
44. Nikodijević, O., Jacobson, K.A., and Daly, J.W., Effects of combinations of methylxanthines and adenosine analogs on locomotor activity in control and chronic caffeine-treated mice, *Drug Develop. Res.*, 30, 104, 1993.
45. Nikodijević, O., Sarges, R., Daly, J.W., and Jacobson, K.A., Behavioral effects of A_1- and A_2-selective adenosine agonists and antagonists: Evidence for synergism and antagonism, *J. Pharmacol. Exp. Therap.*, 259, 286, 1991.
46. Ferré, S. and Fuxe, K., Dopamine denervation leads to an increase in the membrane interaction between adenosine A_2 and dopamine D_2 receptors in the neostriatum, *Brain Res.*, 594, 124, 1992.
47. Garrett, B.E. and Holtzman, S.G., Caffeine cross-tolerance to selective dopamine D_1 and D_2 receptor agonists but not to their synergistic interaction, *Eur. J. Pharmacol.*, 262, 65, 1994.
48. Garrett, B.E. and Holtzman, S.G., D_1 and D_2 dopamine receptor antagonists block caffeine-induced stimulation of locomotor activity in rats, *Pharmacol. Biochem. Behav.*, 47, 89, 1994.
49. Garrett, B.E. and Holtzman, S.G., Does adenosine receptor blockage mediate caffeine-induced rotational behavior? *J. Pharmacol. Exp. Therap.*, 274, 207, 1995.
50. Porter, N.M., Radulovacki, M., and Green, R.D., Desensitization of adenosine and dopamine receptors in rat brain after treatment with adenosine analogs, *J. Pharmacol. Exp. Therap.*, 244, 218, 1988.
51. White, J.M., Behavioral interactions between nicotine and caffeine, *Pharmacol. Biochem. Behav.*, 29, 63, 1988.
52. Cohen, C., Welzl, H., and Bättig, K., Effects of nicotine, caffeine and their combination on locomotor activity in rats, *Pharmacol. Biochem. Behav.*, 40, 121, 1991.
53. Marks, M.J., Grady, S.R., and Collins, A.C., Down-regulation of nicotinic receptor function after chronic nicotine infusion, *J. Pharmacol. Exp. Therap.*, 266, 1268, 1993.
54. Daly, J.W., Shi, D., Wong, V., and Nikodijević, O., Chronic effects of ethanol on central adenosine function of mice, *Brain Res.*, 650, 153, 1994.

55. Jacobson, K.A., Nikodijević, O., Padgett, W.L., Gallo-Rodriguez, C., Maillard, M., and Daly, J.W., 8-(3-Chlorostyryl)caffeine (CSC) is a selective A$_2$-adenosine antagonist *in vitro* and *in vivo*, *FEBS Lett.*, 323, 141, 1993.
56. Snyder, S.H., Katims, J.J., Annau, Z., Bruns, R.F., and Daly, J.W., Adenosine receptors and behavioral actions of methylxanthines, *Proc. Natl. Acad. Sci. U.S.A.*, 78, 3260, 1981.
57. Katims, J.J., Annau, Z., and Snyder, S.H., Interactions in the behavioral effects of methylxanthines and adenosine derivatives, *J. Pharmacol. Exp. Therap.*, 227, 167, 1993.
58. Durcan, M. and Morgan, P.F., Evidence for A$_2$ receptor involvement in the hypomotility effects of adenosine analogs in mice, *Eur. J. Pharmacol.*, 168, 285, 1989.
59. Kaplan, G.B., Greenblatt, D.J., Kent, M.A., Cotreau, M.M., Arcelin, G., and Shader, R.I., Caffeine-induced behavioral stimulation is dose-dependent and associated with A$_1$ adenosine receptor occupancy, *Neuropsychopharmacology*, 6, 145, 1992.
60. Kaplan, G.B., Greenblatt, D.J., Kent, M.A., and Cotreau-Bibbo, M.M., Caffeine treatment and withdrawal in mice: Relationships between dosage, concentrations, locomotor activity and A$_1$ adenosine receptor binding, *J. Pharmacol. Exper. Therap.*, 266, 1563, 1993.
61. Coffin, V.L., Taylor, J.A., Phillis, J.W., Altman, H.J., and Barraco, R.A., Behavioral interaction of adenosine and methylxanthines on central purinergic systems, *Neuroscience Lett.*, 47, 91, 1984.
62. Phillis, J.W., Barraco, R.A., Delong, R.E., and Washington, D.O., Behavioral characteristics of centrally administered adenosine analogs, *Pharmacol. Biochem. Behav.*, 24, 263, 1986.
63. Svenningsson, P. and Fredholm, B.B., Caffeine mimics the effect of dopamine D$_2$/$_3$ receptor agonist on the expression of immediate early genes in globus pallidus, *Neuropharmacology*, 36, 1009, 1997.
64. Nehlig, A., Caffeine, brain energy metabolism and blood flow: A basis for understanding the behavioral effects of the methylxanthines, *Pharmacopsychoecologia*, 7, 97, 1994.

chapter two

Caffeine in the modulation of brain function

J. Patrick Myers, David A. Johnson, and Devon E. McVey

Contents

I. Introduction 17
II. Mechanisms of action: the classical concepts 18
   A. Mobilization of calcium 18
   B. Inhibition of phosphodiesterase..18
   C. Antagonism of adenosine receptors 19
   D. Antagonism of benzodiazepine receptors 19
III. Neurotransmitter modulation...19
   A. Catecholamines 20
   B. Serotonin 21
   C. Acetylcholine 21
   D. Amino acids..21
   E. Benzodiazepine receptors..22
IV. The relationship between neurochemical and behavioral
   effects of caffeine..22
V. Summary 24
References..25

I. Introduction

Caffeine is well known for its stimulant properties and is found in foods, beverages, and as a constituent of drug formulations throughout the world. The mechanisms in the brain by which caffeine produces its effects, however, have not been understood until the last decade, with the advent of powerful research tools that have permitted investigators to study the biochemical

0-8493-1166-7/99/$0.00+$.50
© 1999 by CRC Press LLC

and neuropharmacological properties of this drug. The intent of this article is to review the scientific literature which is related to the role of caffeine as a modulator of neurotransmission in general, and then to focus specifically on how caffeine affects cholinergic pathways. Finally, a link will be made between the changes in a caffeine-mediated cholinergic neurotransmission and behavioral effects of the drug.

II. Mechanisms of action: the classical concepts

Elucidation of caffeine's mechanism of action has been attempted in many types of cells, including skeletal muscle cells,[1] cardiac muscle, hepatocytes, and tracheal cells.[2] The present section will focus on how these proposed mechanisms might affect the release of ACh from neuronal cells.

A. Mobilization of calcium

Several mechanisms of action have been proposed for the stimulant properties of caffeine. The earliest-discovered hypothesis involves the mobilization of intracellular calcium. At high concentrations, caffeine interferes with the uptake and storage of calcium from the sarcoplasmic reticulum in striated muscle,[1] and promotes the translocation of calcium through the plasma membrane.[3] Such a mechanism could affect neurotransmission, because synaptic transmission depends on the influx of calcium into nerve endings. Therefore, neuropharmacologists investigated the direct effects of caffeine on voltage-dependent calcium ion channels. Hughes et al.[4] demonstrated that caffeine directly blocks voltage-dependent calcium ion channels. Further, Mironov and Usachev[5] observed that methylxanthine in the range of 6 to 30 mM produced calcium release from intracellular stores, which also affected calcium ion channels. Other authors have noted that concentrations as low as 250 µM of caffeine will produce shifts in intracellular calcium concentration.[6,7] However, following the ingestion of 1 to 3 cups of coffee, circulating plasma concentrations of caffeine are usually lower than 100 µM. Moreover, toxic effects of caffeine are observed at 200 µM, and lethal concentrations are in the range of 500 µM.[7,8]

B. Inhibition of phosphodiesterase

Methylxanthines are structurally related to the substrates of phosphodiesterases, and competitively inhibit these enzymes in a dose-dependent fashion. Phosphodiesterases are responsible for the metabolism of cAMP. Thus caffeine, via inhibition of the breakdown of cAMP, would potentiate the many effects mediated by this second messenger, including the phosphorylation of enzymes involved in neurotransmission. However, Burg and Warner[9] showed that chronic treatment of caffeine at 25 mg/kg/day did not increase the concentration of cAMP in mice and does not decrease the activity of phosphodiesterase *in vivo*. Other studies have demonstrated that inhibition requires millimolar concentrations of caffeine, a concentration range that is never found *in vivo*.[10] Since the minimal *in vitro* concentration of caffeine which is necessary to produce changes in intracellular calcium and inhibition of phosphodiesterases is well within the toxic range *in vivo*, it seems unlikely that these mechanisms could be responsible for the effects produced by caffeine at therapeutic concentrations.

C. Antagonism of adenosine receptors

The antagonism of adenosine receptors by caffeine is currently the most widely accepted mechanism of its action. The elucidation of this mechanism stemmed from a discovery by Sattin and Rall,[11] who noticed that theophylline often reduced the accumulation of cAMP in cerebral slices rather than increasing it, as would be expected from a phosphodiesterase inhibitor. Furthermore, it was noted that adenosine itself produces effects opposite to those of caffeine, and it was later determined that methylxanthines act as competitive antagonists at adenosine receptors at concentrations well within the therapeutic range (less than 100 µg, which can be attained by drinking 1 to 3 cups of coffee).

In support of the mechanism of adenosine antagonism, Pedata et al.,[12] using rat cortical slices, could inhibit electrically evoked ACh release in a dose-dependent fashion by adding adenosine; this effect was enhanced by dipyridamole (10 µM), a blocker of adenosine uptake. Caffeine (50 µM) facilitated ACh release in the same model, and its effects were blocked by dipyridamole. Moreover, adenosine inhibited the release of neurotransmitters from presynaptic neurons and modified the response to neurotransmitters postsynaptically.[13] Thus, caffeine could enhance the release of neurotransmitters and also potentiate the response of cells to neurotransmitter activation.

D. Antagonism of benzodiazepine receptors

The most recently proposed mechanism of action of caffeine is through its interaction with benzodiazepine binding sites. This mechanism was suggested because caffeine antagonizes or modifies the effects of benzodiazepines on both animal and human behavior.[14] However, caffeine is a much more potent antagonist of adenosine than it is of benzodiazepine receptor, and the inhibition of benzodiazepines may even be mediated by adenosine receptors. Further, like the first two mentioned mechanisms, the inhibition of benzodiazepine receptors occurs only at toxic plasma levels of caffeine.[15]

III. Neurotransmitter modulation

The role of caffeine in modulating neurotransmission can be characterized by examining its effects on: presynaptic adenosine receptors, release and turn-over of neurotransmitters, and postsynaptic receptors. Adenosine has been shown to play an important role in the inhibition of the release of neurotransmitters, including norepinephrine, dopamine, serotonin, acetylcholine, glutamate, and GABA.[7,16] Release inhibition is mediated by activation of presynaptic adenosine receptors. As antagonists of adenosine at both high-affinity (A1) and low-affinity (A2) receptors, caffeine and other methylxanthines have demonstrated the ability to increase the release and functional turn-over of these neurotransmitters by blocking activation of these receptors. In further support for the action of methylxanthines at adenosine receptors, chronic administration of caffeine or theophylline by various routes increased the number of adenosine receptors in mouse brain.[17-19] However, the affinity of the receptors for adenosine is mostly unchanged by chronic administration of caffeine.[17,19] Some studies have shown an increased sensitivity to adenosine in cerebral slices, although an increase in receptor density does not always correspond to a functional change observed in the animal.[20]

A. Catecholamines

Early studies of methylxanthines found that in doses ranging from 2.5 to 100 mg/kg, methylxanthines, including caffeine, did not show any effect on cerebral norepinephrine concentrations.[21-23] However, methylxanthines have been shown to increase the rate of synthesis and turn-over of norepinephrine.[22-24] Methylxanthines have also increased the release of catecholamines from rat brain *in vitro*.[25] This corroborates evidence that methylxanthines increase the spontaneous electrical activity of adrenergic neurons,[26] accompanied by increases in synthesis and turn-over of monoamines.[27] In addition to affecting release and turn-over of catecholamines, methylxanthines have been shown to increase the number of α-adrenoreceptors and to decrease the number of β-adrenoreceptors in the rat.[28] This could most likely be attributed to alterations in the release of norepinephrine caused by antagonism of presynaptic adenosine receptors.

The effects of caffeine on dopamine neurotransmission are less clear. Administration of caffeine increases dopamine concentration in the cerebrum of rats.[29] However, release, uptake, and turn-over of dopamine have been increased, decreased, or unaffected by administration of caffeine as measured in whole brain, gross regional dissections, synaptosomes, tissue slices, or by *in vivo* electrochemistry.[26,30-35] Methylxanthines, including caffeine, have been shown to affect the rate of catecholamine synthesis, with the rate being increased at 30 minutes, but decreased 2 hours after administration for doses of 50 or 100 mg/kg.[36] Also, administration of caffeine decreases the concentration of dihydroxyphenylacetic acid, a dopamine metabolite, in some dopaminergic structures in the rat brain, while increasing it in other areas.[37] Finally, caffeine has been shown to potentiate the effects of amphetamines (release of dopamine),[38] cocaine (blocks reuptake of dopamine),[39] and apomorphine (direct dopamine agonist).[40] Since the action of these drugs increases the synaptic availability of dopamine, the potentiation of their effects is consistent with an enhanced release of dopamine from presynaptic neurons elicited by caffeine and mediated through adenosine receptors.

B. Serotonin

Cerebral concentrations of serotonin, 5-hydroxyindoleacetic acid, and tryptophan are increased when naturally occurring caffeine is added to the diet of rats,[41] or when caffeine is administered IP at doses of 10 to 100 mg/kg.[42-44] Caffeine has also been shown to increase the *in vitro* concentration of serotonin in the brain stem, cerebrum, and cerebellum of the rat.[45,46] Other studies, however, have demonstrated that the postsynaptic availability of serotonin is decreased by administration of caffeine.[47] Decreased availability of serotonin postsynaptically affects central functions under serotoninergic influence, including sleep mechanisms, motor function, and regulation of cerebral blood vessels.[12,48-57]

C. Acetylcholine

The effects of caffeine on brain acetylcholine have not been studied as thoroughly as those of other neurotransmitters. The effects of caffeine on acetylcholine in electrically stimulated brain slices varies as a function of caffeine concentration and frequency of stimulation.[12] Additionally, caffeine and theophylline have been shown to increase acetylcholine outflow from the cerebral cortex of anesthetized rats with the IP administration of 15 and 30 mg/kg.[58] Intracerebral injection of theophylline has also been shown to increase acetylcholine turn-over in the hippocampus of rats.[52] More recent studies have shown the concentration of acetylcholine to be increased in the striatum of anesthetized rats after IP administration of combinations of caffeine and choline at doses of 50 mg/kg caffeine and 120 mg/kg choline.[59] Carter et al.,[60] following oral administration of caffeine (3 to 30 mg/kg), found a significant rise in extracellular levels of acetylcholine in the hippocampus of conscious rats. This increase could be blocked when the microdialysis probe was perfused with tetrodotoxin or the A_1-receptor agonist, N6-cyclopentyladenosine, but not the A_2-receptor agonist CGS 21680. This study, therefore, suggested that orally administered caffeine produced its effect via blockade of A_1-receptors. In support of these findings, Shi et al.[61] found that chronic ingestion of caffeine by mice leads to a significant increase in A_1-adenosine receptors. Others, however, have found that chronic consumption of high doses of caffeine reduces the effect of caffeine on cortical acetylcholine release, an effect which is independent of adenosine.[54]

D. Amino acids

Mouse whole brain concentrations of glutamate are increased, and GABA and glycine are decreased after administration of caffeine at 0.5 mg/ml for 1 week followed by 1 mg/ml for 2 weeks via drinking water.[62] In rats, administration of gradually increasing concentrations of caffeine in drinking water showed increased whole brain concentrations of taurine, histidine, ornithine, and aspartate, unchanged tyrosine concentrations, and decreased GABA and glutamate concentrations.[63,64]

E. Benzodiazepine receptors

Caffeine at 5 to 40 mg/kg IP or 600 mg/kg in food has been shown to increase,[65-67] decrease,[68] or cause no change[34,69] in the number of benzodiazepine receptors in the brain. Moreover, caffeine increases the binding of benzodiazepines to their receptor site *in vivo*,[70,71] but decreases binding in cultured neurons.[72] However, the number of benzodiazepine receptors is affected by stress, and different types of stress can either increase or decrease benzodiazepine receptor density. Also, the addition of caffeine to the diet did not affect benzodiazepine receptor density, while daily administration of caffeine by the intraperitoneal route, which is stressful, did.[73,74] Therefore, the evidence that there is a direct affect by caffeine on benzodiazepine receptors is relatively weak. Finally, the concentrations of caffeine needed to antagonize benzodiazepine receptors is five to ten times higher than those needed to antagonize adenosine receptors.[75,76] This may explain the lack of effect with the administration of caffeine in the food, but may also provide evidence that these effects are also mediated by adenosine.[77]

IV. The relationship between neurochemical and behavioral effects of caffeine

Caffeine can affect several types of behavior. Studies have investigated these effects on spontaneous motor activity, learning, memory, and mental performance, vigilance, endurance, social behavior, aggressiveness, and mood, anxiety, and sleep (see Reference 78 for a review). The broad findings are that caffeine tends to produce a more rapid and clearer flow of thought, to allay drowsiness and fatigue, sustain intellectual affect and a more perfect sensory association of ideas, and to produce a keener appreciation of sensory stimuli and reduce reaction time.[13] It is often difficult, however, to measure with consistency the effects of caffeine on performance because these effects have been found to be both task and situation specific[79] and biphasic.[12]

One approach to understanding the mechanisms underlying the behavioral effects of caffeine is to correlate changes in behavior with changes in neurochemical pathways and levels of specific neurotransmitters. Since caffeine is associated with enhanced cognition, and some aspects of cognition, such as memory, are closely linked to the specific neurotransmitter acetylcholine (ACh), a series of investigations were undertaken to assess alterations in brain ACh induced by caffeine, individually and in combination with other drugs, and to correlate these changes with changes in short-term memory function.

We administered caffeine, with and without the acetylcholine precursor choline, to awake rats and then monitored changes in extracellular ACh in the hippocampus utilizing *in vivo* microdialysis techniques coupled to high performance liquid chromatography with electrochemical detection. The results, summarized in Figure 2.1, demonstrated that caffeine (10 mg/kg) alone produced no significant change in extracellular ACh in the basal ganglia. However, when administered in combination with choline (120 mg/kg), there was a significant synergistic effect between the two drugs, resulting in enhanced ACh concentrations. This synergistic effect may have been the result of enhanced ACh synthesis mediated by the increased availability of choline coupled to disinhibition of ACh release mediated via the blockade of presynaptic adenosine receptors by caffeine.

To correlate the observed neurochemical changes produced by the combination of caffeine and choline on brain ACh concentrations, with potential changes in memory function, additional studies were performed. Rats were treated with the amnestic agent scopolamine (1 mg/kg, IP), and with several doses of caffeine, choline, or combinations of both drugs. Following drug treatment the animals were placed in a standard passive avoidance light-dark box and exposed to a mild shock (1mA; 1sec duration) upon entering the dark side of the box. The following day, the animals were again placed in the box. Control animals, which were not administered any drugs, would not crossover to the dark side of the box indicating that they remembered the shock from the previous day. The animals that were administered scopolamine without caffeine or choline, however, failed to remember the shock of the previous day and rapidly moved to the dark side of the box. Animals that were administered scopolamine with caffeine exhibited a behavior that was not significantly different from those animals that received scopolamine alone, indicating that they also failed to remember the shock of the previous day. It was concluded, therefore, that with this paradigm at least, caffeine alone did not reverse scopolamine-induced amnesia. When amnestic rats were administered caffeine and choline in combination, however, on the following day, they exhibited a significant delay in crossing to the dark side of the box. This response indicated some positive effect by the drug combination on memory retention for the shock of the previous day (Figure 2.2).

The results of the behavioral study, therefore, paralleled those of the neurochemical study since when caffeine and choline were administered individually, there were lower ACh levels as well as less memory retention of an aversive event than when the drugs were administered in combination. It could be suggested, therefore, that caffeine alone in these studies did not enhance memory because there was little change in brain ACh in response to caffeine administration. However, caffeine in combination with choline was effective in elevating brain ACh, thereby reversing the drug-induced amnesia. These two studies demonstrate, therefore, the power of combining both neuropharmacological as well as behavioral investigations to determine the mechanism of action of caffeine on central nervous system function.

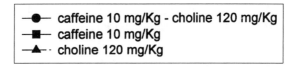

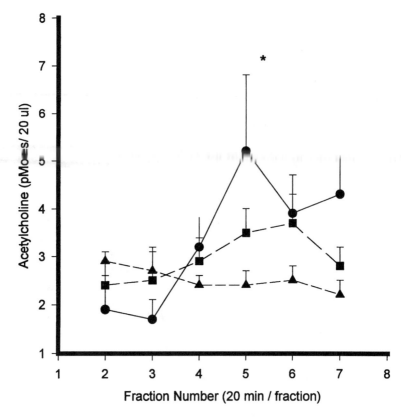

Figure 2.1 The increase in extracellular acetylcholine concentration in the hippocampus of the brains of awake rats following administration of caffeine 10 mg/kg, choline 120 mg/kg, or caffeine 50 mg/kg with choline 120 mg/kg. * indicates significant difference from pre-drug baseline at $p<0.05$.

V. Summary

The effects of caffeine on neurotransmission in brain are complex. The mechanism of action of caffeine in modulating neurotransmission is currently understood to be antagonism of inhibitory presynaptic adenosine receptors. This antagonism generally results in enhanced efflux of neurotransmitter; however, there may also be effects by caffeine on postsynaptic response, as well as neurotransmitter turnover and synthesis. Caffeine affects a wide range of neurotransmitters including catecholamines, acetylcholine, serotonin, and amino acids. Finally, experiments have been able to correlate the neuropharmacological effects of caffeine with its effects on behavior.

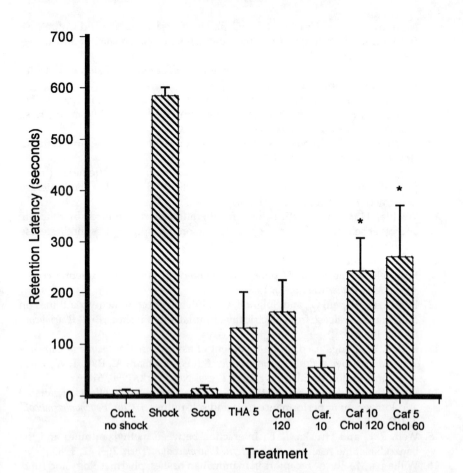

Figure 2.2 The crossover retention latency in scopolamine induced amnestic rats following administration of saline, scopolamine (1 mg/kg), tacrine (5 mg/kg), caffeine (10 mg/kg), choline (120 mg/kg), and combined caffeine with choline (caffeine 10 mg/kg and choline 120 mg/kg; (caffeine 5 mg/kg and choline 60 mg/kg). * indicates significant difference from control at $p < 0.05$.

References

1. Bianchi, C. P., The effect of caffeine on radiocalcium movement in frog sartorius, *J. Gen. Physiol.*, 44, 845, 1961.
2. Guthrie, J. R. and Nayler, W. G., Interaction between caffeine and adenosine on calcium exchangeability in mammalian atria. *Arch. Intern. Pharmacodyn. Ther.*, 170, 323, 1967.
3. Bianchi, C. P., Cellular pharmacology of contraction of skeletal muscle, in *Cellular Pharmacology of Excitable Tissues*, Narahashi, T., Ed., Charles C. Thomas, Springfield, 1975, 485.
4. Hughes, A. D., Herring, S., and Bolton, T. B., The action of caffeine on inward barium current through voltage dependent calcium channels in single rabbit ear artery cells, *Pfluegers Arch.*, 416, 462, 1990.

5. Mironov, S. L. and Usachev, J. M., Caffeine affects Ca⁺ uptake and Ca⁺ release from intracellular stores: fura-2 measurements in isolated snail neurons, *Neurosci. Lett.*, 123, 200, 1991.
6. Sandow, A. and Brust, M., Caffeine potentiation of twitch tension of sartorius muscle, *Biochemistry*, 345, 232, 1996.
7. Rall, T. W., Central nervous system stimulants: the xanthines, in *The Pharmacological Basis of Therapeutics*, 6th ed., Goodman, L. S. and Gilman, A., Eds., MacMillan, New York, 1980, 592.
8. Fredholm, B. B., On the mechanism of action of theophylline and caffeine, *Acta Med. Scand.*, 217, 149, 1985.
9. Burg, A. W. and Warner, E., Effect of orally administered caffeine and theophylline on tissue concentrations of 3′,5′-cyclic AMP and phosphodiesterase (Abstract), *Fed. Proc.*, 34, 332, 1975.
10. Wachtel, H., Characteristic behavioral alterations in rats induced by rolipram and other selective adenosine cyclic 3′, 5′- monophosphate phosphodiesterase inhibitors, *Psychopharmacology*, 77, 309, 1982.
11. Sattin, A. and Rall T. W., The effect of adenosine and adenine nucleotides on the cyclic adenosine 3′, 5′ phosphates content of guinea pig cerebral cortex slices, *Mol. Pharmacol.*, 6, 13, 1970.
12. Pedata, G., Pepeu, G., and Spignoli, G., Biphasic effect of methylxanthines on acetylcholine release from electrically-stimulated brain slices, *Br. J. Pharmacol.*, 83, 69, 1984.
13. Rall, T. W., Drugs used in the treatment of asthma, in, *Goodman and Gilman's The Pharmacological Basis of Therapeutics*, 8th ed., Gilman, A., Rall, T. W., Nies, A. S., and Taylor, P., Eds., Pergamon Press, New York, 1990, 618.
14. File, S. E., Bond, A. J., and Lister, R. G., Interaction between effects of caffeine and lorazepam in performance tests and self-ratings, *J. Clin. Psychopharmacol.*, 2, 102, 1982.
15. Weir, R. L. and Hruska, R. E., Interaction between methylxanthines and the benzodiazepine receptor, *Arch. Intern. Pharmacodyn. Ther.*, 265, 42, 1983.
16. Williams, M., Purine receptors in mammalian tissues: pharmacology and functional significance, *Annu. Rev. Pharmacol. Toxicol.*, 27, 315, 1987.
17. Daval, J. L., Deckert, J., Weiss, S. R. B., Post, R. M., and Marangos, P. J., Upregulation of adenosine A1 receptors and forskolin binding sites following chronic caffeine treatment with caffeine of carbamazepine: a quantitative autoradiographic study, *Epilepsia*, 30, 26, 1989.
18. Fastbom, J. and Fredholm, B. B., Effects of long-term theophylline treatment on adenosine A_1-receptors in rat brain: autoradiographic evidence for increased receptor number and altered coupling to G-proteins, *Brain Res.*, 507, 195, 1990.
19. Lupica, C. R., Jarvis, M. F., and Berman, R. F., Chronic theophylline treatment *in vivo* increases high affinity adenosine A_1 receptor binding and sensitivity to exogenous adenosine in the *in vitro* hippocampal slice, *Brain Res.*, 542, 55, 1991.
20. Matthew, R. J. and Wilson, W. H., Caffeine induced changes in cerebral circulation, *Stroke*, 16, 814, 1985.
21. Karasawa, T., Furukawa, K., Yoshida, K., and Shimizu, M., Effect of theophylline on monoamine metabolism in the rat brain, *Eur. J. Pharmacol.*, 37, 97, 1976.

22. Schlosberg, A. J., Fernstrom, J. D., Kopczynski, M. C., Cusack, B. M., and Gillis, M. A., Acute effects of caffeine injection on neutral amino acids and brain monoamine level in rats, *Life Sci.*, 29, 173, 1981.
23. Chou, D. T., Cuzzone, H., Springstead, J., Ali, R., and Hirsh, K., Differential effects of caffeine on regional brain biogenic amines in rats, *Soc. Neurosci. Abstr.*, 5, 551, 1979.
24. Costill, D. L., Dalsky, G., and Fink, W., Effects of caffeine ingestion on metabolism and exercise performance, *Med.Sci. Sports Exer.*, 10, 155, 1978.
25. Atkinson, J. and Enslen, M., Self-administration of caffeine by the rat, *Arzneimittelforsch*, 26, 2059, 1976.
26. Grant, S. J. and Redmond, Jr., D. E., Methylxanthine activation of noradrenergic unit activity and reversal by clonidine, *Eur. J. Pharmacol.*, 85, 105, 1982.
27. Fernstrom, M. H., Bazil, C. W., and Fernstrom, J. D., Caffeine injection raises brain tryptophan level, but does not stimulate the rate of serotonin synthesis in rat brain, *Life Sci.*, 35, 1241, 1984.
28. Fredholm, B. B., Jonzon, B., and Lendgren, E., Changes in noradrenaline release and in beta receptor number in rat hippocampus following long term treatment with theophylline or L-PIA, *Acta Physiol. Scand.*, 122, 55, 1984.
29. Stern, K. N., Chait, L. D., and Johanson, C. E., Reinforcing and subjective effects of caffeine in norma human volunteers, *Psychopharmacology*, 98, 81, 1989.
30. Berkowitz, B. A., Tarver, J. H., and Spector, S., Release of norepinephrine in the central nervous system by theophylline and caffeine, *Eur. J. Pharmacol.*, 10, 64, 1970.
31. Eitan, A. and Hershkowitz, M., The effects of dibutyril cyclic AMP, theophylline and papaverine on the release of ^{3}H-catecholamines from rat brain striatal and cortical synaptosomes, *Eur. J. Pharmacol.*, 46, 323, 1977.
32. Cardinali, D. P., Effects of pentoxiphylline and theophylline on biogenic amine metabolism in the rat brain, *Eur. J. Pharmacol.*, 47, 239, 1978.
33. Michaelis, M. L., Michaelis, E. K., and Myers, S. L., Adenosine modulation of synaptosomal dopamine release, *Life Sci.*, 24, 2083, 1979.
34. Goldberg, M. R., Curatolo, P. W., and Robertson, D., Caffeine down regulates B- receptors in rat forebrain, *Neurosci. Lett.*, 31, 47, 1982.
35. Galloway, M. P. and Roth, R. H., Clonidine prevents methylxanthine stimulation of norepinephrine metabolism in rat brain, *J. Neurochem.*, 40, 246, 1983.
36. Atuk, N. O., Blaydes, M. C., Westervelt, Jr., F. B., and Wood, J. E., Jr., Effect of aminophylline on urinary excretion of epinephrine and norepinephrine in man, *Circulation*, 35, 745, 1967.
37. Govoni, S., Petkov, V. V., and Montefusco, O., Differential effects of caffeine on dihydroxyphenylacetic acid concentrations in various rat brain dopaminergic structures, *J. Pharm. Pharmacol.*, 36, 458, 1984.
38. White, B. C. and Keller III, G. E., Caffeine pretreatment: enhancement and attenuation of d-amphetamine-induced activity, *Pharmacol. Biochem. Behav.*, 20, 383, 1984.
39. Misra, A. L., Vadlamani, N. L., and Pontani, R. B., Effect of caffeine on cocaine locomotor stimulant activity in rats, *Pharmacol. Biochem. Behav.*, 24, 761, 1986.
40. Klawans, H. L., Moses III, H., and Beaulieu, D. M., The influence of caffeine on *d*-amphetamine-induced activity, *Life Sci.*, 14, 1493, 1974.
41. Yokogoshi, H., Tani, S., and Amano, N., The effects of caffeine and caffeine-containing beverages on the disposition of brain serotonin in rats, *Agric., Biol., Chem.*, 51, 3281, 1987.

42. Haleem, D. J., Yasmeen, A., Haleem, M. A., and Zafar, A., 24-Hour withdrawal following repeated administration of caffeine attenuates brain serotonin but not tryptophan in rat brain: implications for caffeine-induced depression, *Life Sci.*, 57, PL285, 1995.

43. Fernstrom, J. D. and Fernstrom, M. H., Effects of caffeine on monoamine neurotransmitters in the central and peripheral nervous system, in *Caffeine: Perspectives from Recent Research*, Dews, P. B., Ed., Springer, Heidelberg, 1984, 107.

44. Yokogoshi, H., Phenylalanine inhibits caffeine-induced increase in brain serotonin concentrations in rats, *Agric. Biol. Chem.*, 52, 3173, 1988.

45. Berkowitz, B. A. and Spector, S., The effect of caffeine and theophylline on the disposition of brain serotonin in the rat, *Eur. J. Pharmacol.*, 16, 322, 1971.

46. Stromberg, U. and Waldeck, B., Behavioural and biochemical interaction between caffeine and L-DOPA, *J. Pharm. Pharmacol.*, 25, 302, 1973.

47. Hirsh, K., Central nervous system pharmacology of the dietary methylxanthines, in *The Methylxanthines Beverages and Foods: Chemistry, Consumption, and Health Effects* Spiller, G. A. and Alan, R., Eds., Alan R. Liss, New York, 1984, 235.

48. Jouvet, M., Biogenic amines and the states of sleep, *Science*, 163, 32, 1969.

49. Gumulka, W., Samanin, R., Valzelli, L., and Console, S., Behavioral and biochemical effects following the stimulation of the nucleus raphe dorsalis in rats, *J. Neurochem.*, 18, 533, 1971.

50. Holman, R. B., Glen, R., Elliot, B. S., and Barchas, J. D., Neuroregulators and sleep mechanisms, *Annu. Rev. Med.*, 26, 499, 1975.

51. Warbritton, J. D., Stewart, R. M., and Baldessarini, R. J., Increased sensitivity to intracerebroventricular infusion of serotonin and deaminated indoles after lesioning rat with dihydroxytryptamine, *Brain Res.*, 177, 355, 1980.

52. Murray, T. F., Blaker, W. D., Cheney, D. L, and Costa, E., Inhibition of acetylcholine turnover rate in rat hippocampus and cortex by intraventricular injection of adenosine analogs, *J. Pharmacol. Exp. Therap.*, 222, 550, 1982.

53. Pedata, F., Pepeu, G., and Spignoli, G., Effects of methylxanthines on acetylcholine release from electrically stimulated cortical slices (Abstract), *Br. J. Pharmacol.*, 80 Suppl., 471P, 1983.

54. Corradetti, R., Pedata, F., Pepeu, G., and Vannucchi, M. G., Chronic caffeine treatment reduces caffeine but not adenosine effects on cortical acetylcholine release, *Br. J. Pharmacol.*, 88, 671, 1986.

55. Reith, M. E. A., Sershen, H., and Lajtha, A., Effects of caffeine on monoaminergic systems in mouse brain, *Acta Biochim. Biophysich. Hungaria*, 22, 149, 1987.

56. Hadfield, M. G. and Milio, C., Caffeine and regional brain monoamine utilization in mice. *Life Sci.*, 45, 2637, 1989.

57. Kirch, D. G., Taylor, T. R., Gerhardt, G. A., Benowitz, N. L., Stephen, C., and Wyatt, R. J., Effect of chronic caffeine administration of monoamine and monoamine metabolite concentrations in the rat brain, *Neuropharmacology*, 29, 599, 1980.

58. Phillis, J. W., Siemens, R. K., and Wu, P. H., Effects of diazepam on adenosine and acetylcholine release from rat cerebral cortex: further evidence for a purinergic mechanism in the action of diazepam, *Br. J. Pharmacol.*, 70, 1085, 1980.

59. Johnson, D. A., Ulus, I. H., and Wurtman, R. J., Caffeine potentiates the enhancement of striatal acetylcholine release, *Life Sci.*, 51, 1597, 1992.

60. Carter, A. J., O'Connor, W. T., Carter, M. J., and Ungerstedt, U., Caffeine enhances acetylcholine release in the hippocampus *in vivo* by a selective interaction with Adenosine A1 receptors, *J. Pharmacol. Exp. Therap.*, 273, 637, 1995.

61. Shi, D., Nikodijevic, O., Jacobson, K. A., and Daly, J. W., Effects of chronic caffeine on adenosine, dopamine and acetylcholine systems in mice, *Arch. Internation. Pharmacol. Ther.*, 328, 261, 1994.

62. Lajtha, I. J., Banay-Schwartz, M., and Lajtha, A., The effect of caffeine on some mouse brain free amino acid levels, *Neurochem. Res.*, 14, 317, 1989.

63. Portoles, M., Minana, M. D., Jorda, A., and Grisolia, S., Caffeine-induced changes in the composition of the free amino acid pool of the cerebral cortex, *Neurochem. Res.*, 10, 887, 1985.

64. Colombatto, S., Fasulo, L., Mondardini, A., Malabaila, A., and Grillo, M. A., Effect of caffeine of ornithine metabolism in rat brain, liver and kidney, *Ital. J. Biochem.*, 28, 75, 1989.

65. Marangos, P. J., Paul, S. M., and Goodwin, F. K., Putative endogenous ligands for the benzodiazepine receptor, *Life Sci.*, 25, 1093, 1979.

66. Wu, P. H. and Coffin, V. L., Up-regulation of brain [^{3}H]diazepam binding sites in chronic caffeine-treated rats, *Brain Res.*, 294, 186, 1984.

67. Boulenger, J. P., Salem, N., Marangos, P. J., and Uhde, T. W., Plasma adenosine levels: measurement in humans and relationship to the anxiogenic effects of caffeine, *Psychiat. Res.*, 21, 247, 1987.

68. Daval, J. L. and Vert, P., Effect of chronic caffeine exposure to methylxanthines on Diazepam cerebral binding in female rats and their offspring, *Develop. Brain Res.*, 27, 175, 1986.

69. Hunter, R. E., Barrera, C. M., Dohanich, G. P., and Dunlap, W. P., Effects of uric acid and caffeine of A_1 adenosine receptor binding in developing rat brain, *Pharmacol. Biochem. Behav.*, 35, 791, 1990.

70. Koe, B. K., Kondratas, E., and Russo, L. L., [^{3}H]Ro15-1788 binding to benzodiazepine receptors in mouse brain *in vivo*: marked enhancement by GABA agonists and other CNS drugs, *Eur. J. Pharmacol.*, 142, 373, 1987.

71. Wu, P. H. and Phillis, J.W., Up-regulation of brain [^{3}H] diazepam binding sites in chronic caffeine treated rats, *Gen. Pharmacol.*, 17, 501, 1988.

72. Kaplan, G. B., Greenblatt, D. J., Leduc, B. W., Thompson, M. L., and Shader, R. I., Relationship of plasma and brain concentration of caffeine and metabolites to benzodiazepine receptor binding and locomotor activity, *J. Pharmacol. Exp. Therap.*, 248, 1078, 1989.

73. Boulenger, J. P., Marangos, P. J., Zander, K. J., and Hanson, J., Stress and caffeine: effects on central adenosine receptors, *Clin. Neuropharmacol.*, 9, 79, 1986.

74. Lopez, F., Miller, L. G., Greenblatt, D. J., Kaplan, G. B., and Shader, R. I., Interaction of caffeine with the $GABA_A$ complex: alterations in receptor function but not ligand binding, *Eur. J. Pharmacol.*, 172, 453, 1989.

75. Snyder, S. H., Katims, J. J., Annau, Z., Bruns, R. F., and Daly, J. W., Adenosine receptors and behavioral actions of methylxanthines, *Proc. Natl. Acad. Sci. U.S.A.*, 78, 3260, 1981.

76. Nehlig, A., Daval, J. L., Pereira De Vasconcelos, A., and Boyet, S., Caffeine-diazepam interaction and local glucose utilization in the conscious rat, *Brain Res.*, 419, 272, 1987.

77. Pole, P., Bonetti, E. P., Pieri, L., Cumin, R., Angioi, R. M., Mohler, H., and Haefely, W. E., Caffeine antagonizes several central effects of Diazepam, *Life Sci.*, 28, 2265, 1981.
78. Nehlig, A., Daval, J. L., and Debry, G., Caffeine and the central nervous system: Mechanism of action, biochemical, metabolic, and psychostimulant effects, *Brain Res. Rev.*, 17, 139, 1992.
79. Gupta, U., Effects of caffeine on recognition, *Pharmacol. Biochem. Behav.*, 44, 393, 1993.

chapter three

Cerebral energy metabolism and blood flow: useful tools for the understanding of the behavioral effects of caffeine

Astrid Nehlig

Contents

Abstract...32
I. Introduction 32
II. Caffeine and cerebral energy metabolism 33
   A. Effects of caffeine on the nigrostriatal dopaminergic system
      and on locomotor activity 33
   B. Effects of caffeine on the noradrenergic and serotoninergic
      cell groupings and on the sleep–wake cycle...........................35
   C. Effects of caffeine on the mesolimbic dopaminergic system
      and psychostimulant effects 35
   D. Effects of chronic caffeine exposure on cerebral
      energy metabolism 37
III. Caffeine and cerebral blood flow...38
   A. General effects of caffeine on cerebral blood flow..............38
   B. Vasoconstrictive effects of caffeine, headache and
      caffeine withdrawal..41
IV. Coupling between cerebral blood flow and metabolism....................41
V. Caffeine, cerebral blood flow and pathology.............................42
VI. Conclusion..42
Acknowledgments...43
References...43

Key words: Caffeine. Methylxanthines. Cerebral blood flow. Cerebral energy metabolism. Behavior.

Abstract

The effects of caffeine on cerebral energy metabolism and blood flow, although not studied in great detail, represent a powerful approach to the understanding of the behavioral effects of pharmacological agents. In adult animals, an acute high dose of caffeine (10 mg/kg) induces a generalized increase in the rates of cerebral glucose utilization. Low doses of caffeine representative of the daily human consumption (1 to 2.5 mg/kg) specifically increase rates of glucose utilization in the caudate nucleus, the raphe nuclei, and the locus coeruleus, structures regulating locomotor activity and the sleep–wake cycle, respectively. Conversely, the shell of the nucleus accumbens belonging to the brain circuitry of addiction and reward is only activated at high doses of caffeine (10 mg/kg) and in a nonspecific manner, rendering the addictive potential of caffeine quite unlikely.

Caffeine and theophylline increase cerebrovascular resistance and decrease cerebral blood flow. The vasoconstrictive properties of caffeine have been shown in both humans and animals, and have often been used for the treatment of painful cerebrovasodilation associated with vascular headaches. Caffeine decreases cerebral blood flow mostly in regions where energy metabolism is simultaneously markedly increased. Thus, caffeine resets the level of coupling between cerebral blood flow and metabolism by inducing a relative hypoperfusion at a given metabolic level. Thus, methylxanthines are able to change the regulatory mechanism that couples cerebral blood flow and metabolism levels. This effect may be mediated by adenosine which represents one of the main factors involved in the dynamic coupling between cerebral blood flow and metabolism, i.e., between delivery and use of oxygen and metabolic substrates.

I. Introduction

Caffeine is the most widely used psychoactive substance in the world. It has well-known stimulant effects on the central nervous system that are reflected by increased alertness, wakefulness, and motor activity as well as stimulated neuronal activity (for a review, see References 1 and 2). Many of the effects of caffeine and other methylxanthines on the central nervous system can be related to their antagonism at the level of adenosine receptors.[1,3,4] Therefore, the effects of methylxanthines could be of particular importance with respect to the key role of adenosine in coupling cerebral blood flow to energy metabolism.[5,6] Indeed, cerebral energy metabolism (i.e., oxygen and glucose utilization) and blood flow are closely interrelated[7-9] such that changes in the activity of the brain lead to parallel changes in cerebral energy metabolism and blood flow.[7,10,11]

In the present chapter, we will review the effects of caffeine on cerebral blood flow and glucose utilization and try to correlate these data with the well-known effects of the methylxanthines on behavior. Moreover, we will discuss the effects of caffeine and theophylline on cerebral blood flow in pathological conditions.

II. Caffeine and cerebral energy metabolism

The effects of methylxanthines on cerebral metabolic rates for glucose (LCM-Rglcs) described in the present review have been explored by means of the quantitative autoradiographic [^{14}C]2-deoxyglucose method of Sokoloff et al.,[12] which allows the simultaneous visualization of functional activity in discrete areas of the brain of conscious animals. This technique permits the identification of neuronal pathways affected by a pharmacological agent and is very useful for relating behavioral effects to the central action of a drug.

A. Effects of caffeine on the nigrostriatal dopaminergic system and on locomotor activity

The nigrostriatal dopaminergic system originates in neurons located in the substantia nigra, mainly in the pars compacta and to a lesser extent in the pars reticulata. These neurons project via the median forebrain bundle and the lateral hypothalamus to the globus pallidus and terminate in the caudate nucleus.[13,14] This system is involved in the control of locomotion.

The stimulant effects of caffeine on LCMRglcs in the structures mediating locomotor activity have been shown previously at quite high doses of caffeine, i.e., after the acute injection of 10 mg/kg caffeine or the continuous perfusion of the methylxanthine at a rate of 0.30 mg/kg/min. These studies all reported increases over control values in the rates of energy metabolism in the dopaminergic substantia nigra, both the pars reticulata and compacta and in structures of the extrapyramidal motor system (caudate nucleus, globus pallidus, sensorimotor and cerebellar cortex) as well as in numerous thalamic motor and sensory relay nuclei.[15-19]

Our more recent data show that the caudate nucleus is very sensitive to the effects of caffeine since LCMRglc is already significantly activated in this structure after the administration of the lowest dose of caffeine, 1 mg/kg to adult male rats (Figure 3.1). The functional activity of this nucleus is further increased at 2.5 mg/kg, about 40% over control levels, and remains activated at the two higher doses of the methylxanthine. In the two other structures of the nigrostriatal dopaminergic system (the substantia nigra pars compacta and the globus pallidus), LCMRglc increases after the injection of 2.5 to 10 mg/kg of caffeine, while the sensorimotor cortex shows a significant increase in LCMRglcs only after 5 mg/kg of caffeine. The high sensitivity of the caudate nucleus to caffeine is confirmed by the fact that a direct ionto-phoretic administration of the methylxanthine can also modify the spontaneous electrical activity of neurons in the caudate nucleus of the rat. This

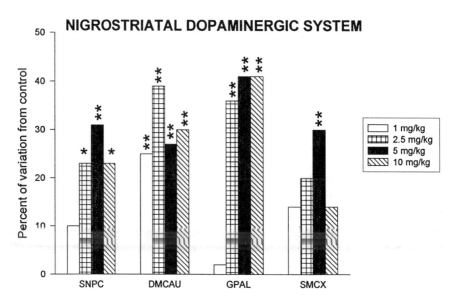

Figure 3.1 Effects of the acute administration of increasing doses of caffeine on LCMRglcs in selected regions of the nigrostriatal dopaminergic system. Data are presented as percent of variation from control values. The animals received an acute i.v. injection of 1 to 10 mg/kg caffeine at 15 min before the onset of the [14C]2-deoxyglucose procedure. For experimental details, see Nehlig et al.[17] SNPC: substantia nigra pars compacta; DMCAU: dorsomedial caudate nucleus; GPAL: globus pallidus; SMCX: sensorimotor cortex. * p< 0.05; ** p< 0.01, statistically significant differences from controls (Dunnett's t-test for multiple comparisons).

activation is further confirmed by the caffeine-induced dopamine release in the caudate nucleus recently shown by microdialysis.[20,21] Thus, the methylxanthine is able to activate the nigrostriatal pathway by stimulating dopamine release from the nigrostriatal nerve endings.[22]

There is a good correlation between caffeine-induced functional activation of the structures belonging to the nigrostriatal pathway and the well-known stimulant effects of the methylxanthine on locomotor activity. Indeed, the activation of LCMRglcs occurring in the structures belonging to the nigrostriatal pathway at low doses of caffeine (Figure 3.1), as well as the increase in functional activity in most areas of the extrapyramidal motor system recorded at high doses of the methylxanthine,[15-19] reflect the stimulant effects of caffeine on locomotor activity and general behavior (for a review, see References 1 and 2). This effect is dose-dependent, and the minimal dose of caffeine necessary to affect locomotion is 1.5 mg/kg, which correlates well with the increase in LCMRglc in the caudate nucleus after 1 mg/kg of the methylxanthine. The stimulant effect of caffeine on locomotion has been shown to increase with doses ranging from 10 to 20 mg/kg and decrease with doses higher than 40 mg/kg (for a review, see References 1 and 2). The effect of caffeine on the nigrostriatal dopaminergic pathway has also been shown indirectly since 10 to 50 mg/kg of the methylxanthine are able to antagonize the akinesia induced by catecholamine depletion in mice.[23]

B. Effects of caffeine on the noradrenergic and serotoninergic cell groupings and on the sleep–wake cycle

Previous studies using high doses of caffeine, i.e., the acute injection of 10 mg/kg caffeine or the continuous perfusion of methylxanthine at a rate of 0.30 mg/kg/min reported increases in energy metabolism in structures known to control the sleep–wake cycle, such as the mesencephalic reticular formation, locus coeruleus, and raphe nuclei.[1,15-19] Our present data show that the serotoninergic cell groupings, the medial and dorsal raphe nuclei, as well as the noradrenergic cell grouping, the locus coeruleus, are very sensitive to caffeine. In these three structures, LCMRglcs increase at all doses of caffeine used (1 to 10 mg/kg) and are already activated at the lowest one, 1 mg/kg (Figure 3.2). These data correlate well with the known sensitivity of sleep functions to the absorption of caffeine.[24] In the cat, 10 mg/kg of caffeine produce an activation of the cortical EEG similar to the activity induced by the direct stimulation of the reticular formation, a structure which plays a central role in vigilance and awakening.[25] In the rat, caffeine stimulates spontaneous electrical activity in neurons of the reticular formation, and this effect appears at low doses, 1 to 2.5 mg/kg i.v. This response is dose-dependent, and the duration of activation increases with the dose.[26,27] Simultaneous to the activation of the cortex, caffeine lowers electrical activity in the thalamus, also at very low doses, 0.1 to 0.5 mg/kg i.v. The medial thalamus appears, therefore, to be an important site for the arousal induced by caffeine.[28,29] Moreover, caffeine can simultaneously activate the reticular formation and inhibit the medial thalamus, which means that the methylxanthine excites all levels of brain activity even by reinforcing some inhibitory influences. Finally, caffeine reduces serotonin availability at postsynaptic receptor sites,[30] which elicits a reduction in the sedative effect of the amine on activity and has repercussions on sleep mechanisms and motor function.[31,32]

C. Effects of caffeine on the mesolimbic dopaminergic system and psychostimulant effects

The possible dependence on caffeine has been considered by several groups for over a decade.[33-37] The mesolimbic dopaminergic system plays a critical role in drug dependence.[38] The mesolimbic dopaminergic system originates in the ventral tegmental area, projects to the nucleus accumbens, and terminates in the medial prefrontal cortex. The nucleus accumbens that plays a central role in the mechanism of drug dependence is functionally and morphologically divided into a core and a shell part. The medioventral shell part is related to the limbic "extended amygdala," assumed to play a role in emotional, motivational, and reward functions, whereas the laterodorsal core

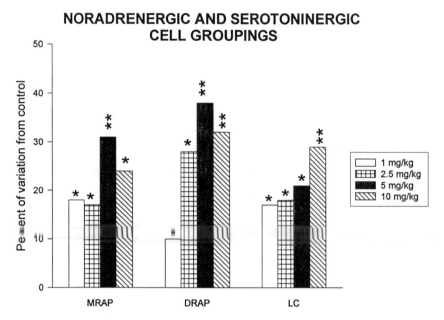

Figure 3.2 Effects of the acute administration of increasing doses of caffeine on LCMRglcs in the noradrenergic and serotoninergic cell groupings. Data are presented as percent of variation from control values. The animals received an acute i.v. injection of 1 to 10 mg/kg caffeine at 15 min before the onset of the [¹⁴C]2-deoxyglucose procedure. For experimental details, see Nehlig et al.[17] MRAP: medial raphe, DRAP: dorsal raphe, LC: locus coeruleus. * p< 0.05, ** p< 0.01, statistically significant differences from controls (Dunnett's t-test for multiple comparisons).

part regulates somatomotor functions.[39] The specificity of cocaine, amphetamine, morphine, alcohol, Δ⁹tetrahydrocannabinol, and also nicotine is to selectively activate the dopaminergic neurotransmission in the shell of the nucleus accumbens,[40-42] a property that has been related to the strong addictive properties of these drugs.[38,43] Conversely, caffeine at doses ranging from 0.5 to 5 mg/kg that reflect the level of human consumption does not trigger any release of dopamine in the shell of the nucleus accumbens.[44]

Previous studies using high doses of caffeine, i.e., the acute injection of 10 mg/kg caffeine or the continuous perfusion of methylxanthine at a rate of 0.30 mg/kg/min reported increases in energy metabolism in the ventral tegmental area, the nucleus accumbens, and the medial prefrontal cortex.[1,15-19] However, none of them discriminated between the shell and the core part of the nucleus accumbens. Our recent studies show that the increase in LCMRglcs recorded in the structures of the mesolimbic dopaminergic system are of lower amplitude (Figure 3.3) than those recorded in the other brain regions studied (Figures 3.1 and 3.2). Moreover, the significant activation of functional activity appears only at quite high doses, 5 mg/kg for the area of origin, the ventral tegmental area, and 10 mg/kg for the two subdivisions

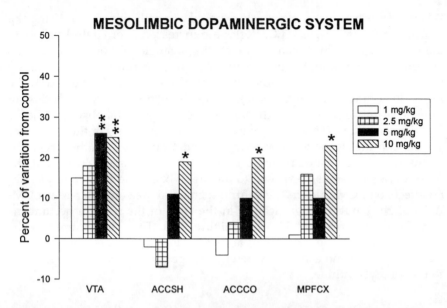

Figure 3.3   Effects of the acute administration of increasing doses of caffeine on LCMRglcs in selected regions of the mesolimbic dopaminergic system. Data are presented as percent of variation from control values. The animals received an acute i.v. injection of 1 to 10 mg/kg caffeine at 15 min before the onset of the [^{14}C]2-deoxyglucose procedure. For experimental details, see Nehlig et al.[17] VTA: ventral tegmental area, ACCSH, nucleus accumbens, shell, ACCCO: nucleus accumbens, core, MPFCX: medial prefrontal cortex. * $p< 0.05$, ** $p< 0.01$, statistically significant differences from controls (Dunnett's t-test for multiple comparisons).

of the nucleus accumbens and the medial prefrontal cortex (Figure 3.3). These data show that at the doses consumed daily by most people (2 to 2.5 mg/kg), caffeine does not activate the brain circuitry of dependence and reward. Moreover, the activation of functional activity in the shell of the nucleus accumbens occurs only at high doses of caffeine (10 mg/kg, i.e., about 4 to 5 times the average daily human consumption) at which the methylxanthine also activates the core part of the nucleus and induces widespread nonspecific metabolic increases in a majority of brain regions.[1,15-19] These widespread effects of high doses of caffeine on brain functional activity are also likely to reflect the numerous side and adverse effects of the ingestion of the methylxanthine. Conversely, the effects of amphetamine, cocaine, and nicotine on the neural substrates underlying addiction are rather specific and occur at doses that do not usually lead to the activation of many other brain regions.[41,45,46]

D. Effects of chronic caffeine exposure on cerebral energy metabolism

An acute administration of 10 mg/kg of caffeine increases LCMRglcs in numerous brain regions,[18] mainly those presented in Figures 3.4 and 3.5.

When the rats have been previously exposed to a daily injection of 10 mg/kg of caffeine for two weeks, LCMRglcs are increased over control values to the same extent as in the animals receiving a single injection of 10 mg/kg of caffeine in the caudate nucleus involved in the control of motor activity (Figures 3.4a and 3.5), the medial and dorsal raphe nuclei involved in the regulation of the sleep (Figure 3.4c). In those structures every single dose of caffeine seems to be able to activate rates of energy metabolism and hence functional activity. These data confirm the high sensitivity of the locomotor system and the sleep–wake cycle to caffeine. Conversely, in the structures of the mesolimbic dopaminergic system, the acute injection of caffeine given to rats previously exposed to the methylxanthine for two weeks has no effects on LCMRglcs in the ventral tegmental area and increases LCM-Rglcs slightly but not significantly in the shell of the nucleus accumbens and the medial prefrontal cortex (Figure 3.4c). Thus, in contrast to the former structures, the circuitry of addiction and reward becomes less sensitive to caffeine and is not activated any longer in rats chronically exposed to the methylxanthine.

III. Caffeine and cerebral blood flow

The effects of methylxanthines on rates of local cerebral blood flow (LCBF) described in the present review have been explored by means of the quantitative autoradiographic [14C]iodoantipyrine method of Sakurada et al.,[47] which allows the simultaneous visualization of functional activity in discrete areas of the brain of conscious animals. The [14C]iodoantipyrine technique as the [14C]2-deoxyglucose method permits the identification of neuronal pathways affected by a pharmacological agent.

A. General effects of caffeine on cerebral blood flow

Methylxanthines such as caffeine or theophylline induce vasodilation, except in the central nervous system where they raise cerebrovascular resistance; this actually contributes to a reduction in cerebral blood flow. The cerebral vasoconstrictive properties of methylxanthines have been demonstrated in humans[48-51] and animals.[15,16,52,53] In rats, as shown in Figure 3.5, caffeine induces a decrease in the rates of LCBF, mainly in the areas where it increases metabolism, i.e., in monoaminergic cell groupings, motor and limbic systems, and in the thalamus.[15,16,52] In humans, cerebral blood flow is decreased by 20 to 30% after an intravenous dose of 250 mg of caffeine with no interregional differences.[48,49,51,54] The value of cerebral blood flow before caffeine exposure strongly affects the extent of decrease in cerebral blood flow induced by the methylxanthine, suggesting that some regulatory mechanism may prevent decreases below some minimum tolerable level, which could be 30 ml/100 g/min.[48] The decrease of cerebral blood flow is often accompanied by a significant increase in subjective anxiety ratings in response to caffeine.[48,52]

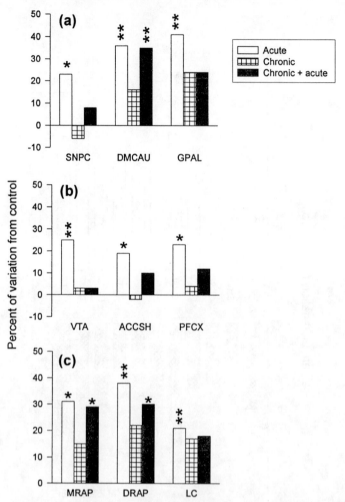

Figure 3.4 Effects of the chronic and acute administration of 10 mg/kg of caffeine on LCMRglcs in selected regions of the nigrostriatal dopaminergic system (a), the mesolimbic dopaminergic system (b), and the noradrenergic and serotoninergic cell groupings (c). Data are presented as percent of variation from control values. The animals received either a two-week daily saline treatment followed by an acute i.v. injection of 10 mg/kg caffeine on the day of the experiment (acute group) or a two-week daily treatment with 10 mg/kg of caffeine followed on the day of the experiment either by an acute saline i.v. injection (chronic group) or an acute 10 mg/kg caffeine i.v. injection (chronic + acute group). Control animals received a daily saline injection for two weeks, followed by an acute saline injection on the day of the experiment. The acute injection was performed at 5 to 6 h after the last chronic injection. The [^{14}C]2-deoxyglucose procedure was initiated at 15 min after the acute saline or caffeine injection. For experimental details, see Nehlig et al.[18] SNPC: substantia nigra pars compacta, DMCAU: dorsomedial caudate nucleus, GPAL: globus pallidus, VTA: ventral tegmental area, ACCSH, nucleus accumbens, shell, PFCX: medial prefrontal cortex, MRAP: medial raphe, DRAP: dorsal raphe, LC: locus coeruleus.* $p < 0.05$, ** $p < 0.01$, statistically significant differences from controls (Dunnett's t-test for multiple comparisons).

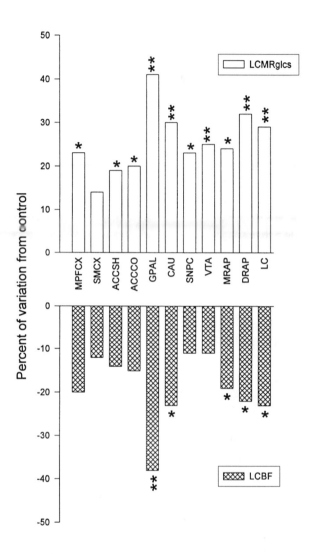

Figure 3.5 Effects of the acute administration of 10 mg/kg of caffeine on LCMRglcs and rates of local cerebral blood flow (LCBF) in selected regions of the rat brain. Data are presented as percent of variation from control values. The animals received an acute i.v. injection of 10 mg/kg caffeine at 15 and 35 min before the onset of the [^{14}C]2-deoxyglucose or [^{14}C]iodoantipyrine procedure, respectively. For experimental details, see Nehlig et al.[18,52] MPFCX: medial prefrontal cortex, SMCX: sensorimotor cortex, ACCSH, nucleus accumbens, shell, ACCCO, nucleus accumbens, core, GPAL: globus pallidus, CAU: caudate nucleus, SNPC: substantia nigra pars compacta, VTA: ventral tegmental area, MRAP: medial raphe, DRAP: dorsal raphe, LC: locus coeruleus. * $p < 0.05$, ** $p < 0.01$, statistically significant differences from controls (Dunnett's t-test for multiple comparisons).

B. Vasoconstrictive effects of caffeine, headache and caffeine withdrawal

The cerebral vasoconstrictive properties of caffeine have been used regularly for the treatment of painful cerebral vasodilations associated with vascular headaches. Indeed, several efficient anti-headache medications contain caffeine. The methylxanthine has also been shown to be able to relieve pain significantly and in a dose-dependent way. This anti-headache effect of caffeine is similar to that of acetaminophen (which is frequently associated with caffeine in analgesic preparations) and is also independent on the effects of caffeine on mood or on previous exposure to caffeine.[55]

One of the most commonly reported symptoms of caffeine withdrawal is headache that is easily relieved by caffeine consumption (for a review, see References 1 and 2). Weekend attacks in migraine patients appear to be linked to caffeine withdrawal.[56] There is a strong positive correlation between caffeine consumption, fasting, and headaches before and after surgical procedures. For every increase in the usual daily consumption of 100 mg of caffeine (about a cup of coffee), the risk of headache immediately before and after surgery is increased by 12 and 16%, respectively, and also correlates with the duration of fasting.[57,58] The risk of headaches is reduced in individuals who drink caffeine or get substitutive caffeine tablets on the day of the surgery.[59-61] Therefore, it was advised by three studies that the numerous healthy patients who drink caffeine-containing beverages daily and are undergoing minor surgical procedures should be permitted to ingest preoperative caffeine.[58,60,61] Moreover, there is a relationship between caffeine withdrawal, the development of headaches, and changes in cerebral blood flow. The cerebral blood flow velocities are increased during withdrawal headaches, mainly in frontal regions, but only in high caffeine consumers (more than 6 cups daily for at least 3 years). In low caffeine consumers (less than 3 cups daily for a minimum of 3 years), there are no differences between the resting, post-placebo, and post-caffeine cerebral blood flow levels.[54] The administration of caffeine to subjects with withdrawal symptoms leads to a significant decrease in cerebral blood flow levels within 30 min after intake and a return to baseline values after 2 h.[56] This evolution is similar in the high and low caffeine-consumer groups.[54] The significant reduction of cerebral blood flow after the acute administration of caffeine in both low and heavy caffeine consumers argues against the development of tolerance of cerebral blood flow to caffeine.

IV. Coupling between cerebral blood flow and metabolism

In most situations, cerebral blood flow and glucose utilization are closely coupled in all cerebral regions, so that modifications in cerebral activity elicit parallel changes in cerebral glucose utilization and blood flow.[5,8,9,62] In general, changes in cerebral blood flow are the consequence of variations in cerebral energy metabolism.[7,10,11] Contrary to the majority of pharmacological agents to which man is frequently exposed, and as shown in Figure 3.5, caffeine has the property to induce cerebral hypoperfusion accompanied by a simultaneous increase in glucose utilization[15-19,52]; in other words, caffeine resets the level of coupling between cerebral blood flow and energy metabolism. Methylxanthines thus seem to modify the regulating mechanism between cerebral blood flow and cerebral metabolism. Although this mechanism is not yet fully understood, adenosine, with which methylxanthines compete, is known to be one of the modulators of the regulation of the relationship of blood flow to metabolism in the central nervous system.[5,6,62] Indeed, theobromine, a weaker adenosine antagonist than caffeine or theophylline, has only minor effects on cerebral blood flow and metabolism, while propentophylline, an adenosine uptake blocker, induces the reverse effect, i.e., an increase in cerebral blood flow and a decrease in glucose utilization. Thus, several xanthine derivatives can reset the relationship between metabolism and blood flow in the brain.[66]

V. Caffeine, cerebral blood flow and pathology

In adult rats, rabbits, and dogs, theophylline has been shown to significantly attenuate, reduce the duration, or even block the increase in cerebral blood flow recorded during moderate and severe hypoxia,[64-67] whereas dipyridamole and papaverine, inhibitors of adenosine uptake, have the opposite effects.[68] Conversely, theophylline produces no significant effect on hypercapnia-induced increase in cerebral blood flow.[64,69] These results confirm the specificity of action of theophylline and other methylxanthines on hypoxia-induced adenosine release.[64-69] Likewise, theophylline significantly reduces the hyperemia observed during seizures in adult animals and enhances brain damage. These data strongly suggest that adenosine is partly responsible for the increase in cerebral blood flow recorded during seizures and that it has neuroprotective effects.[70]

Finally, the acute administration of caffeine in rats accelerates ischemic damage consecutive to stroke, while chronic administration of the methylxanthine protects the brain against ischemic damage, probably through the increase of the number of adenosine receptors.[71] In man, chronic consumption of caffeine is inversely related to the risk of fatal and nonfatal stroke.[72] The advice is that one should drink enough coffee to increase the number of central adenosine receptors, but also be able to stop the intake of drinks containing the methylxanthine when a stroke happens, in order to prevent caffeine from antagonizing the beneficial effects of adenosine at the cerebral level.[73]

VI. Conclusion

In conclusion, it appears that, in adult humans and animals, methylxanthines increase cerebral energy metabolism and decrease cerebral blood flow. These substances are thus able to reset the level of coupling between cerebral blood flow and energy metabolism, inducing a relative hypoperfusion of the brain at a constant metabolic level, which may be deleterious in some pathological situations. Moreover, these data stress the important regulatory role of adenosine in the mechanism of coupling between cerebral blood flow and energy metabolism. They also show the high sensitivity of the areas controlling locomotor activity and the sleep–wake cycle to low concentrations of caffeine, while the mesolimbic dopaminergic system involved in addiction and reward appears to be activated only at high doses of the methylxanthine. Since these doses simultaneously activate a large number of brain regions, they are also likely to induce the side and adverse effects well described after the ingestion of high doses of caffeine in humans,[74,75] rendering the risk of addiction to caffeine quite unlikely.

Acknowledgments

This work was supported by grants from the Institut National de la Santé et de la Recherche Médicale and the Institute for Scientific Information on Coffee (I.S.I.C), Paris, France.

References

1. Nehlig, A., Daval, J. L., and Debry, G., Caffeine and the central nervous system: mechanisms of action, biochemical, metabolic and psychostimulant effects, *Brain Res. Rev.*, 17, 139, 1992.
2. Nehlig, A. and Debry, G., Effects of coffee on the central nervous system, in *Coffee and Health*, Debry, G., Ed., John Libbey, Paris, 1994, 157.
3. Fredholm, B. B., On the mechanism of action of caffeine and theophylline, *Acta Med. Scand.*, 217, 149, 1985.
4. Daly, J. W., Mechanism of action of caffeine, in *Coffee, Caffeine and Health*, Garattini, S., Ed., Raven Press, New York, 1993, 97.
5. Kuschinsky, W., Coupling of function, metabolism and blood flow in the brain, *News Physiol. Sci.*, 2, 217, 1987.
6. Winn, H. R., Rubio, G. R., and Berne, R. M., The role of adenosine in the regulation of cerebral blood flow, *J. Cereb. Blood Flow Metab.*, 1, 239, 1981.
7. McCulloch, J., Kelly, P. A. T., and Ford, I., Effect of apomorphine on the relationship between local cerebral glucose utilization and local cerebral blood flow (with an appendix on its statistical analysis), *J. Cereb. Blood Flow Metab.*, 2, 487, 1982.
8. Raichle, M. E., Grubb, R. L., Gado, M. H., Eichling, J. O., and Ter-Pogossian, M. M., *In vivo* correlations between regional cerebral blood flow and oxygen utilization in man, *Acta Neurol. Scand.*, 56, 240, 1977.
9. Sokoloff, L., Relationship among local functional activity, energy metabolism and blood flow in the central nervous system, *Fed. Proc.*, 40, 2311, 1981.
10. Grome, J. J. and Harper, A. M., The effects of quipazone, a putative serotonin agonist on local cerebral blood flow and glucose utilization in the rat, and pial vascular diameter in the cat, *J. Cereb. Blood Flow Metab.*, 3, Suppl. 1, S302, 1983.

11. Kelly, P. A. T. and McCulloch, J., The effect of GABAergic agonist muscimol upon the relationship between cerebral blood flow and glucose utilization, *Brain Res.*, 258, 338, 1983.
12. Sokoloff, L., Reivich, M., Kennedy, C., Des Rosiers, M. H., Patlak, C. S., Pettigrew, K. D., Sakurada, O., and Shinohara, M., The [^{14}C]2-deoxyglucose method for the measurement of local cerebral glucose utilization: theory, procedure and normal values in conscious and anesthetized albino rats, *J. Neurochem.*, 28, 897, 1977.
13. Arluison, M., Agid, Y., and Javoy, F., Dopaminergic nerve endings in the neostriatum of the rat. I. Identification by intracerebral injections of 6-hydroxydopamine, *Neuroscience*, 3, 657, 1978.
14. Ungerstedt, U., Stereotaxic mapping of the monoamine pathways in the rat brain, *Acta Physiol. Scand.*, Suppl. 367, 69, 1971.
15. Grome, J. J. and Stefanovich, V., Differential effects of xanthine derivatives on local cerebral blood flow and glucose utilization in the conscious rat, in *Adenosine. Receptors and Modulation of Cell Function, Stefanovich, V., Rudolphi, K.,* and Schubert, P., Eds., IRL Press Ltd., Oxford, 1985, 453.
16. Grome, J. J. and Stefanovich, V., Differential effects of methylxanthines on local cerebral blood flow and glucose utilization in the conscious rat, *Naunyn–Schmiedeberg's Arch. Pharmacol.*, 333, 172, 1987.
17. Nehlig, A., Lucignani, G., Kadekaro, M., Porrino, L. J., and Sokoloff, L., Effects of acute administration of caffeine on local cerebral glucose utilization in the rat, *Eur. J. Pharmacol.*, 101, 91, 1984.
18. Nehlig, A., Daval, J. L., Boyet, S., and Vert, P., Comparative effects of acute and chronic administration of caffeine on local cerebral glucose utilization in the conscious rat, *Eur. J. Pharmacol.*, 129, 93, 1986.
19. Nehlig, A., Pereira de Vasconcelos, A., Collignon, A., and Boyet, S., Comparative effects of caffeine and L-phenylisopropyladenosine on local cerebral glucose utilization in the rat, *Eur. J. Pharmacol.*, 157, 1, 1988.
20. Okada, M., Mizuno, K., and Kaneko, S., Adenosine A1 and A2 receptors modulate extracellular dopamine levels in rat striatum, *Neurosci. Lett.*, 212, 53, 1996.
21. Okada, M., Kiryu, K., Kawata, Y., Mizuno, K., Wada, K., Tasaki, H., and Kaneko, S., Determination of the effects of caffeine and carbamazepine on striatal dopamine release by *in vivo* microdialysis, *Eur. J. Pharmacol.*, 321, 181, 1997.
22. Hirsh, K., Forde, J., and Chou, D. T., Effects of caffeine and amphetamine SO$_4$ on single unit activity in the caudate nucleus (abstract), *Soc. Neurosci. Abstr.*, 8, 898, 1982.
23. Popoli, P., Caporali, M. G., and Scotti de Carolis, A., Akinesia due to catecholamine depletion in mice is prevented by caffeine: Further evidence for an involvement of adenosinergic system in the control of motility, *J. Pharm. Pharmacol.*, 43, 280, 1991.
24. Snel, J., Coffee and caffeine: Sleep and wakefulness, in *Coffee, Caffeine and Health*, Garattini S., Ed., Raven Press, New York, 1993, 255.
25. Jouvet, M., Benoit, O., Marsallon, A., and Courjon, J., Action de la caféine sur l'activité électrique cérébrale, *C. R. Soc. Biol.*, 151, 1542, 1957.
26. Forde, J. H. and Hirsh, R. K., Caffeine effects on reticular formation neurons in the decrebrate cat (abstract), *Soc. Neurosci. Abstr.*, 2, 867, 1976.
27. Hirsh, K., Forde, J., and Pinzone, M., Caffeine effects on spontaneous activity of reticular formation neurons (abstract), *Soc. Neurosci. Abstr.*, 257, 1974.

28. Chou, D. T., Forde, J. H., and Hirsh, K. R., Unit activity in medial thalamus: comparative effects of caffeine and amphetamine, *J. Pharmacol. Exp. Therap.*, 213, 580, 1980.

29. Foote, W. E., Holmes, P., Prichard, A., Hatcher, C., and Mordes, J., Neurophysiological and pharmacodynamic studies on caffeine and on interactions between caffeine and nicotinic acid in the rat, *Neuropharmacology,* 17, 7, 1978.

30. Hirsh, K., Central nervous system pharmacology of the dietary methylxanthines, in *The Methylxanthines Beverages and Food: Chemistry, Consumption and Health Effects,* Spiller, G. A., Ed., Alan Liss, New York, 1984, 235.

31. Gerson, S. C. and Baldessarini, R. J., Motor effects of serotonin in the central nervous system, *Life Sci.*, 27, 1435, 1980.

32. Jouvet, M., Biogenic amines and the states of sleep, *Science*, 163, 32, 1969.

33. Griffiths, R. R. and Woodson, P. P., Caffeine physical dependence: a review of human and laboratory animal studies, *Psychopharmacology,* 94, 437, 1988.

34. Griffiths, R. R. and Mumford, G. K., Caffeine reinforcement, discrimination, tolerance and physical dependence in laboratory animals and humans, in *Handbook of Experimental Pharmacology,* Vol. 118, Schuster, C. R. and Kuhar, M. J., Eds., Springer Verlag, Heidelberg, 1996, 315.

35. Heishman, S. J. and Henningfield, J. E., Stimulus functions of caffeine in humans: Relation to dependence potential, *Neurosci. Biobehav. Rev.*, 16, 273, 1992.

36. Holtzman, S. G., Caffeine as a model drug of abuse, *Trends Pharmacol. Sci.*, 11, 355, 1990.

37. Strain, E. C., Mumford, G. K., Silverman, K., and Griffiths, R. R., Caffeine dependence syndrome. Evidence from case histories and experimental evaluations, *JAMA*, 272, 1043, 1994.

38. Self, D. W. and Nestler, E. J., Molecular mechanisms of drug reinforcement and addiction, *Annu. Rev. Neurosci.*, 18, 463, 1995.

39. Heimer, L., Zahm, D. S., Churchill, L., Kalivas, P. W., and Wohltmann, C., Specificity in the projection patterns of accumbal core and shell in the rat, *Neuroscience*, 41, 89, 1991.

40. Pontieri, F. E., Tanda, G., and DiChiara, G., Intravenous cocaine, morphine, and amphetamine preferentially increase extracellular dopamine in the "shell" as compared with the "core" of the rat nucleus accumbens, *Proc. Natl. Acad. Sci. U.S.A.*, 92, 12304, 1995.

41. Pontieri, F. E., Tanda, G., Orzi, F., and DiChiara, G., Effects of nicotine on the nucleus accumbens and similarity to those of addictive drugs, *Nature*, 382, 255, 1996.

42. Tanda, G., Pontieri, F. E., and DiChiara, G., Cannabinnoid and heroin activation of mesolimbic dopamine transmission by a common μ_1 opioid receptor mechanism, *Science*, 276, 2048, 1997.

43. Koob, G. F., Drugs of abuse: anatomy, pharmacology and function of reward pathways, *Trends Pharmacol. Sci.*, 13, 177, 1992.

44. Tanda, G., Loddo, P., Frau, R., Acquas E., and DiChiara, G., Effect of intravenous caffeine on limbic and cortical dopamine transmission in the rat: a microdialysis study, Proceedings of the *6th International Symposium on Adenosine and Adenine Nucleotides,* Ferrara, Italy, May 19-24, 1998.

45. Porrino, L. J., Domer, F. R., Crane, A. M., and Sokoloff, L., Selective alterations in cerebral metabolism within the mesocorticolimbic dopaminergic system produced by acute cocaine administration in rats, *Neuropsychopharmacology,* 1, 109, 1988.

46. Stein, E. A. and Fuller, S. A., Selective effects of cocaine on regional cerebral blood flow in the rat, *J. Pharmacol. Exp. Therap.*, 262, 327, 1992.

47. Sakurada, O., Kennedy, C., Jehle, J., Brown, J.D., Carbin, G.L., and Sokoloff, L., Measurement of local cerebral blood flow with [^{14}C]iodoantipyrine, *Am. J. Physiol.*, 234, H59, 1978.

48. Cameron, O. G., Modell, J. G., and Hariharan, M., Caffeine and human cerebral blood flow: a positron emission tomography study, *Life Sci.*, 47, 1141, 1990.

49. Mathew, R. J., Barr, D. L., and Weinman, M. L., Caffeine and cerebral blood flow, *Br. J. Psychiat.*, 143, 604, 1983.

50. Mathew, R. J. and Wilson, W. H., Caffeine induced changes in cerebral circulation, *Stroke*, 16, 814, 1987.

51. Mathew, R. J. and Wilson, W. H., Behavioral and cerebrovascular effects of caffeine in patients with anxiety disorders, *Acta Psychiatr. Scand.*, 82, 17, 1990.

52. Nehlig, A., Pereira de Vasconcelos, A., Dumont, I., and Boyet, S., Effects of caffeine, L-phenylisopropyladenosine and their combination on local cerebral blood flow in the rat, *Eur. J. Pharmacol.*, 179, 271, 1990.

53. Puiroud, S., Pinard, E., and Seylaz, J., Dynamic cerebral and systemic circulatory effects of adenosine, theophylline and dipyridamole, *Brain Res.*, 453, 287, 1988.

54. Mathew, R. J. and Wilson, W. H., Caffeine consumption, withdrawal and cerebral blood flow, *Headache*, 25, 305, 1985.

55. Ward, N., Whitney, C., Avery, D., and Dunner, D., The analgesic effect of caffeine in headache, *Pain*, 44, 151, 1991.

56. Couturier, E. G. M., Hering, R., and Steiner, T. J., Weekend attacks in migraine patients: caused by caffeine withdrawal?, *Cephalalgia*, 12, 99, 1992.

57. Fennelly, M., Galletly, D. C., and Purdie, G. I., Is caffeine withdrawal the mechanism of postoperative headache?, *Anesth. Analg.*, 72, 449, 1991.

58. Nikolajsen, L., Larsen, K. M., and Kierkegaard, O., Effect of previous frequency of headache, duration of fasting and caffeine abstinence on perioperative headache, *Br. J. Anaesth.*, 72, 295, 1994.

59. Hampl, K. F., Schneider, M. C., Ruttimann, U., Ummenhofer, U., and Drewe, J., Perioperative administration of caffeine tablets for prevention of postoperative headaches, *Can. J. Anaesth.*, 42, 789, 1995.

60. Weber, J. G., Ereth, M. H., and Danielson, D. R., Perioperative ingestion of caffeine and postoperative headache, *Mayo Clin. Proc.*, 68, 842, 1993.

61. Weber, J. G., Klindworth, J. T., Arnold, J. J., Danielson, D. R., and Ereth, M. K., Prophylactic intravenous administration of caffeine and recovery after ambulatory surgical procedure, *Mayo Clin. Proc.*, 72, 621, 1997.

62. Kuschinsky, W., Regulation of cerebral blood flow: An overview, in *Neurophysiological Basis of Cerebral Blood Flow Control: An Introduction*, Mraovitch S. and Sercombe, R., Eds., John Libbey, London, 1996, 245.

63. Kuschinsky, W., Adenosine and cerebral blood flow, in *Role of Adenosine in Cerebral Metabolism and Blood Flow*, Stefanovich, V. and Okyayuz-Baklouti, I., Eds., VNU Science Press, Utrecht, 1987, 95.

64. Hoffman, W. E., Albrecht, R. F., and Miletich, D. J., The role of adenosine in CBF increases during hypoxia in young vs. aged rats, *Stroke*, 15, 124, 1984.

65. Morii, S., Ngai, A. C., Ko, K. R., and Winn, H. R., Role of adenosine in regulation of cerebral blood flow: effects of theophylline during normoxia and hypoxia, *Am. J. Physiol.*, 253, H165, 1987.

66. Pinard, E., Puiroud, S., and Seylaz, J., Role of adenosine in cerebral hypoxic hyperemia in the unanesthetized rabbit, *Brain Res.*, 481, 124, 1989.
67. Phillis, J. W., Preston, G., and DeLong, R. E., Effects of anoxia on cerebral blood flow in the rat brain: evidence for a role of adenosine in autoregulation, *J. Cereb. Blood Flow Metab.*, 4, 586, 1984.
68. Phillis, J. W., DeLong, R. E., and Towner, J. K., Adenosine deaminase inhibitors enhance cerebral anoxic hyperemia in the rat, *J. Cereb. Blood Flow Metab.*, 5, 295, 1985.
69. Emerson, T. E., Jr. and Raymond, R. M., Involvement of adenosine in cerebral hypoxic hyperemia in the dog, *Am. J. Physiol.*, 241, H134, 1981.
70. Pinard, E., Riche, D., Puiroud, S., and Seylaz, J., Theophylline reduces cerebral hyperemia and enhances brain damage induced by seizures, *Brain Res.*, 511, 303, 1990.
71. Sutherland, G. R., Peeling, J., Lesiuk, H. J., Brownstone, R. M., Rydzy, M., Saunders, J. K., and Geiger, J. D., The effects of caffeine on ischemic neuronal injury as determined by magnetic resonance imaging and histopathology, *Neuroscience*, 42, 171, 1991.
72. Grobbee, D. E., Rimm, E. B., Giovannucci, E., Colditz, G., Stampfer, M., and Willett, W., Coffee, caffeine, and cardiovascular disease in men, *N. Engl. J. Med.*, 323, 1026, 1990.
73. Longstreth, W. T. and Nelson, L. M., Caffeine and stroke (Letter). *Stroke*, 23, 117, 1992.
74. Curatolo, P. W. and Robertson, D., The health consequences of caffeine, *Ann. Int. Med.*, 98, 641, 1983.
75. James, J. E., *Caffeine and Health*, Academic Press, London, 1991.

chapter four

Caffeine effects on locomotor and reward behavior

Ernest N. Damianopoulos and Robert J. Carey

Contents

I. Introduction: conditioning, sensitization, and reward effects............49
II. Methods 52
    A. Subjects, apparatus, and materials 52
    B. Design and procedure..53
III. Experimental findings and discussion 56
    A. Locomotor stimulant effects and sensitization/tolerance 56
    B. Conditioning...56
    C. Biochemical assays 58
    D. Reward/aversion and place preference..59
IV. Conclusion...69
Acknowledgment...72
References..72

I. Introduction: conditioning, sensitization, and reward effects

Caffeine is a widely used drug consumed primarily in coffee and other caffeine-containing beverages, seemingly for its stimulant/arousal effects.[1,2] The literature has suggested that adenosine receptor antagonism mediates these stimulatory effects.[3-6] More recent evidence indicates that the adenosine antagonism initiated by caffeine results in indirect changes in dopamine neurotransmission.[3,7-10] Specifically, a negative coupling has been reported between the hippocampal A1 adenosine × D1 dopamine receptors and between the striatal A2 adenosine × D2 dopamine receptors. That is, A1 adenosine receptors exert an inhibitory influence over the D1 receptors in the hippocampus, and the antagonism of these adenosine receptors has been cited as a mechanism for the arousal effects of caffeine.[7] A similar A2 × D2 receptor interaction has been suggested as a mechanism for caffeine's locomotor stimulant effects.[3,8] Such observations lead to the conclusion that caffeine-induced adenosine antagonism may exert effects on locomotor behavior functionally similar to those induced by dopaminergic psychostimulants.

A number of studies, however, have generated a body of literature indicating important differences between caffeine and the dopaminergic psychostimulants. Drugs such as cocaine and amphetamine have a high abuse liability.[11-14] Positive reward effects induced by these drugs have been identified both as initiating mechanisms by which drug addictions develop and also as mechanisms sustaining addictive behavior associated with drug abuse.[12,13] In addition, conditioning has been cited as playing an important role in the development of addictive behavior as a mechanism linking reward and stimulant effects as well as abstinence/withdrawal effects to external environment stimuli associated with drug taking.[15-16] For these drugs, the stimulant effects on locomotor behavior parallel their reward effects in terms of sensitization and conditioning to associated contextual stimuli.[17-20]

As a psychostimulant, caffeine can induce locomotor stimulant effects like the dopaminergic psychostimulants.[21-22] Unlike dopaminergic drugs, however, caffeine does not have substantial drug abuse liability.[23] Indeed, caffeine appears to have largely negative reinforcement effects resulting from the alleviation of withdrawal symptoms consequent to periods of abstinence.[1,2] Nonetheless, reward effects for caffeine have been observed, primarily with choice/preference operant behavior paradigms[23] and, in one study,[24] with the conditioned place preference/conditioned place aversion (CPP/CPA) paradigm. While the reward effects of dopaminergic agonists appear linked to their psychomotor stimulant effects,[25] in the case of caffeine, limited evidence suggests that this might not be the case.[24,26]

There have been relatively few studies directly comparing the effects of repeated caffeine vs. repeated dopaminergic psychostimulant treatment on locomotor behavior.[26] In view of this lack of direct behavioral comparisons between caffeine and the dopaminergic psychostimulants and given the newly developed evidence that caffeine may act on behavior through the dopaminergic receptors, in our first study, we directly compared the locomotor stimulant effects of repeated caffeine treatment with those of cocaine. We used the locomotion stimulant effect of cocaine (10 mg/kg i.p.) as the standard in the selection of an appropriate caffeine treatment dosage. It is well-established that repeated cocaine treatments at 10 mg/kg induce conditioned locomotor stimulant effects, as well as sensitization, and reward effects as expressed in conditioned place preference (CPP).[27] In the first study, we were interested in determining whether caffeine, when equated in locomotor stimulant efficacy to the 10 mg/kg cocaine dosage, would induce similar levels of conditioned locomotion and sensitization/tolerance based on the assumption of a commonality of action on dopaminergic neurotransmission.

In a preliminary study, a group of rats was administered 10 mg/kg i.p. cocaine, and the locomotor stimulant effects were assessed in a 20-min open-field test. Additional groups were administered caffeine at 5, 10, 20, and 30 mg/kg i.p. in order to determine a caffeine dose that would induce a statistically equivalent stimulant effect to that of the 10 mg/kg i.p. cocaine. We found that a 10 mg/kg caffeine dose induced a locomotor stimulant effect that was equivalent to that of 10 mg/kg cocaine. With the initial locomotion stimulant effects of caffeine and cocaine equated behaviorally, we then compared the effects of the 10 mg/kg caffeine to the 10 mg/kg cocaine treatment on conditioned locomotion and sensitization.

In the second phase of experimentation, caffeine reward effects were investigated. In this study, we used the conditioned place preference/conditioned place aversion (CPP/CPA) paradigm[27] to assess caffeine reward and aversive effects. Embedded within the CPP/CPA paradigm is a pairing treatment protocol in which a target drug is paired with one test environment and a vehicle treatment with another test environment. The underlying idea of the CPP/CPA paradigm is that a drug which induces positive reward effects will have these effects associated with the drug environment. Following a series of pairings of drug treatment with test compartment placement, as well as a control vehicle treatment with vehicle test compartment placement, the animals are given a choice test to determine compartment preference. The expectation is that animals would prefer the drug-associated compartment (i.e., significantly above 50% or chance level) if the drug has positive reward effects; or, conversely, the vehicle compartment, if the drug treatment has aversive effects.

A number of studies have shown that dopaminergic drugs such as cocaine[28,29] and opioid drugs such as morphine[30-32] induce a preference for the environment paired with the drug treatment. In the case of caffeine, however, the behavioral drug effects appear to be more complicated.[24,33] That is, the withdrawal effects of caffeine can be aversive.[2] As a consequence, the alternation between caffeine and vehicle treatment which typically occurs in the conventional CPP/CPA protocol could result in the vehicle treatment environment being associated with the withdrawal effects of caffeine. As a consequence, the animals could display a preference for the caffeine-associated environment based, not upon positive caffeine reward effects, but upon aversion to the vehicle environment due to withdrawal effects of caffeine associated with the vehicle environment.

In order to circumvent this latter complication, we used a paired vs. unpaired caffeine treatment protocol. The paired protocol group(s) received the caffeine in the caffeine-associated environment. The unpaired group(s) received saline in the same test environment but caffeine 30 min later in the home cage. Thus, the unpaired group did not have the caffeine associated with the test environment; but it had equivalent caffeine exposure. On the next day, both the paired and unpaired caffeine groups are administered the vehicle treatment and tested in another environment. Since the two groups are equated for caffeine treatment, both should express similar withdrawal aversion effects. Thus, if withdrawal aversion effects are critical for CPP/CPA in caffeine-treated animals, then the two groups would be equivalent. On the other hand, if caffeine induced a positive reward effect and no withdrawal effects, then the paired group alone would exhibit a CPP effect.

In addition to the paired–unpaired treatment protocol, we measured locomotor behavior during CPP/CPA testing in order to directly assess for caffeine the relationship between locomotor stimulation and reward/aversion. Furthermore, we extended the conventional CPP/CPA treatment protocol of four paired and four unpaired treatments (followed by a test for CPP/CPA) by repeating this cycle two more times in order to detect transient as well as late-developing CPP/CPA effects. In conducting this CPP/CPA study with caffeine, we used two doses which represent stimulant effects at the ascending and descending portions of the inverted U-shaped dose–response function for caffeine-induced locomotor stimulant effects in rats.[22] The selected doses were 10 mg/kg i.p. and 50 mg/kg i.p. This chapter presents the results of these two sets of experiments followed by a discussion of the similarities and differences between caffeine and dopaminergic drugs with high abuse liability, such as cocaine.

II. Methods

A. Subjects, apparatus, and materials

The subjects were male Sprague–Dawley rats (Taconic Farms, Germantown, NY), 6 months in age, weighing, on the average, 500 g at the start of the experiment. The animals, upon arrival from the supplier, were weighed and handled and then housed in individual $25 \times 17 \times 17$ cm wire mesh cages on a rack in a climate-controlled room (22°C). Daily handling and weighing was continued for six more times and then, in the next three days, the animals were administered saline i.p. injections. In the final phase of the initial adaptation treatment, the animals were administered two 10-min saline trials in the open-field test apparatus of each experiment and the results, along with the animal's weight, were used to form matched groups in each study. Throughout, the animals were maintained on a 12-h dark/light cycle and were tested only during the light cycle. The experimental protocols were approved by the animal care committees of both the SUNY Health Science Center and by the VAMC at Syracuse, NY.

In the first study, locomotor behavior was measured by a video-image analyzing system (Coulbourn Instruments, Lehigh Valley, PA) in two separate compartments. One was a white 60-cm square compartment with 40-cm side panels, while the other was a white round compartment of 70-cm diameter with a 40-cm-high wall enclosure. Interior area was equivalent. The floor of each compartment was lined by a white polysorb pan liner which was changed after every trial. The test compartments were in a darkened room illuminated by 2 overhead 12 v projection lamps (with a red filter) placed above and next to the two video cameras 50 cm above the center of each test compartment. Also, a white-noise generator/speaker was placed in the same location above the test compartment. The animal's head area was blackened with water-soluble ink and this was the only body part tracked by the image analyzer. On-line analog camera images of the freely moving animal in each compartment were digitized and converted into summated scores and printed at each 5-min interval. The printers were placed in a separate room. To directly observe specific behavioral responses, the animal's behavior was also videotaped during the first and last drug treatment tests and during the two interspersed post-treatment tests for conditioning.

In the second study, behavior was monitored in a CPP/CPA apparatus consisting of two black test compartments: a square compartment (53.17 cm interior dimensions with 40.00 cm high walls) and a cylindrical compartment (60.00 cm diameter and a 40.00 cm high wall with an enclosed area approximately equal to that of the square compartment). The two compartments were connected by a small rectangular middle runway (10 × 20 cm × 40 cm), which remained closed except during CPP/CPA testing. The apparatus was in a light-attenuated room and illuminated from above with red-light illumination as in the first study. Two overhead mounted cameras placed over the center of each compartment and two Video-Path Analyzers, Model E61-21 (Coulbourn Instruments, Lehigh, PA) tracked the animal's body in the test compartment and measured spontaneous and drug-induced locomotor behavior, expressed as distance traversed as well as time spent in each compartment. Two speakers were placed between the two cameras 50 cm directly above the connecting runway to produce an 80 dB ambient white noise during testing.

Caffeine (Sigma Chemical Co. St. Louis, MO) was dissolved in warm sterile distilled H_2O to a concentration of 10 mg/ml and 50 mg/ml and administered by i.p. injection at doses of 1 ml/kg. Cocaine hydrochloride (Malinckrodt Specialty Chemical, St. Louis, MO) was dissolved in sterile distilled H_2O to a concentration of 10 mg/ml and administered by i.p. injections at doses of 1 ml/kg.

B. Design and procedure

1. Sensitization and conditioning experiment

Three matched groups (n = 7) in the first study were administered repeated saline, caffeine, or cocaine treatments depending on group assignment. The first group received saline in the square compartment and then saline in the round compartment; the second group received saline in the square compartment and then caffeine (10 mg/kg) in the round compartment; the third group received saline in the square compartment and cocaine (10 mg/kg) in the round compartment. In all groups, the saline treatment was administered immediately prior to placement into the square compartment. After a 20-min test session in this compartment, the animals were administered saline, caffeine, or cocaine, depending upon group assignment. These treatments were administered immediately prior to a 20-min test session in the round compartment. An ambient 80 dB white noise was turned-on immediately prior to test compartment placement and turned-off upon removal. The saline, caffeine, and cocaine treatments were administered once/wk for a total of eight drug treatments. Saline tests for conditioning were conducted 1 week after every fourth drug treatment.

2. *Reward/aversion experiment*

In the second study, the evaluation of locomotor stimulant and reward/aversion effects was carried out using a Pavlovian drug conditioning protocol embedded within a CPP/CPA paradigm.[17,31] Animals were assigned to five (n = 6) matched groups (matched on the total locomotion distance in the two compartments); either to a paired (P10) or to an unpaired (UP10) caffeine treatment group, which received the moderate 10 mg/kg caffeine dose. There were also a paired (P50) and an unpaired (UP50) treatment group, which received the high 50 mg/kg caffeine dose. In addition to these four caffeine groups, there was a vehicle treatment group consisting of 6 matched animals which received either the paired (n = 3) or unpaired treatment (n = 3) protocol as in the other groups, but with a vehicle injection in place of caffeine.

In the paired treatment protocol, the animals received 10 mg/kg or 50 mg/kg caffeine, depending on group assignment, 20 min before a 20-min placement into one of the two test compartments. Animals in the unpaired treatment protocol received saline 20 min before testing but also caffeine 30 min after removal from the test compartment. The vehicle control animals received vehicle injection 20 min before (paired group) or 30 min after removal from the test compartment (unpaired group). These protocols were repeated with vehicle treatment on the next day in all the animals, but the animals were placed into the second of the two test compartments 20 min after injection. Ambient 80 dB noise was turned-on immediately prior to test compartment placement and turned-off upon removal. Assignment of treatment compartment to caffeine or vehicle administration was counterbalanced in each group. Half of the animals were assigned to one sequence, while the other half had the reverse sequence. The caffeine and vehicle treatments in the paired and unpaired treatment groups were administered for 8 consecutive days (4 caffeine and 4 vehicle treatments) with one 20-min trial per day. After a 24-h withdrawal, the animals were administered a nondrug CPP/CPA test. Each animal in the CPP/CPA test was placed into the connecting runway with equal access to each of the drug and vehicle treatment compartments for a 20-min session. There were three treatment and CPP/CPA testing cycles.

3. Biochemical assay procedures

The experimental treatment animals were sacrificed following completion of behavioral testing. The animal was placed into a plastic restraining cone (Braintree Products, Braintree, MA) and sacrificed by guillotine decapitation. A bilateral limbic tissue sample, which included the nucleus accumbens, olfactory tubercle on the overlying pyriform cortex, was dissected under magnification.

Ex vivo measurements were made on dopamine (DA) and serotonin (5-HT) and their respective metabolites, DOPAC for dopamine and 5-HIAA for serotonin. Immediately after dissection, the samples of brain tissue were weighed, placed in tubes containing 0.5 ml of 0.1 M perchloric acid and 4.5 µl of 10 µg/ml dihydroxybenzylamine (DHBA) as an internal standard, and then homogenized and centrifuged. The resulting supernatant was filtered through 0.2 µm pore filters and the extracts stored at –70°C until the HPLC–EC analysis, which was completed within 24 to 72 h. The tissue samples were analyzed for dopamine, DA (3-hydroxytyramine), the dopamine metabolite, DOPAC (3,4-dihydroxyphenyl-acetic acid), serotonin, 5-HT (5-hydroxytryptamine), and the serotonin metabolite, 5-HIAA (5-hydroxyindole-3-acetic acid). For the catecholamine and indoleamine analyses in brain tissue, a BAS biophase column [C18 reverse phase (4.6 × 250 mm 5 µm)] was used. The buffer used was 0.15 M monocholoroacetic acid, pH 3.1, 2 mM EDTA, 0.86 mM SOS (sodium octyl sulfate). This was added to 35 ml acetonitrile (3.5%) to make 1 L. This solution was then filtered and degassed, and 18 ml (1.8%) tetrahydrofuran (THF) was added. The mobile phase flow rate is 1.2 ml/min and a BAS 4B EC detector is set at 0.8 V. Trunk blood at sacrifice was also collected and centrifuged for 15 min at 2,500 rpm in order to obtain the plasma component. A solid phase extraction procedure was followed in preparing the plasma sample for corticosterone analysis. The extraction column was a C18 3 ml (500 mg) column. Under vacuum, the column was conditioned with 2 × 3 methanol followed by 2 × 3 ml HPLC-grade H_2O. Before the column could dry, 0.5 to 2.0 ml of plasma (depending upon availability of samples) was passed through it and immediately followed with a 2 ml HPLC-grade H_2O/acetonitrile wash (80:20) for corticosterone. Next, the column was air-dried for 3 min. Finally, the sample was eluted with 2 × 0.5 ml methanol for corticosterone. The corticosterone plasma samples were injected into a BAS phase II C18 reverse phase column (4.6 × 250 mm, 5 µm) with mobile phase of 60% MeOH, 40% H_2O run at a flow rate of 1.0 ml/min. A BAS variable wavelength UV detector was used with the setting at 254 nm.

4. Statistical analysis

A multivariate ANOVA (analysis of variance) was used to analyze the behavioral data to determine treatment group effects, repeated measurement effects as well as within-session effects. Subsequently, 1-way ANOVAs as well as *post hoc* tests for specific group differences were used (e.g., Duncan's multiple range test) to analyze locomotor behavior. P <.05 was used as the criterion for statistical significance.

III. Experimental findings and discussion

A. Locomotor stimulant effects and sensitization/tolerance

The drug test results of the first study on Day 1 and Day 8 (the first and eighth caffeine and cocaine drug treatments) are shown in Figure 4.1 to indicate the locomotor stimulant and sensitization/tolerance effects of the intermittent drug treatment schedule. In the square or nondrug compartment, as shown in the top panel of Figure 4.1, there were no group differences in total locomotion distance among the three treatment groups as determined by a 2-way ANOVA (NS-F-tests). This outcome indicates that there were no chronic intersession drug effects or that group equivalence under nondrug conditions was maintained throughout the experiment immediately prior to the caffeine and cocaine treatments. In the drug compartment, however, as shown in the bottom panel of Figure 4.1, both caffeine and cocaine increased locomotion distance compared to the saline-treated control animals. A two-way ANOVA revealed a significant group effect (F $(2/42)$ = 21.4, p<.01); no day of treatment effect (F $1/42$) = 1.6, p>.05); and a significant interaction effect (F $(1/42)$ = 3.4, p<.05). Subsequent pair-wise group comparisons indicated that both the caffeine and cocaine groups had significantly higher mean locomotion distance scores compared to the saline group on both Test 1 and Test 8 (p<.05). There was, however, no change in the caffeine locomotion stimulant effect on Test 8 compared to Test 1 (p>.05); whereas, for cocaine, distance scores increased on Test 8 compared to Test 1 (p<.05). Thus, the repeated once/wk intermittent drug treatment schedule induced a reliable locomotor stimulant effect throughout in both drug treatment groups, but neither tolerance nor sensitization was observed for the caffeine treatment, while, by the eighth day, cocaine sensitization had occurred.

B. Conditioning

The results of the saline tests for conditioning in terms of locomotion distance were pooled and analyzed by 1-way ANOVAs. The conditioning test results in the drug compartment are presented in Figure 4.2. The 1-way ANOVA revealed that there were no significant differences between the saline, caffeine, and cocaine groups in the nondrug compartment (NS-F-tests). The same analysis of the results in the drug compartment, however, revealed a significant group effect (F $(2/40)$ = 4.8, p<.01). Subsequent pair-wise comparisons indicated no significant difference between the saline and caffeine groups (p>.05), but the cocaine group had a significantly higher mean level of locomotion distance compared to the saline group (p<.05). Thus, while the caffeine dose used was sufficient to induce a locomotion stimulant effect,

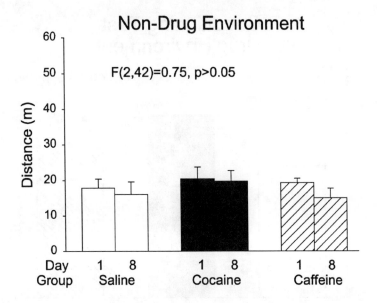

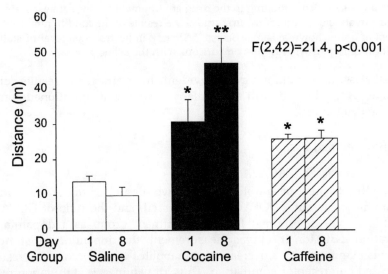

Figure 4.1 Means and SEMs of locomotion in meters of distance traversed in the non-drug and drug environments for the saline, cocaine, and caffeine treatment groups of the first study on the first and eighth drug treatments. *Asterisks indicate p<05 level of statistical significance in pair-wise group comparisons with the saline group. **Asterisks indicate p<05 level of statistical significance in within-group comparisons of the first vs. the eighth drug treatment.

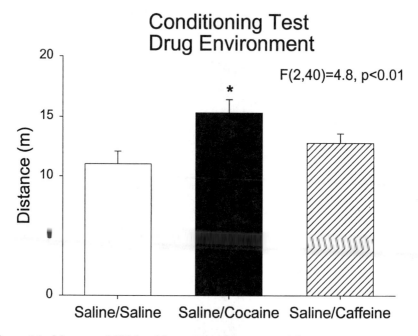

Figure 4.2 Means and SEMs of locomotion in meters of distance traversed during the saline tests for conditioning in the drug environment for the saline, cocaine, and caffeine treatment groups of the first study. The results of the conditioning tests were pooled and are shown as single scores. *Asterisk indicates p<05 level of statistical significance in pair-wise group comparisons with the saline group.

nonetheless, no conditioning was evident. In contrast, the results for the cocaine group showed both a stimulant as well as a conditioned context-specific drug effect.

C. Biochemical assays

1. Limbic DOPAC/DA ratio

The limbic tissue concentrations (µg/g wet-tissue) of dopamine and the dopamine metabolite, DOPAC, were assessed, and the ratio of DOPAC to dopamine was determined for each treatment group as a dopaminergic activation indicator. In *ex vivo* measurement, the dopaminergic stimulant effect is manifested by a decrease in metabolites which occurs as a result of increased autoreceptor stimulation. Thus, dopaminergic stimulation results in a decreased DOPAC/DA ratio. The results are presented in the top panel of Figure 4.3. These results were analyzed by 1-way ANOVA. Group effects were significant F (2/39) = 13.5, p<.01. *Post hoc* comparison indicated that the caffeine and cocaine groups had significantly lower DOPAC/DA ratios compared to the saline group (p<.05) but that the two drug groups were statistically equivalent. Thus, caffeine and cocaine had an equal impact on the limbic dopaminergic system, indicating increased activation. It is important to note that these groups were equivalent in terms of DA concentration (p>.05).

2. Limbic 5-HIAA/5-HT ratio

The limbic tissue concentrations (μg/g wet-tissue) of serotonin (5-HT) and the serotonin metabolite, 5-HIAA, were assessed, and the ratio of 5-HIAA to 5-HT was determined for each treatment group as a serotoninergic activation indicator. The results are presented in the middle panel of Figure 4.3. These results were also analyzed by a 1-way ANOVA, as were the DOPAC/DA results. Group effects were again significant (F (2/39) = 4.7, p<.05) but the *post hoc* comparison of group means indicated that the cocaine group had a lower ratio compared to both the caffeine and saline groups (p<.05). There was no statistically significant difference between the latter groups. Thus, cocaine, but not caffeine, had an impact on the limbic serotoninergic system, indicating increased activation. Again, it should be noted, that all three groups were equivalent in terms of 5-HT concentration (p>.05).

3. Serum corticosterone

Serum corticosterone concentration (μg/ml) was determined as an index of drug-induced effects. The results are presented in the bottom panel of Figure 4.3. These results were also analyzed by a 1-way ANOVA. Group effects were again significant (F (2/39) = 7.7, p<.01. *Post hoc* comparison of the group means indicated that corticosterone was equally elevated in the caffeine and cocaine groups compared to that of the saline group (p<.05). These results indicate that caffeine and cocaine had a significant impact on the stress-related hormone of corticosterone, with the drug groups showing increased serum corticosterone.

D. Reward/aversion and place preference

1. Caffeine stimulant effects on locomotor behavior and sensitization/tolerance

The stimulant as well as the sensitization/tolerance effects of the 10 mg/kg caffeine treatments during the three 8-day treatment cycles of the second study are shown in Figures 4.4 to 4.6. The top and bottom panels of Figure 4.4 present the locomotion distance results for the paired and unpaired treatment groups of the 10 mg/kg caffeine dose. As can be seen from the top panel, the paired, P10, group shows a stable locomotor stimulant effect across the three 8-day treatment cycles. This was confirmed by three separate 1-way ANOVAs, one for each treatment cycle, comparing locomotion in the drug vs. vehicle treatment compartment: F (1/10) = 49.36, p<.001; F 1/10) = 32.80, p<.01 and F (1/10) = 22.35, p<.001, respectively. The same analysis for the unpaired, UP10, group, as shown in the bottom panel of Figure 4.4, indicated that the activity levels in the two compartments did not differ statistically (NS-F-tests).

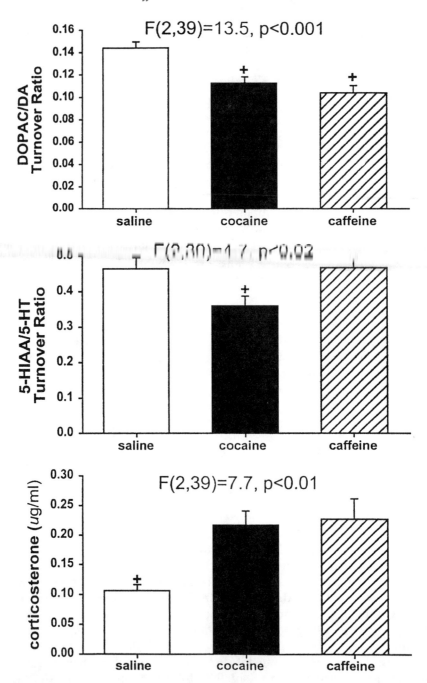

Figure 4.3 Means and SEMs of DOPAC/DA ratios in the limbic area sampled for the saline, cocaine, and caffeine treatment groups with DOPAC and DA measured as μg/g wet-tissue (top panel); limbic 5-HIAA/5-HT ratios measured as μg/g wet-tissue (middle panel); and, serum corticosterone concentration bottom panel. + indicates p<05 level of statistical significance in pair-wise group comparisons with the saline group.

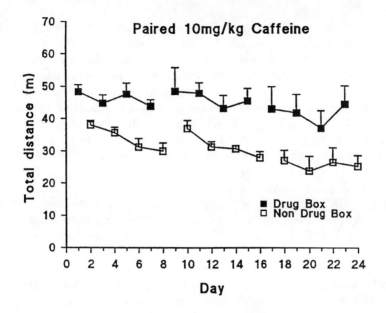

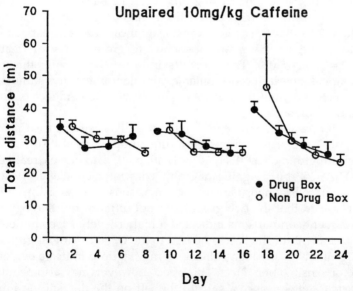

Figure 4.4 Means and SEMs of locomotion in meters of distance traversed in the drug box and in the vehicle box for the moderate dose (10 mg/kg) caffeine treatment groups of the second study. The upper panel shows the results for the paired (P10) caffeine treatment group, while the bottom panel shows the results for the unpaired (UP10) caffeine treatment group. The means and SEMS of total distance traveled in each test compartment are shown in three cycles of 8-day treatments. See text for corresponding statistical analyses. (From Carey, R.J. and Damianopoulos, E.N., *Pharmacopsychoecologia*, 7, 137, 1994. With permission from the Pharmacopsychoecological Association.)

The results for the high dose 50 mg/kg caffeine treatment are presented in Figure 4.5. The P50 group (top panel), in contrast to the P10 group, shows a session effect in that the locomotion activity levels in the drug compartment were related to treatment cycle. During the first cycle, there was no statistically significant drug effect (NS-F-test). In the second and third treatment cycles, however, the caffeine treatment increased locomotion. This was confirmed by a statistically significant locomotor stimulant effect in the 1-way ANOVAs for the second and third cycles: ($F (1/10) = 19.14$, $p<.001$ and ($F (1/10) = 20.67$, $p<.001$). In contrast, the activity levels of the UP50 control group (bottom panel) remained the same in both compartments (NS-F-tests), as well as those of the vehicle treatment group, as shown in Figure 4.6 (NS-F-tests). Thus, there were no progressive changes in *baseline* locomotion activity levels either as a result of drug treatment *per se* (as shown by the results of the UP50 group) or as a result of repeated compartment exposure (as shown by the results of the vehicle group).

The within-subject analysis above is sensitive to the treatment variables in that the subject variable remains constant. In order to assess overall treatment effects as well as across-session effects, a between-group analysis of the locomotion activity levels was performed comparing the paired vs. the unpaired treatment groups across the three treatment cycles. A 2-way ANOVA of the P10 vs. the UP10 groups indicated a significant group effect, with the P10 group showing a statistically higher level of locomotor activity ($F (1/10) = 77.38$, $p<.001$); there was no significant day of treatment effect ($F (23/110) = 1.01$, $p>.05$) or an interaction of group × day of treatment ($F (23/110) = .96$, $p>.05$). These results indicate that, while the moderate 10 mg/kg caffeine induced a reliable stimulant effect, the stimulant effect was stable throughout the treatment period with no evidence of either sensitization or tolerance.

The comparable analysis for the P50 vs. UP50 caffeine treatment groups in the drug compartment revealed a different pattern of results. There was a significant group effect ($F (1/10) = 11.49$, $p<.01$); no day of treatment effect ($F (23/110) = .57$; but a significant group × day interaction effect ($F (23/110) = 4.10$, $p<.01$). Subsequent pair-wise comparisons of days 1 through 8 (first cycle) revealed that the two groups did not differ at this stage ($p>.05$), but all subsequent comparisons indicated a higher level of locomotion on each day of treatment for the P50 compared to the UP50 treatment group ($p<.01$). These results with the high caffeine 50 mg/kg dose indicate evidence for a locomotor sensitization effect. This effect, however, may also be interpreted as a *tolerance* effect, since a shift to the left on the descending limb of the inverted U-shaped function of activation could yield the same kind of behavioral result.

Taken together, the results presented above show that the moderate 10 mg/kg caffeine dose treatments did generate a reliable locomotor stimulant effect in the drug environment, and this was demonstrated by both a within-group analysis comparing the same animals in the drug vs. the vehicle compartments, as well as by a between-group analysis comparing the

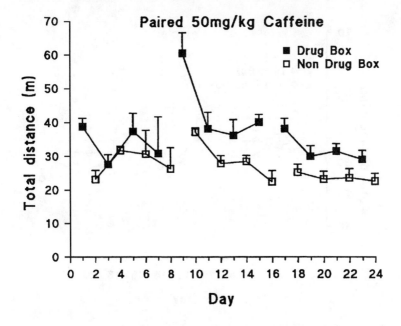

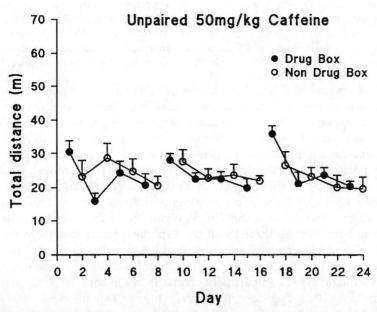

Figure 4.5 Means and SEMs of locomotion in meters of distance traversed in the drug box and in the vehicle box for the high dose (50 mg/kg) caffeine treatment groups of the second study. The upper panel shows the results for the paired (P50) caffeine treatment group, while the bottom panel shows the results for the unpaired (UP50) caffeine treatment group. The means and SEMS of total distance traveled in each test compartment are shown in three cycles of 8-day treatments. See text for corresponding statistical analyses. (From Carey, R.J. and Damianopoulos, E.N., *Pharmacopsychoecologia,* 7, 137, 1994. With permission from the Pharmacopsychoecological Association.)

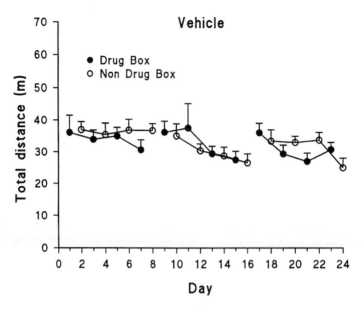

Figure 4.6 Means and SEMs of locomotion in meters of distance traversed in the drug box and in the vehicle box for the vehicle reference control treatment group of the second study. The Means and SEMS of total distance traveled in each test compartment are shown in three cycles of 8-day treatments. See text for corresponding statistical analyses. (From Carey, R.J. and Damianopoulos, E.N., *Pharmacopsychoecologia*, 7, 137, 1994. With permission from the Pharmacopsychoecological Association.)

two drug treatment groups, only one of which had the caffeine treatment in the drug-associated test compartment. Moreover, there was no evidence of sensitization or tolerance for the moderate 10 mg/kg caffeine dose, because there was no statistically significant progressive change across days of treatment. For the interpretation of this result as well as for the CPP/CPA results, it is important to note that the unpaired treatment group showed no differential levels of locomotion activity in the drug vs. vehicle test compartments, as demonstrated by the within-group comparisons of activity. This last result suggests that there was a stimulus equivalence of the two compartments which did not change throughout the experiment and that there were no intersession drug effects. The inference of baseline stability is further supported by the unchanging vehicle control group treatment baselines in the drug vs. the nondrug compartment comparisons. In contrast to these results with the low 10 mg/kg dose, the high 50 mg/kg caffeine dose showed no locomotor stimulation effects in the first treatment cycle followed by a reliable increase in locomotion activity in the second and third cycles. Behaviorally, this is a sensitization effect; however, this characterization is simply descriptive and fails to consider that the 50 mg/kg dosage is on the descending arm of the dose–response function. From this perspective, a sensitization effect would be expected to induce a decrease in locomotion. Thus, the increase in locomotion fits well with a shift to the left in the dose–response function, or, tolerance.

2. Place preference

The results of the CPP/CPA tests are presented in Figures 4.7 to 4.9. In the analysis of these results, both locomotion distance and time spent in each compartment were evaluated. Locomotion, but more specifically, time spent in each compartment would reveal place preference and, by inference, caffeine reward/aversion effects.

a. Conditioned locomotion. Two derived scores were calculated for each animal based on the locomotion results of the three CPP/CPA tests after the fourth, eighth, and twelfth caffeine treatments, respectively. One derived score was a difference score calculated as total locomotion in the drug compartment minus total locomotion in the nondrug compartment (DB-NDB). The other derived score was calculated as a ratio score of total locomotion in the drug compartment divided by total locomotion in the drug and nondrug compartments (DB/DB+NDB). In this analysis, difference scores can be distributed from negative to positive values on the ordinate, as shown in Figure 4.7, expressing an aversion (negative hedonic effect) or a preference (positive hedonic effect), respectively, for the drug associated compartment. Similarly, ratio scores, as shown in Figure 4.7, can be distributed from below to above .50 on the ordinate. A ratio score below .50 represents an aversion for the drug-associated test compartment, while a ratio score above .50 represents a preference. A ratio score of .50 represents hedonic neutrality. As can be seen from the two top panels of Figure 4.7, the difference scores and the ratio scores for the 10 mg/kg caffeine dose show no preference for the drug compartment compared to the vehicle treatment compartment in terms of increased locomotion activity either by the animals in the paired P10 or by the animals in the unpaired UP10 group. Moreover, from these results it can also be determined that no context-specific conditioning appears to have occurred with the moderate caffeine dose in this study as well.

In contrast to these results, the animals in the P50 group, as shown in the top panels of Figure 4.8, revealed increased locomotion activity in the drug vs. the vehicle compartment, but this effect reached statistical significance only after completion of the third treatment cycle (p>.05). The unpaired treatment UP50 group results, as shown in the same panel of Figure 4.8, reflect no differential increase either in the difference or ratio scores. The vehicle control animals, shown in Figure 4.9, also, reflect no differential effects (p>.05).

b. Time spent in each compartment. Difference scores as well as ratio scores were similarly calculated for total time spent in each compartment for all the animals. As shown in the bottom panels of Figure 4.7, again,

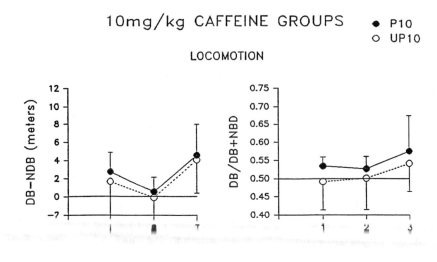

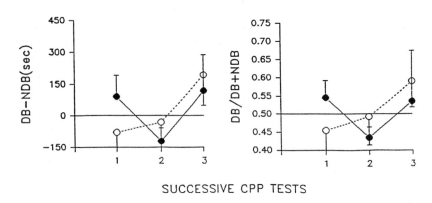

Figure 4.7 Means and SEMs of locomotion in meters of distance traversed (upper panels) and time spent in each box (lower panels) during the three post-treatment CPP tests for the paired 10 mg/kg caffeine treatment group and for the unpaired 10 mg/kg caffeine treatment group of the second study. Results in the left panels indicate difference scores between the DRUG and NON DRUG BOXES (DB-NDB), while the results in the right panels show ratio scores for the DRUG BOX vs. the DRUG BOX + NON DRUG BOX (DB/DB+NBD) in the three 20-min non drug CPP placement tests after each treatment cycle. (From Carey, R.J. and Damianopoulos, E.N., *Pharmacopsychoecologia*, 7, 137, 1994. With permission from the Pharmacopsychoecological Association.)

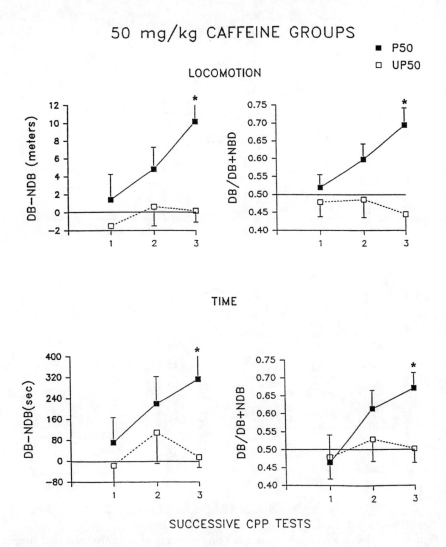

Figure 4.8 Means and SEMs of locomotion in meters of distance traversed (upper panels) and time spent in each box (lower panels) during the three post-treatment CPP tests for the paired 50 mg/kg caffeine treatment group and for the unpaired 50 mg/kg caffeine treatment group of the second study. Results in the left panels indicate difference scores between the DRUG and NON DRUG BOXES (DB-NDB), while the results in the right panels show ratio scores for the DRUG BOX vs. the DRUG BOX + NON DRUG BOX (DB/DB+NDB) in the three 20-min non drug CPP placement tests after each treatment cycle. *Asterisk indicates a significant group difference as determined by t-test for independent means (p<.01). (From Carey, R.J. and Damianopoulos, E.N., *Pharmacopsychoecologia,* 7, 137, 1994. With permission from the Pharmacopsychoecological Association.)

VEHICLE GROUP

LOCOMOTION

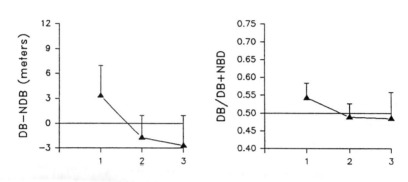

TIME

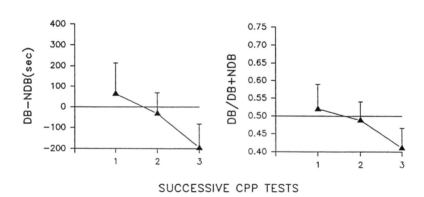

SUCCESSIVE CPP TESTS

Figure 4.9 Means and SEMs of locomotion in meters of distance traversed (upper panels) and time spent in each box (lower panels) during the three post-treatment CPP tests for the vehicle treatment group of the second study. Results in the left panels indicate difference scores between the DRUG and NON DRUG BOXES (DB-NDB), while the results in the right panels indicate ratio scores for the DRUG BOX vs. the DRUG BOX + NON DRUG BOX (DB/DB+NDB) in the three 20-min non drug CPP placement tests after each treatment cycle. (From Carey, R.J. and Damianopoulos, E.N., *Pharmacopsychoecologia, 7*, 137, 1994. With permission from the Pharmacopsychoecological Association.)

neither the difference scores (bottom left panel) nor the ratio scores (bottom right panel) reflect a preference for the drug-associated compartment (NS-F-tests) in the 10 mg/kg caffeine treatment groups. Thus, no shift occurred from the initial hedonic neutrality for the two test compartments in the 10 mg/kg caffeine groups during the three caffeine treatment cycles. Importantly, for these paired caffeine 10 mg/kg animals, locomotor stimulant effects had occurred, as shown in Figure 4.4, but neither a reward/preference for the drug-associated compartment was evident nor conditioning to the contextual stimuli. Thus, a dissociation of locomotor stimulant effects from reward/preference effects appears to be the case for the caffeine 10 mg/kg treatment; since, as shown in Figure 4.4, caffeine had induced an increased level of locomotor activity but, as shown in Figure 4.7, there was no preference for the drug-associated compartment.

In contrast to these results with the 10 mg/kg caffeine treatment, and as can be seen from the bottom panels of Figure 4.8, a clear preference emerged in the animals of the high dose P50 caffeine treatment group for the drug-associated compartment compared to the UP treatment group; but the increased time in the drug-associated compartment reached statistical significance only at the end of the third treatment cycle ($p<.05$). This result, as well as the increase in activity level shown in Figure 4.8, suggests that the high 50 mg/kg caffeine dose had a dual behavioral effect: a conditioned stimulant and a reward/reinforcement effect. For the high caffeine dose, the locomotor stimulant effects were not dissociated from the reward/preference effects.

For the vehicle treatment animals, as can be seen from Figure 4.9, especially from the ratio score analysis (bottom right hand panel), compartment preference remained neutral as their ratio scores remained at about .50 (NS-F-tests). These results indicate that there were no shifts from compartment preference neutrality during the course of treatment, and compartment equivalence was maintained throughout the study.

IV. Conclusion

The results of the first caffeine study provide evidence suggesting that the repeated caffeine treatment effects on conditioning and sensitization/tolerance of locomotor stimulant effects differ from those observed with the dopaminergic agonist, cocaine. However, it must be noted that not all dopaminergic psychostimulants induce conditioned hyperlocomotion effects. For example, several studies of the direct-acting dopaminergic agonist, apomorphine, with dosages sufficient to induce both locomotor stimulation as well as sensitization, did not result in context-specific conditioning of these effects.[35-37] Thus, it would appear, based on the pattern of results obtained in the present investigation as well as on earlier studies with dopaminergic agonists, that dopaminergic activation, while it may be essential for inducing a locomotor stimulant response (or a UR in Pavlovian conditioning terminology,[34,38-40] is not sufficient (and it may not even be necessary) for conditioning to occur.

Perhaps the biochemical results of the present experiment as well as those of our previous studies[41,42] and those of others provide[43] a clue as to what other neurotransmission systems are needed besides dopaminergic activation for context-specific conditioning to occur. Serotoninergic involvement in terms of increased activation in the limbic system, as shown by the cocaine results of the present study, and in the medial prefrontal cortex, as shown by the results of our previous study,[41] may mediate context-specific conditioning of locomotor stimulant effects. Thus, future studies with serotoninergic agonist and antagonist manipulations may confirm this hypothesized role of serotoninergic limbic and medial prefrontal cortex neurotransmission in the development of context-specific conditioning of locomotor stimulant responses.

While chronic caffeine treatment can lead to tolerance[22,44,45] this effect is perhaps treatment schedule-dependent as well as context-dependent, as it was not observed in the present results with the moderate 10 mg/kg caffeine dose. The once/wk treatment in the first study and the once/48 h treatment in the second study (with interspersed vehicle treatments in a separate test compartment) produced a stable caffeine stimulant effect for an extended period. In connection with the first study, the stable stimulant effect demonstrated the occurrence of an undiminished drug UR on each drug treatment, showing UCS (unconditioned stimulus) adequacy in the selected drugs, one of the necessary elements for Pavlovian conditioning[34,38,39]; while in the second study the stable stimulant effect showed that the stimulant response was not diminished by tolerance.

While tolerance was not observed, the question of why sensitization did not occur can be raised. One possibility is based on the absence of a Pavlovian conditioned caffeine stimulant effect. Pert et al.[46] hypothesized that context-specific behavioral sensitization includes an algebraic summation of the drug response with the conditioned drug response. The absence of conditioning of the locomotor stimulant drug effect would imply that no locomotor sensitization can develop. The experience from our laboratory and those of others is that behavioral sensitization due to repeated drug treatment can be dissociated from conditioned locomotor stimulant effects.[28,40] Dopaminergic drug conditioning of locomotor behavior has been observed with[47] and without sensitization,[48] pointing to an independence of mediating central processes. The possibility yet to be tested, however, is that the stimulus control of the conditioned response is not by the contextual stimuli alone, but by a compound of the interoceptive drug cue plus the contextual stimuli.[49] Accordingly, in saline tests for conditioning, the interoceptive drug cue would be absent, and it can happen under these circumstances that the context is only a partial component of the conditioned stimulus and, therefore, has insufficient stimulus control over the behavioral drug response. Even when extended to include interoceptive drug cue conditioning, the Pert et al.[46] hypothesis still cannot account for instances when conditioning does occur but without sensitization.[48] Our findings with the moderate caffeine dose indicate another type of an outcome, one in which stimulant effects are induced but neither sensitization nor conditioning occur.

The results with the high 50 mg/kg caffeine dose may suggest evidence for a sensitization effect; but as analyzed earlier, based on the inverted U-shaped function of the locomotor stimulant effects for caffeine, tolerance would be a more appropriate description. However, conditioning of the caffeine stimulant effects on locomotion was observed. Thus, the increased locomotion and increased time spent in the drug-associated compartment during the interspersed CPP/CPA tests suggest that both conditioning to contextual stimuli as well as a reward/reinforcement effect had occurred with the high 50 mg/kg dosage. With regard to the latter, there is, however, one complication. According to the analysis of Scoles and Siegel,[32] a preference for one compartment may actually reflect an aversion to the other compartment (in the two-compartment CPP/CPA paradigm) due to either a drug-induced aversion to the drug compartment or due to withdrawal aversion leading to an aversion to the vehicle compartment. Caffeine has been known to induce aversive effects;[24,50] but, in the present study, preference shown for the drug-associated compartment discounts the first possibility; i.e., high dose caffeine-induced aversion. The other possibility, i.e., withdrawal aversion associated with the vehicle compartment, can also be discounted by the neutral responding of the unpaired UP50 group relative to the two compartments. If withdrawal aversion associated to the vehicle compartment during the nondrug test trials had occurred, these animals would have shown increased time spent in the drug compartment.

As already indicated, caffeine has indirect effects on the dopaminergic system, and the locomotor stimulant effects of adenosine antagonism may be mediated by a striatal A2 adenosine × D2 dopamine receptor interaction. This possibility is further supported by the anatomical finding of colocalized striatal adenosine and dopamine receptors wherein the adenosine receptors exert a downward modulatory effect on the dopamine receptors.[7] In spite of this commonality, a caffeine dosage sufficient to induce a locomotor stimulant effect, which constitutes the observable drug UR in the Pavlovian paradigm, did not result in context-specific conditioning. The high caffeine dose, however, yielded evidence of conditioning after a long treatment regimen though as yet to be determined mechanisms. In spite of the common end-point D2 receptor stimulation by caffeine relative to dopaminergic psychostimulants, it appears that dopamine transmission is not essential for the induction and/or expression of conditioned drug-induced locomotor stimulant effects. Thus, the focus on dopamine may be misdirected. Other transmission systems which mediate the development of conditioning with dopaminergic psychostimulants are implicated. That is, dopaminergic psychostimulant drugs, but not moderate doses of caffeine, impinge upon the serotonin system.[41-43] It would appear likely that it is this additional serotoninergic agonist effect upon the serotoninergic system which may be critical for the development and/or expression of conditioned locomotor behavior and/or sensitization. Possibly, the high doses of caffeine are able to activate the serotoninergic system. Thus, the results of the present study raise a number of intriguing issues in psychopharmacology and indicate that some widely accepted assumptions regarding the behavioral impact of caffeine need to be re-examined.

Acknowledgment

This research was supported by NIDA grant RO1DA05366-11.

References

1. Richardson, N. J., Rogers, P. J., and Elliman, N. A., Conditioned flavor preferences reinforced by caffeine consumed after lunch, *Physiol. Behav.*, 60, 257, 1996.

2. Rogers, P. J., Richardson, N. J., and Elliman, N. A., Overnight caffeine abstinence and negative reinforcement of preference for caffeine containing drinks, *Psychopharmacology*, 120, 457, 1995.

3. Garrett, B. E. and Holtzman, S.G., Does adenosine receptor blockade mediate caffeine-induced rotational behavior? *J. Pharmacol. Exp. Therap.*, 274, 207, 1995.

4. Howell, L. L., Coffin, V. L., and Spealman, R. D., Behavioral and physiological effects of xanthines in non-human primates, *Psychopharmacology*, 129, 1, 1997.

5. Katims, J. J., Murphy, K. M., and Snyder S. H., Xanthine stimulants and adenosine, in *Stimulants: Neurochemical, Behavioral, and Clinical Perspectives*, Creese, I. Ed., Raven, New York, 1983, 63.

6. Snyder, S. H. J., Katims, S. J. Anau, Z., Bruns, R. F., and Daly, J. W., Adenosine receptors and behavioral actions of methylxanthines, *Proc. Natl. Acad. Sci. U.S.A.*, 78, 3260, 1981.

7. Garrett, B. E. and Griffiths, R. R., The role of dopamine in the behavioral effects of caffeine in animals and humans, *Pharmacol. Biochem. Behav.*, 57, 533, 1997.

8. Morelli, M. Fena, S., Pinna, A., and di Chiara, G., Adenosine D_2 receptors interact negatively with dopamine D_1 and D_2 receptors in unilaterally 6-hydroxydopamine-lesioned rats, *Eur. J. Pharmacol.*, 251, 21, 1994.

9. Pinna, A., di Chiara, G., Wardas, J., and Morelli, M., Blockade of A_{2a} adenosine receptors positively modulates turning behavior induced by D_1 agonists in dopamine denervated rats, *Eur. J. Neurosci.*, 8, 1176, 1996.

10. Ferre, S., Popoli, P., Tinner-Staines, B., and Fuxe, K., Adenosine A_1 receptor-dopamine D_1 receptor interaction in the rat limbic system: modulation of dopamine D_1 receptor antagonist binding sites, *Neurosci. Lett.*, 208, 109, 1996.

11. Foltin, R. W. and Fischman, M. W., Effects of methadone or buprenorphine maintenance on the subjective and reinforcing effects of intravenous cocaine in humans, *J. Pharmacol. Exp. Therap.*, 278, 1153, 1996.

12. Koob, G. F., Drugs of abuse: Anatomy, pharmacology and functional reward pathways, *TiPS*, 13, 177, 1992.

13. Siegel, S., Pharmacological conditioning and drug effects, in *Psychoactive Drugs*, Goudie, A. J. and Emmett-Oglesby, M. W., Eds., The Human Press, New York, 1989, 115.

14. Woolverton, W. L. and Johnson, K., Neurobiology of cocaine abuse, *TiPS*, 131, 193, 1992.

15. Gold, M. S., Byron, C.A., Dackis, C.A., and Sweeny, D. R., Paraphernalia-induced cocaine craving, *Soc. Neurosci. Abstr.*, 12, 936, 1986.

16. McClellan, A., Childress, A., Ehrmann, R., and O'Brien, C., Opiate and cocaine related stimuli elicit craving and physiological responses in drug-abuse patients. Presented at the annual meeting of the *Society for Stimulus Properties of Drugs,* Cape Cod, MA, June 28, 1988.
17. Damianopoulos, E. N. and Carey, R. J., A new method to assess Pavlovian conditioning of psychostimulant drug effects, *J. Neurosci. Meth.,* 53, 7, 1994.
18. Damianopoulos, E. N. and Carey, R. J., Pavlovian conditioning of CNS drugs effects: A critical review and new experimental design, *Rev. Neurosci.,* 3, 65, 1992.
19. Kalivas, P. W., Interactions between dopamine and excitatory amino acids in behavioral sensitization to psychostimulants, *Drugs Alcoh. Depend.,* 37, 95, 1995.
20. Kalivas, P. W., Striplin, C. D., Steketee, J. D., Klitenick, M. A., and Duffy, P., Cellular mechanisms of behavioral sensitization in drugs of abuse (Review), *Ann. N Y Acad. Sci.,* 654, 128, 1992.
21. Choi, O.H., Shamin, M. T., Padgett W. T., and Daly, J. W., Caffeine and theophylline analogues: Correlation of behavioral effects with activity as adenosine receptor antagonists and as phosphodiesterase inhibitors, *Life Sci.,* 43, 397, 1988.
22. Holtzman, S. G. and Finn, I., Tolerance to behavioral effects of caffeine in rats, *Pharmacol. Biochem. Behav.,* 29, 411, 1988.
23. Liguori, A. and Hughes, J. R., Caffeine self-administration in humans: 2. A within-subjects comparison of coffee and cola vehicles, *Exp. Clin. Psychopharmacol.,* 5, 295, 1997.
24. Brockwell, N. T., Eikelboom, R., and Beninger, R. J., Caffeine-induced place and taste conditioning: Production of dose-dependent preference and aversion, *Pharmacol. Biochem. Behav.,* 38, 513, 1991.
25. Wise, R. A., Neurobiology of drug addiction, *Curr. Opin. Neurobiol.,* 6, 243, 1996.
26. Herz, R. S. and Beninger R. J., Comparison of the ability of (+)-amphetamine and caffeine to produce environment-specific conditioning, *Psychopharmacology,* 92, 365, 1987.
27. Hoffman, D. C., The use of place conditioning in studying the neuropharmacology of drug reinforcement, *Brain Res. Bull.,* 23, 378, 1989.
28. Ali, I. and Kelly, M. E., Buspirone fails to effect cocaine-induced conditioned place preference in the mouse, *Pharmacol. Biochem. Behav.,* 58, 311, 1997.
29. Spyraki, C., Nomikos, G., and Varonos, D. D., Intravenous cocaine-induced place preference: attenuation by haloperidol, *Behav. Brain Res.,* 26, 57, 1987.
30. Popik, P. and Danysz, W., Inhibition of reinforcing effects of morphine and naloxone-precipitated opioid withdrawal by N-methyl-D-aspartate receptor antagonist, memantine, *J. Pharmacol. Exp. Therap.,* 280, 854, 1997.
31. Piepponen, T. P., Kivastik, T., Katajamaki, J., Zharkovsky, A., and Antee, L., Involvement of opioid mu 1 receptors in morphine-induced conditioned place preference in rats, *Pharmacol. Biochem. Behav.,* 58, 275, 1997.
32. Scoles, M.T. and Siegel, S., A potential role of saline trials in morphine-induced place-preference conditioning, *Pharmacol. Biochem. Behav.,* 25, 1169, 1986.
33. Bhattacharya, S. K., Satyan, K. S., and Chakrabarti, A., Anxiogenic action of caffeine: An experimental study in rats, *J. Psychopharmacol.,* 11, 219, 1997.

34. Stewart, J. and Eikelboom, R., Conditioned drug effects, in *Handbook of Psychopharmacology*, Iversen, L. L., Iversen, I. D., and Snyder, S. H., Eds., Plenum Press, New York, 1987, 10.

35. Damianopoulos, E. N. and Carey, R. J., Apomorphine sensitization effects: Evidence for environment contingent behavioral reorganization processes, *Pharmacol. Biochem. Behav.*, 45, 653, 1993.

36. Mattingly, B. A., Gotsick, J. E., and Marin, C., Locomotor activity and stereotypy in rats following repeated apomorphine treatments at 1-, 3- or 7-day intervals, *Pharmacol. Biochem. Behav.*, 31, 871, 1989.

37. Mattingly, B. A., Gotsick, J. E., and Salamanca, L., Latent sensitization to apomorphine following repeated low doses, *Behav. Neurosci.*, 102, 553, 1988.

38. Pavlov, I. P., *Conditioned Reflex*, Oxford University Press, London, 1927.

39. Pavlov, I. P., *Lectures on Conditioned Reflexes*, International, New York, 1928.

40. Stewart, J. and Vezina, P., Conditioning and behavioral sensitization, in *Sensitization in the Nervous System*, Barnes, C. R. and Kalivas, P. W., Eds., Telford Press, Caldwell, NJ, 1988, 207.

41. Carey, R. J. and Damianopoulos, E. N., Conditioned induced hyperactivity: an association with increased medial pre-frontal cortex serotonin, *Behav. Brain Res.*, 62, 177, 1994.

42. Damianopoulos, E. N. and Carey, R. J., Evidence for N-methyl-D-aspartate receptor mediation of cocaine induced corticosterone release and cocaine conditioned stimulant effects, *Behav. Brain Res.*, 68, 219, 1995.

43. Levy, A. D., Li, Q. A., Kerr, J. E., Rittenhouse, P. A., Milonas, G., Cabrera, T. M., Battaglia, G., Alvarez Sanz, M. C., and Van De Kar, L. D., Cocaine-induced elevation of plasma adrenocorticotropin hormone and corticosterone is mediated by serotoninergic neurons, *J. Pharmacol. Exp. Therap.*, 259, 495, 1991.

44. Lau, C. E. and Falk, J. L., Dose-dependent surmountability of locomotor activity in caffeine tolerance, *Pharmacol. Biochem. Behav.*, 52, 139, 1995.

45. Howell, L. L. and Lundrum, A. M., Effects of chronic caffeine administration on respiration and schedule controlled behavior in rhesus monkeys, *J. Pharmacol. Exp. Therap.*, 283, 190, 1997.

46. Pert, A., Post, R., and Weiss, S. R. B., Conditioning as a critical determinant of sensitization induced by psychomotor stimulants, in *Neurobiology of Learning and Memory. NIDA Research Monograph, 97*, Erinoff, L., Ed., 1990, 208.

47. Carey, R. J. and Gui, J., Cocaine sensitization can accelerate the onset of peak cocaine behavioral effects, *Pharmacol. Biochem. Behav.*, 60, 395, 1997.

48. Martin-Iverson, M. T. and Fawcett, S. L., Pavlovian conditioning of psychomotor stimulant-induced behaviours: has convenience led us astray? *Behav. Pharmacol.*, 7, 26, 1996.

49. Carey, R. J. and Gui, J., Cocaine conditioning and cocaine sensitization: what is the relationship? *Behav. Brain Res.*, 92, 67, 1998.

50. Steigerwald, E. S., Rusniak, R. W., Eckel, D. L., and O'Regan, M. H., Aversive conditioning properties of caffeine in rats, *Pharmacol. Biochem. Behav.*, 31, 579, 1988.

chapter five

Behavioral effects of caffeine coadministered with nicotine, benzodiazepines, and alcohol

Jason M. White

Contents

  I. Caffeine and nicotine 76
  II. Caffeine and benzodiazepines...78
  III. Caffeine and alcohol..81
  IV. Conclusion..82
References..83

The widespread use of caffeine provides considerable opportunity for interactions between this methylxanthine and other drugs. Such interactions may simply be an inevitable consequence of the use of multiple drugs by individuals. There are also occasions when the combined drug effect may be actively sought or when one drug is used to modify the effects of another. An example of the latter is the use of caffeine to reverse alcohol-induced intoxication.

Drug interactions are inherently difficult to study. The outcome of any interaction will depend on the doses of the drugs under study: there may be an additive effect at one dose combination and an antagonistic effect at another, for example. The effect measured is also important. Again, the qualitative nature of the interaction may vary, depending on the particular response under investigation. The organism's prior experience with each drug may also be a modifying factor. Decreases in drug effect are most commonly observed, but sensitization may occur with some drugs. This may be of particular relevance to caffeine, because most users have a history of daily use over many years. The tolerance developed through the history may change the nature of the interaction with other drugs.

Having noted the complexities of analyzing drug interactions, it must be acknowledged that there are relatively few studies examining behavioral changes when caffeine is coadministered with other centrally acting drugs. Three of the more widely used drugs/drug classes have been chosen for review: nicotine, benzodiazepines, and alcohol. Given the frequency with which these are coadministered with caffeine, the lack of data appears somewhat surprising. The review will cover both human and animal studies in which behavioral outcomes have been recorded and will examine possible mechanisms that may help explain the results found.

I. Caffeine and nicotine

Like caffeine, nicotine is classed as a stimulant. Its behavioral effects are due to a number of actions, including agonist activity at central nicotinic cholinergic receptors.[1] At moderate doses, nicotine can increase rates of schedule-controlled responding, an effect that is blocked by the nicotinic receptor antagonist mecamylamine.[2] Nicotine is administered in the form of tobacco smoke, other tobacco products, nicotine chewing gum, and transdermal patches. Studies of coffee drinking and cigarette smoking have shown a moderately strong positive relationship between the two.[3] This could be explained by individual factors, such as a predisposition to use stimulants, but may possibly reflect an interaction of some kind between the two drugs.

In a study of schedule-controlled behavior in rats, White[4] showed that a dose of caffeine that moderately increased response rate (3.0 mg/kg) had an additive effect with nicotine. Thus, the rate-increasing effects of caffeine and nicotine were greater than the rate-increasing effects of either drug alone. In contrast, higher doses of caffeine that did not increase response rates diminished the stimulant effects of nicotine. The disruption to the temporal pattern of responding was greatest at the highest doses of the two drugs. In contrast to their combined effects on positively reinforced behavior, nicotine failed to alter the effects of caffeine on avoidance behavior in mice.[5]

Lee et al.[6] also observed additive effects of caffeine and nicotine. They measured the locomotor activity of rats and noted increases with caffeine (6 mg/kg) and nicotine (0.5 mg/kg). The increase that occurred with coadministration of these two was approximately equal to the sum of the effects of each drug alone. Cohen et al.[7] used a somewhat different dose range in a study of the locomotor activity of rats in a tunnel maze. Caffeine blocked nicotine-induced decreases in activity following acute exposure. With chronic nicotine exposure, caffeine administration resulted in an enhanced stimulant effect. A greater than additive effect of nicotine combined with caffeine was observed by Sansone et al.[5] Using rats, doses of 0.5 mg/kg nicotine and 2.5 to 10 mg/kg caffeine had little effect on locomotor activity. However, nicotine combined with 5 and 10 mg/kg caffeine produced significant increases.

In a different species and employing a different activity measure, Kurib-ara et al.[8] also observed enhanced effects when nicotine was coadministered with caffeine. However, the change in activity produced by each drug was a decrease rather than an increase. Caffeine suppressed the wheel-running of mice during the dark phase of the 24-hour cycle. Nicotine had a similar effect and greater suppression was evident when the two were coadminis-tered. The effect appeared less than additive.

Unpublished data by Rumbold and White suggest that enhancement of the nicotine discriminative stimulus may occur following administration of a stimulant dose of caffeine (3.0 mg/kg) to rats trained to discriminate nic-otine from saline. At this dose, caffeine alone produced a partial nicotine-like discriminative effect, but also enhanced the effects of low nicotine doses. Together, these animal studies suggest that the effect of caffeine–nicotine combinations is generally greater than the effect of either drug alone, par-ticularly when the doses of caffeine used are those that have stimulant effects.

There is little information on the behavioral effects of nicotine–caffeine combinations in humans. Rose and Behm[9] asked subjects to report their arousal levels following administration of nicotine and caffeine. Nicotine was administered as tobacco smoke and caffeine as coffee, with appropriate placebos for each. Under the conditions of this study, neither caffeine nor nicotine had a significant effect on arousal. However, the combination decreased arousal from the baseline level. The authors suggested that caffeine may have induced a high arousal state, in which condition nicotine has a depressant effect. There is no direct evidence in support of this hypothesis.

Kerr et al.[10] found facilitatory effects of both nicotine and caffeine on both memory and motor function using a variety of psychomotor tasks. While there were differences across tasks, combining nicotine and caffeine did not appear to produce a greater effect than administering either drug alone.

In a survey of prison inmates conducted by Hughes and Boland,[11] high caffeine consumption was associated with negative mood states. In those subjects who consumed large amounts of caffeine and also smoked tobacco, the association was not present. Unfortunately, the nature of the study pre-cludes conclusions about whether these differences reflect drug effects or the characteristics of different drug users. The effects of caffeine on rates of cigarette smoking have been examined in several studies,[12a,b] but the results are inconclusive. In contrast, the elevated consumption of caffeine by ciga-rette smokers may be explained by increased caffeine tolerance in smokers as a result of the enzyme aryl hydrocarbon hydroxylase.[13]

The results from most animal studies show additive effects of combined nicotine–caffeine administration, but this is not the case with human studies. To date, there is little information from humans, and the effects of each drug alone vary across experiments. In part, this is because each study uses only a single dose of each drug. The more robust conclusions of animal studies are generally based on a range of doses of each compound.

II. Caffeine and benzodiazepines

Benzodiazepines are widely used for the treatment of anxiety and insomnia as well as a variety of other disorders. In most countries they are available on medical prescription. Problems of dependence and abuse have been recognized for some time, although the dependence potential of these drugs remains a matter of contention.[14] In addition to their anxiolytic and hypnotic effects, benzodiazepines produce sedation, muscle relaxation, and cognitive and psychomotor impairment.

A number of studies using animals have shown that caffeine and benzodiazepines share a mutually antagonistic relationship. In a study of locomotor activity in mice, Kaplan et al.[15] found increases following caffeine administration and decreases following alprazolam administration. Coadministration of the two drugs revealed that caffeine and alprazolam each blocked the behavioral effects of one another. Alternating mice, Zwyrbora et al.[16] observed a reduction in the locomotor stimulatory effects of caffeine by doses of diazepam that had no effect when administered alone. The same group also found that caffeine reduced the impairment in avoidance behavior produced by diazepam.[17] Additionally, caffeine has been shown to block a diazepam-induced reduction in skeletal muscle tone.[18]

This antagonistic relationship may hold for certain behavioral effects only. De Angelis et al.[19] found that the effects of caffeine were reversed by desmethyldiazepam and chlor-desmethyldiazepam in the hole-board and spontaneous motor activity tests in mice. In contrast, benzodiazepine coadministration enhanced the increase in open-field exploratory behavior induced by caffeine. It should be noted that in this latter test some benzodiazepine doses produced increased exploratory activity when administered alone. Similarly, caffeine reversed the depressant effects of diazepam on walking and rearing in the open field test, but failed to attenuate the decrease in ambulation.[20]

In two very detailed studies of the effects of caffeine and alprazolam on the performance of rats on differential reinforcement of low-rate (DRL) behavior, Lau and Wang[21] and Lau et al.[22] observed additive effects of the two drugs. Both drugs disrupted DRL performance when administered alone. The effects of combinations were then predicted based on these findings. It was shown that there was no significant variation from a model of simple additivity. Importantly, there was no evidence of antagonism.

These same authors earlier studied another type of behavior that is impaired by both caffeine and benzodiazepines: fine motor control. In studies of acute[23] and chronic[24] administration of caffeine and midazolam to rats, the combination produced a greater degree of impairment than either drug alone. Together, the findings of this group clearly demonstrate that the benzodiazepines and caffeine do not have a simple antagonistic relationship, particularly where their effects are similar when administered alone.

Several studies have been concerned with anxiolytic/anxiogenic effects of caffeine–benzodiazepine combinations. Using the social interaction test of anxiety in rats, Baldwin and File[25] found that the anxiogenic effects of caffeine could be reversed by chlordiazepoxide. Tang et al.[26] used the consumption of 1.5% NaCl solution by rehydrating rats as a measure of anxiolysis. Caffeine increased consumption of the solution at low doses (an anxiolytic effect), but decreased it at higher doses. Clonazepam had a clear anxiolytic effect which was attenuated by caffeine. Using the elevated plus maze, Bhattacharya et al.[27] observed anxiolytic effects following lorazepam and anxiogenic effects when caffeine was administered. Lorazepam clearly attenuated the anxiogenic effects of caffeine. A final model of anxiety is the reversal of suppression induced by a punishing stimulus. Coffin and Spealman[28] showed that the food-reinforced behavior of squirrel monkeys, suppressed by response produced electric shocks, was increased by chlordiazepoxide. Coadministration of caffeine either had no effect or magnified the increase. These findings would appear to contradict those described above using different methods and species, but confirm an earlier report from the same laboratory.[29] They also contrast with the findings from Polc et al.[18] using similar methodology in rats, who found that caffeine blocked the anticonflict effect of diazepam.

Finally, two studies of the discriminative effects of chlordiazepoxide and caffeine showed that each is able to block the discriminative effects of the other. Holloway et al.[30] found that chlordiazepoxide was able to block the discriminative effects of caffeine in a dose-related manner. However, it should be noted that the reversal was incomplete: doses of caffeine large enough to markedly decrease response rates did not completely abolish the caffeine discriminative cue. Pentobarbital failed to alter the caffeine discriminative cue, suggesting that the chlordiazepoxide-induced changes may be specific to benzodiazepines and not all sedative compounds. Gauvin et al.[31] showed a parallel shift to the right in the chlordiazepoxide dose–response curve following caffeine coadministration. However, after comparing the effects of caffeine with those of a benzodiazepine antagonist, the authors concluded that the effects of caffeine were due to perceptual masking rather than receptor interaction.

Studies with humans have generally supported the hypothesis of a mutually antagonistic relationship between caffeine and benzodiazepines. In a brief report, Mattila et al.[32] indicated that caffeine blocked the impairment of cognitive skills, increase in muscle relaxation, and subjective calming produced by 10 mg diazepam. Similar results were reported in a more detailed follow-up study.[33] Using the benzodiazepine triazolam, Mattila et al.[34] found relatively mild sedative effects that were largely reversed by caffeine. Similarly, the alerting effects of caffeine were diminished when triazolam was coadministered. Caffeine, at doses that had little effect when administered alone, has been shown to reverse the effects of both lorazepam and triazolam on learning, performance, and ratings of sedation.[35,36] File et al.[37] administered a battery of tests that measured psychomotor and cognitive performance, mood-state, and other factors. Lorazepam impaired psychomotor and cognitive performance. On some tests (e.g., digit-symbol substitution), caffeine improved performance and reversed the effect of lorazepam. Caffeine also reversed the self-reported anxiolytic effects of lorazepam. A similar battery of tests was used by Roache and Griffiths[38] in a study notable for using a range of doses of both diazepam and caffeine. In a number of tasks, but not all, diazepam and caffeine each blocked the effects of the other. In addition to the type of test, the effects were dependent on the exact doses of each drug.

Somewhat contradictory findings were obtained in two studies using caffeine and diazepam by Ghoneim and colleagues. Loke et al.[39] evaluated the effects of a range of doses of each drug as well as combinations of the two using a number of tests. The major effects of diazepam were on cognitive tasks and mood, while caffeine had effects on relatively few tasks. Only in the symbol cancellation test did caffeine ameliorate the impairment produced by diazepam. Ghoneim et al.[40] also found little evidence of reversal of diazepam-induced impairment by caffeine.

Johnson et al.[41] examined the day-after effects of two benzodiazepines, flurazepam and triazolam. Caffeine alone enhanced early morning alertness, but also reversed drowsiness evident in mornings after benzodiazepine administration.

It is apparent that a number of both animal and human studies have provided evidence of a mutually antagonistic relationship between caffeine and benzodiazepines. Such a relationship is most likely to be found where caffeine has an effect opposite to the benzodiazepine. This will depend on the nature of the task investigated as well as doses of the drugs tested. There are also many instances of the failure of caffeine to reverse the effects of benzodiazepines.

There are a number of possible explanations for the interaction between caffeine and benzodiazepines. Ghoneim et al.[40] examined the hypothesis that the interaction may be pharmacokinetic in nature. They found that coadministration of caffeine with diazepam resulted in a 22% reduction in plasma diazepam levels. Lau and Falk[24] observed some effect of alprazolam on caffeine absorption, but no other pharmacokinetic interaction between the two. The differences observed in each of the studies were of relatively small magnitude and unlikely to account for the mutually antagonistic relationships between caffeine and benzodiazepines.

A number of investigators have examined whether caffeine interacts with the benzodiazepine receptor. Recent evidence suggests that most of the effects of methylxanthines such as caffeine are due to their action as antagonists at adenosine receptors.[42] Nevertheless, some evidence has accumulated showing that caffeine interacts with benzodiazepine receptors. Caffeine has weak antagonist properties at these receptors, but this occurs only at relatively high concentrations.[43] Lopez et al[44] have evidence that caffeine, while not interacting directly with the benzodiazepine receptor site, may alter benzodiazepine function through changes in $GABA_A$ activity. Other studies have failed to find evidence of a common substrate for caffeine and benzodiazepines.[45]

The studies of Lau, Falk, and colleagues on DRL performance and fine motor control in rats would appear to confirm this view. For their measures, caffeine and benzodiazepines both had similar effects, and there was no evidence of one blocking the other. Rather, the relationship was a simple additive one. Such findings argue strongly against a common neural substrate at which caffeine and the benzodiazepines have opposite actions.

III. Caffeine and alcohol

The CNS effects of alcohol are complex, and the behavioral effects it produces are dependent on the concentration of the drug. With respect to general behavioral changes, low to moderate concentrations of alcohol often produce stimulation, while at higher concentrations depressant effects are more prominent. Alcohol enhances the activity of GABA at the $GABA_A$-benzodiazepine receptor complex and is a noncompetitive antagonist at the NMDA receptor.[46]

Animal behavioral studies examining the interaction between caffeine and alcohol have revealed a complex picture. In an early study, Waldeck[47] tested doses of caffeine ranging from 25 to 100 mg/kg in mice together with ethanol (1 to 4 g/kg). Ethanol alone produced moderate increases in activity, as did the two lower doses of caffeine. These doses of caffeine potentiated the locomotor stimulatory effects of low alcohol doses, but blocked the effects of the highest dose. The highest caffeine dose reduced the stimulatory alcohol effects. Thus, the effects of alcohol and caffeine were generally additive at the lower alcohol doses, but somewhat antagonistic at higher doses. Kuribara et al.[16] also observed increased locomotor activity in mice when alcohol (0.4 to 3.2 g/kg) and caffeine (10 mg/kg) were coadministered. The effects were somewhat less than the added effects of each drug administered alone.

A series of studies by Dar and colleagues examined ataxia induced by caffeine and alcohol.[48-50] Interestingly, they observed potentiation of ethanol-induced motor incoordination by caffeine. This was noted following administration of caffeine at doses of 45 and 90 mg/kg[48] and 62.5 mg/kg.[49] Dar[50] tested a range of caffeine doses and the results suggested a biphasic effect of caffeine: antagonism of the ataxic effects of alcohol at doses less than 20 mg/kg and enhancement of the effects at higher doses.

Like Waldeck,[47] Hilakivi[51] had a baseline of increased locomotor activity following ethanol administration. In this experiment, caffeine (60 mg/kg) blocked the effects of alcohol. In the same study, ethanol decreased social interaction — an effect that was reversed by administration of 30 mg/kg caffeine.

Taken together, these findings suggest that the nature of the caffeine–ethanol interaction depends critically on the effect of the ethanol dose, as well as the dose of caffeine. Where ethanol has a depressant effect, low doses of caffeine are able to reverse this change. Conversely, stimulant effects of alcohol are enhanced by low caffeine doses and blocked by high caffeine doses. Further studies using a range of caffeine and alcohol concentrations are needed to confirm these conclusions.

In a more complex study, Elsner et al.[52] examined the behavior of rats in a discrete trial spatial alternation schedule. Microanalysis of the behavior yielded a number of variables, but the authors suggested that they could be divided into those reflecting performance (accuracy, etc.) and those reflecting depression. The effect of caffeine was to potentiate the deleterious effects of alcohol on performance, but to reverse the alcohol-induced depression. According to their criteria, earlier studies had examined depression-related variables.

Avoidance behavior of mice was studied by Kuribara and Tadokoro.[17] Both caffeine and alcohol were able to decrease avoidance performance. However, a dose of caffeine with no effect on performance significantly worsened the alcohol induced decrement. Finally, Schechter[53] showed that caffeine (100 mg/kg) failed to reverse the discriminative effects of ethanol (15 mg/kg). By comparison, amphetamine (4.0 mg/kg) partially reversed these effects.

Human studies of alcohol-caffeine interactions will be described only briefly here; they are discussed in more detail elsewhere in this volume. Most attention has been directed to the disruptive effects of alcohol on skilled performance and the ability of caffeine to reverse these effects. In general, most studies have shown that caffeine reduces the impairment of performance in psychomotor and cognitive tasks produced by alcohol.[10,54,55] However, there are also contradictions in the literature that are not easy to resolve. For example, Nuotto et al.[56] failed to observe any modification of the effects of alcohol (1 g/kg) by doses of caffeine in the range of 200 to 500 mg. Franks et al.[57] also failed to observe any reversal of alcohol effects by caffeine in any test except their measure of reaction time. In apparent contradiction to almost all other studies, Osborne and Rogers[58] found that caffeine (150 mg) potentiated the increase in reaction time resulting from alcohol administration (0.18 g/kg). There are a number of factors that could explain these discrepancies, including variations in methodology, doses, and statistical criteria.

In both animal and human studies where data on blood concentrations of alcohol have been collected, interactions between the two compounds have been characterized as pharmacodynamic rather than pharmacokinetic in nature.[50] Furthermore, a number of effects of caffeine on alcohol-induced behavioral changes have been replicated with other methylxanthines.[59] This has led to the suggestion by Dar and colleagues that adenosine receptors may be the locus for the interaction between caffeine and alcohol. While caffeine is a known adenosine antagonist, the evidence for a direct effect of alcohol on these receptors is less convincing.

IV. Conclusion

Interactions between caffeine and widely used drugs such as nicotine, benzodiazepines, and alcohol constitute a potentially valuable area of research.

However, in no case can the research being carried out or the data be described as definitive or exhaustive. The clearest relationship seems to be between caffeine and benzodiazepines. Both animal and human studies show that caffeine and benzodiazepines frequently have opposite effects, and that combinations result in reduced effects. Where the behavioral effects are similar, the combination produces a greater effect than either drug alone. Thus, it is likely that caffeine and benzodiazepines combine in a simple additive manner, and the interaction cannot be attributed to action at a common receptor site. Animal studies are also consistent in showing an additive relationship between the stimulants caffeine and nicotine. Unfortunately, relatively few human studies have attempted to characterize this interaction, and the information available is not consistent.

The data are most contradictory for alcohol–caffeine interactions. Human studies have generally examined the effects of caffeine on alcohol-induced decrements in performance. While there is some evidence for a reduction in these decrements, there is also evidence that such a change does not occur. Animal studies are difficult to interpret because of the biphasic effects of alcohol. Thus, the effects of caffeine depend largely on the baseline provided by alcohol alone. In at least some studies, however, caffeine is able to reverse alcohol-induced sedation. Further research that uses a range of doses of both alcohol and caffeine is needed to characterize the interaction between the compounds more precisely.

References

1. Benowitz, N. L., Clinical pharmacology of nicotine, *Annu. Rev. Med.*, 32, 181, 1986.
2. White, J. M. and Ganguzza, C. C., Effects of nicotine on schedule-controlled behaviour: role of fixed interval length and modification by mecamylamine and chlorpromazine, *Neuropharmacology,* 24, 75, 1985.
3. Istvan, J. and Matarazzo, J. D., Tobacco, caffeine and alcohol use: a review of their interrelationships, *Psychol. Bull.*, 95, 301, 1984.
4. White, J. M., Behavioral interactions between nicotine and caffeine, *Pharmacol. Biochem. Behav.*, 29, 63, 1988.
5. Sansone, M., Battaglia, M., and Castellano, C., Effect of caffeine and nicotine on avoidance learning in mice: lack of interaction, *J. Pharm. Pharmacol.*, 46, 765, 1994.
6. Lee, E. H. Y., Tsai, M. J., Tang, Y. P., and Chai, C. Y., Differential biochemical mechanisms mediate locomotor stimulation effects on mood, memory, and psychomotor performance, *Psychopharmacology,* 87, 344, 1987.
7. Cohen, C., Welzl, H., and Bättig, K., Effects of nicotine, caffeine and their combination on locomotor activity in rats, *Pharmacol. Biochem. Behav.*, 40, 121, 1991.
8. Kuribara, J., Shonida, M., and Uchihashi, Y., Delayed effects of ethanol, caffeine and nicotine assessed by wheel-running and drinking in mice, *J. Toxicol. Sci.*, 20, 595, 1995.
9. Rose, G. E. and Behm, F. M., Psychophysiological interactions between caffeine and nicotine, *Pharmacol. Biochem. Behav.*, 38, 333, 1991.

10. Kerr, J.S., Sherwood, N., and Hindmarch, I., Separate and combined effects of the social drugs on psychomotor performance, *Psychopharmacology,* 104, 113, 1991.

11. Hughes, G. B. and Boland, F. J., The effects of caffeine and nicotine consumption on mood and somatic variables in a penitentiary inmate population, *Addict. Behav.,* 17, 447, 1992.

12. Marshall, W. R., Epstein, L. H., and Green, F. B., Coffee drinking and cigarette smoking. a. Coffee, caffeine and cigarette smoking behaviour, *Addict. Behav.,* 5, 389; b. Coffee, urinary pH and cigarette smoking behavior, *Addict. Behav.,* 5, 395, 1980.

13. Parsons, W. D. and Neims, A. H., Effects of smoking on caffeine clearance, *Clin. Pharmacol. Therap.,* 24, 40, 1978.

14. Wood, J.H, Katz, J.L, and Winger, G., Benzodiazepines: use, abuse and consequences, *Pharmacol. Rev.,* 44, 151, 1992.

15. Kaplan, G. B., Tai, N. T, Greenblatt, D. J., and Shader, R. I., Separate and combined effects of caffeine and alprazolam on locomotor activity and benzodiazepine receptor binding *in vitro, Psychopharmacology,* 101, 539, 1990.

16. Kuribara, H., Asahi, T., and Takokoro, S., Ethanol enhances, but diazepam and pentobarbital reduce the ambulation-increasing effect of caffeine in mice, *Jap. J. Alcoh. Drug Depend.,* 27, 528, 1992.

17. Kuribara, H. and Tadokoro, S., Caffeine does not effectively ameliorate, but rather may worsen the ethanol intoxication when assessed by discrete avoidance in mice, *Jap. J. Pharmacol.,* 59, 393, 1992.

18. Polc, P., Bonetti, E. P., Pieri, L., Cumin, R., Angioi, R. N., Möhler, H., and Haefely, W. E., Caffeine antagonises several central effects of diazepam, *Life Sci.,* 28, 2265, 1981.

19. De Angelis, L. N., Bertolissi, M., Nardini, G., Traversa, U., and Vertua, R., Interaction of caffeine with benzodiazepines: behavioral effects in mice, *Arch. Internat. Pharmacodyn. Ther.,* 255, 89, 1982.

20. Hughes, R. N., Effects on open-field behavior of diazepam and buspirone alone and in combination with chronic caffeine, *Life Sci.,* 53, 1217, 1993.

21. Lau, C. E. and Wang, J., Alprazolam, caffeine and their interaction: relating DRL performance to pharmacokinetics, *Psychopharmacology,* 126, 115, 1996.

22. Lau, C. E., Wang, Y., and Falk, J. L., Differential reinforcement of low rate performance, pharmacokinetics and pharmacokinetic-pharmacodynamic modeling: independent interaction of alprazolam and caffeine, *J. Pharmacol. Exp. Therap.,* 281, 1013, 1997.

23. Falk, J. L. and Lau, C. E., Synergism by caffeine and by cocaine of the motor control deficit produced by midazolam, *Pharmacol. Biochem. Behav.,* 38, 525, 1991.

24. Lau, C. E. and Falk, J. L., Sustained synergism by chronic caffeine of the motor control deficit produced by midazolam, *Pharmacol. Biochem. Behav.,* 40, 723, 1991.

25. Baldwin, H. A. and File, S. E., Caffeine-induced anxiogenesis: the role of adenosine, benzodiazepine and noradrenergic receptors, *Pharmacol. Biochem. Behav.,* 32, 181, 1989.

26. Tang, M., Kuribara, H., and Falk, J. L., Anxiolytic effect of caffeine and caffeine-clonazepam interaction: evaluation by NaCl solution intake, *Pharmacol. Biochem. Behav.,* 32, 773, 1989.

27. Bhattacharya, S. L, Satyan, K. S., and Chakrabarti, A., Anxiogenic action of caffeine: an experimental study in rats, *J. Psychopharmacol.*, 11, 219, 1997.
28. Coffin, V. I. and Spealman, R. D., Modulation of the behavioral effects of chlordiazepoxide by methylxanthines and analogs of adenosine in squirrel monkeys, *J. Pharmacol. Exp. Therap.*, 235, 724, 1985.
29. Valentine, J. O. and Spealman, R. D., Effects of caffeine and chlordiazepoxide on schedule-controlled responding of squirrel monkeys, *Fed. Proc.*, 42, 1158, 1983.
30. Holloway, F. A., Modrow, H. E., and Michaelis, R. C., Methylxanthine discrimination in the rat: possible benzodiazepine and adenosine mechanisms, *Pharmacol. Biochem. Behav.*, 22, 815, 1985.
31. Gauvin, D. V., Peirce, J. M., and Holloway, F. A., Perceptual masking of the chlordiazepoxide discriminative cue by both caffeine and buspirone, *Pharmacol. Biochem. Behav.*, 47, 153, 1994.
32. Mattila, M. J., Palva, E., and Savolainen, K., Caffeine antagonises diazepam effects in man, *Med. Biol.*, 60, 121, 1992.
33. Mattila, M. J. and Nuotto, E., Caffeine and theophylline counteract diazepam effects in man, *Med. Biol.*, 61, 337, 1983.
34. Mattila, M. E., Mattila, M. J., and Nuotto, E., Caffeine moderately antagonises the effects of triazolam and zopiclone on the psychomotor performance of healthy subjects, *Pharmacol. Toxicol.*, 70, 286, 1992.
35. Rush, C. R., Higgins, S. T., Bickel, W. K., and Hughes, J. R., Acute behavioral effects of lorazepam and caffeine alone and in combination, in humans, *Behav. Pharmacol.*, 5, 245, 1994.
36. Rush, C. R., Higgins, S. T., Hughes, J. R., and Bickel, W. K., Acute behavioral effects of triazolam and caffeine, alone and in combination, in humans, *J. Exp. Clin. Psychopharmacol.*, 2, 211, 1994.
37. File, S. E., Bond, A. J., and Lister, R., Interaction between effects of caffeine and lorazepam in performance tests and self-ratings, *J. Clin. Psychopharmacol.*, 2, 102, 1982.
38. Roache, J. D. and Griffiths, R. R., Interactions of diazepam and caffeine: behavioral and subjective dose effects in humans, *Pharmacol. Biochem. Behav.*, 26, 801, 1987.
39. Loke, W. H., Hinrichs, J. V., and Ghoneim, M. M., Caffeine and diazepam: separate and combined effects on mood, memory, and psychomotor performance, *Psychopharmacology*, 87, 344, 1985.
40. Ghoneim, M. M., Hinrichs, J. V., Chiang, C. K., and Loke, W. H., Pharmacokinetic and pharmacodynamic interactions between caffeine and diazepam, *J. Clin. Psychopharmacol.*, 6, 75, 1986.
41. Johnson, L. C., Spinweber, C. L., and Gomez, S. A., Benzodiazepine and caffeine: effect on daytime sleepiness, performance and mood, *Psychopharmacology*, 101, 160, 1990.
42. Nehlig, A., Daval, J. L., and Debry, G., Caffeine and the central nervous system: mechanisms of action: biochemical, metabolic and psychostimulant effects, *Brain Res. Rev.*, 17, 139, 1992.
43. Snyder, S. H., Katims, J. J., Annau, Z., Bruns, R. F., and Daly, J. W., Adenosine receptors and behavioral actions of methylxanthines, *Proc. Natl. Acad. Sci. U.S.A.*, 78, 3260, 1981.

44. Lopez, F., Miller, L. G., Greenblatt, D. J., Kaplan, G. B., and Shader, R. I., Interaction of caffeine with the GABA$_A$ receptor complex: alterations in receptor function but not ligand binding, *Eur. J. Pharmacol. — Mol. Pharmacol. Sect.*, 172, 453, 1989.
45. Sannerud, C. A., Marley, R. J., Serdikoff, S. L., Alastra, A. J. G., Cohen, C., and Goldberg, S. R., Tolerance to the behavioral effects of chlordiazepoxide: pharmacological and biochemical selectivity, *J. Pharmacol. Exp. Therap.*, 267, 1311, 1993.
46. Tabakoff, B. and Hoffman, P. L., Alcohol: neurobiology, in *Substance Abuse: A Comprehensive Textbook*, Ruiz, P. and Millman, R. B., Eds., Williams & Wilkins, Baltimore, 1993, 152.
47. Waldeck, B., Ethanol and caffeine: Complex interaction with respect to locomotor activity and central catecholamines, *Psychopharmacologia*, 36, 209, 1974.
48. Dar, M. S. and Wooles, W. R., Effect of chronically administered methylxanthines on ethanol induced motor incoordination in mice, *Life Sci.*, 39, 1429, 1986.
49. Dar, M. S., Jones, M., Close, G., Mustafa, S. J., and Wooles, W. R., Behavioural interactions of ethanol and methylxanthines, *Psychopharmacology*, 91, 1, 1987.
50. Dar, M. S., Biphasic effects of centrally and peripherally administered caffeine on ethanol-induced motor incoordination in mice, *J. Pharm. Pharmacol.*, 40, 482, 1988.
51. Hilakivi, L. A., Durcan, M. J., and Lister, R. G., Effects of caffeine on social behaviour, exploration and locomotor activity: Interactions with ethanol, *Life Sci.*, 44, 543, 1989.
52. Elsner, J., Alder, S., and Zbinden, G., Interactions between ethanol and caffeine in operant behavior in rats, *Psychopharmacology*, 96, 194, 1988.
53. Schechter, M. D., Effect of propranolol, d-amphetamine and caffeine on ethanol as a discriminative cue, *Eur. J. Pharmacol.*, 29, 52, 1974.
54. Rush, C.R., Higgins, S. T., Hughes, J. R., Bickel, W. K., and Wiegner, M. S., Acute behavioral and cardiac effects of alcohol and caffeine alone and in combination in humans, *Behav. Pharmacol.*, 4, 562, 1993.
55. Hasenfratz, M., Bungey, A., Dal Pra, G., and Bättig, K., Antagonistic effects of caffeine and alcohol on mental performance parameters, *Pharmacol. Biochem. Behav.*, 46, 463, 1993.
56. Nuotto, E., Mattila, M. J., Seppala, T., and Konno, K., Coffee and caffeine and alcohol effects on psychomotor function, *Clin. Pharmacol. Therap.*, 31, 68, 1982.
57. Franks, H. N., Hagedorn, H., Hensley, V. R., Hensley, W. J., and Starmer, G. A., The effect of caffeine on human performance, alone and in combination with ethanol, *Psychopharmacologia*, 45, 177, 1975.
58. Osborne, D. J. and Rogers, Y., Interactions of alcohol and caffeine on human reaction time, *Aviat. Space Environ. Med.*, 54, 528, 1983.
59. Dar, M. S., Mustafa, S. J., and Wooles, W. R., Possible role of adenosine in the CNS effects of ethanol, *Life Sci.*, 33, 1363, 1983.

chapter six

Caffeine and arousal: a biobehavioral theory of physiological, behavioral, and emotional effects

Barry D. Smith, Kenneth Tola, and Mark Mann

Contents

I. Introduction 88
II. Theoretical models of arousal 89
   A. Dual-interactionism 89
   B. A general arousal model 89
   C. An arousal model for caffeine 92
III. Neural and physiological mechanisms 93
IV. Effects of acute caffeine ingestion 94
   A. Physiological effects 94
   B. Interactions with other arousal agents 96
   C. Effects on behavior and performance 98
   D. Caffeine and cognition 98
   E. Caffeine and effect 105
V. Effects of habitual caffeine exposure 108
   A. Cardiovascular impact 108
   B. Physiological effects 109
   C. Effects on behavior and performance 109
VI. The interaction of habitual and acute ingestion 111
VII. Individual differences 111
VIII. Caffeine and psychopathology 112
   A. Anxiety 112
   B. The anxiety disorders 113

0-8493-1166-7/99/$0.00+$.50
© 1999 by CRC Press LLC

C. Caffeinism 113
D. Depression... 115
E. Schizophrenia 115
IX. Conclusion.. 116
References... 117

I. Introduction

Over 50% of both high school and college students report experience with illicit drugs, and 27% of all Americans abuse drugs at some time in their lives.[1] In fact, 20 million people in the United States reported heavily using and abusing drugs in 1994.[2] Some of these drugs, including cocaine and the amphetamines are stimulants that increase arousal and provide the user with a psychological "high. But cocaine, for example, was used by only 3.6% of the U.S. populace in 1994,[3] as compared with the 80% of U.S. adults who drink coffee or tea daily. The average adult in the U.S. and Canada consumes 4 mg/kg/day,[4] and many exceed 15 mg/kg.[5,6] Similarly, Asians and some Europeans consume large amounts of caffeine in tea. Coffee contains a larger amount of caffeine, averaging 85 mg for ground roasted and 60 mg for instant in 5 oz, while leaf tea (5 oz) averages 30 mg and instant tea averages 20.[7]

In addition to coffee and tea, the psychological effects of caffeine can be obtained from a number of other food sources. Chocolate is a popular and widely consumed source, but the drug is also found in considerable quantities in a number of medications, both prescription and over-the-counter (OTC). Caffeine tablets (e.g., No-Doz®) are sold for those who use the drug to study, drive, or engage in other activities. Less obvious is the caffeine content in analgesics, cold preparations, and anorectants.

The fact that caffeine is present in many different consumables means that investigators studying total consumption must be exceptionally careful to assess all possible sources of the drug in the dietary and pharmacological intakes of their subjects. It also means that in experimental studies where the desire is to dose caffeine relative to the subject's typical intake or to population averages, those consumption levels should be calculated taking all common sources into account. In some kinds of studies, where the concern is not only with caffeine *per se* but with arousal drugs more generally, other pharmacological sources of arousal must also be assessed. Ginseng, for example, has been widely advertised in recent years. It is available OTC and does act to increase arousal. Similarly, theobromine (2,7-dimethylxanthine) is found in cocoa and some other foods and may affect arousal.[8] Theophylline (1,3-dimethylxanthine) is found in some medications, in small amounts in tea, and in trace quantities in cocoa and coffee.[8]

Caffeine is clearly the most widely consumed of all psychotropic drugs.[5,9-11] Although its popularity is typically attributed to its stimulant effects,[11] it is also used because it slows and smoothes habituation[12,13] and because it enhances and sustains attentional focus.[14,15] These desirable effects and the resulting use (and sometimes abuse) of this powerful stimulant are among the reasons that it has been subjected to considerable scientific scrutiny in recent years.[16]

An additional reason for the increased scientific attention has been the concern that caffeine may have detrimental physiological and psychological effects. Physical health concerns have focused primarily on cardiovascular function,[7,17] with early studies suggesting that caffeine consumption may increase the risk of some cardiovascular problems.[18,19] On the psychological side, the drug can produce caffeinism [20] and may contribute to other disorders. We will take up both physical and psychological health issues later in this paper.

II. Theoretical models of arousal

A better understanding of the causes and effects of both acute and habitual caffeine consumption is likely to follow from the development and testing of a theoretical model of arousal that treats caffeine as one of a number of known arousal agents. The senior author has developed a set of models that place arousal within the broader context of a dual-interaction theory of behavior and its biological underpinnings. The arousal theory follows from this more general model and addresses the hypothetical effects of any arousal agent, including caffeine.

A. Dual-interactionism

The dual-interaction model (Figure 6.1) postulates that both physiological functioning and behavior are ultimately products of the characteristics of the person, the situational context, and the interaction of these two determinants. The person, in turn, is a product of developmental factors that are both biological (e.g., genetic) and environmental (e.g., learning, long-term caffeine exposure). The interaction of the two forms a separate source of variance. The person is seen in the model as consisting of both long-term or trait dimensions and short-term or state dimensions and has two important functional components, simply termed *arousal* and *cognition*. As the region-of-intrapersonal interaction concept suggests, these latter cross with the state–trait distinction to permit both state and trait forms of both arousal and cognition. The model places the person in a situational context, and it can be seen in Figure 6.1 that both the person and the situation contribute separately and interactively to both physiological functioning and behavior.

B. A general arousal model

Figure 6.2 presents a multidimensional model of the arousal component of the dual-interaction theory that updates and refines the one presented earlier.[21] Biological and environmental background factors contribute separately and interactively to both chronic and acute arousal. Exposure to such envi-

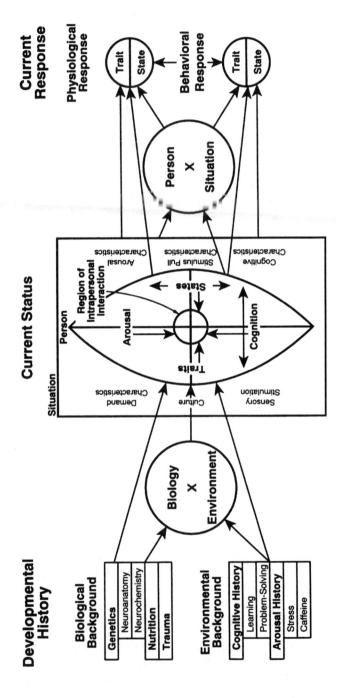

Figure 6.1 The dual-interaction model of physiology and behavior. (From Smith, B. D. Effects of acute and habitual caffeine ingestion on physiology and behavior: Tests of a biobehavioral arousal theory. *Pharmacopsychoecologia*, 151–167, 1994. With permission.)

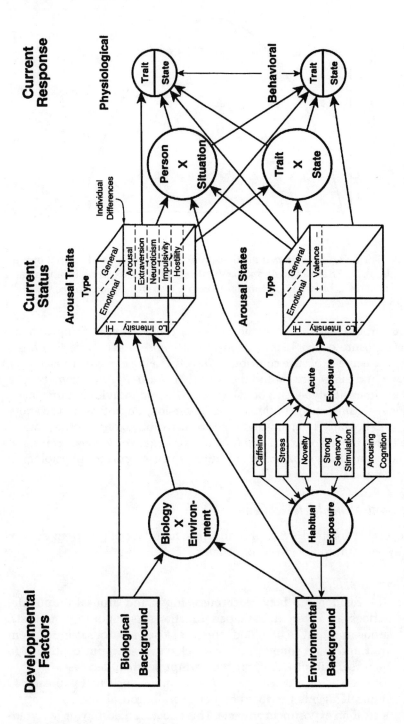

Figure 6.2 The biobehavioral arousal model of arousal. (From Smith, B. D., Effects of acute and habitual caffeine ingestion on physiology and behavior: Tests of a biobehavioral arousal theory. *Pharmacopsychoecologia*, 151–167, 1994. With permission.)

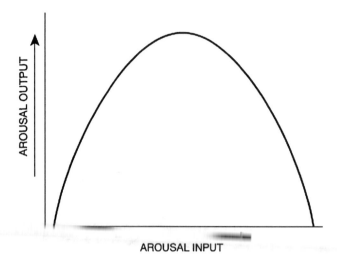

Figure 6.3 The inverted U-shaped relationship of arousal input and arousal output. (From Smith, B. D., Effects of acute and habitual caffeine ingestion on physiology and behavior: Tests of a biobehavioral arousal theory. *Pharmacopsychoecologia*, 151–167, 1994. With permission.)

ronmental arousal agents as caffeine and stress can be chronic, acute, or both. Chronic exposure contributes to arousal traits and thereby affects arousal states, while acute exposure contributes directly to the current arousal state. Both traits and states are multidimensional. The three dimensions of trait arousal are intensity, type (emotional or general), and individual differences, the latter including such continua as extraversion, neuroticism, sensation-seeking, and hostility. Intensity and type are also dimensions of state arousal, and the third dimension is valence (positive or negative). Arousal traits and states then contribute separately and interactively to current physiological functioning and behavior.

C. An arousal model for caffeine

Four general theoretical principles and five more specific hypotheses are especially relevant to the effects of caffeine.

1. General principles
- The relationship between arousal input and arousal output describes an inverted U-shaped function. As Figure 6.3 suggests, arousal output (e.g., blood pressure) rises as arousal input from trait and state sources increases. However, if input continues to increase, output will approach asymptote, then decrease.
- The effects of arousal agents are additive, such that exposure to multiple agents produces higher overall arousal and stronger effects than exposure to any one. This effect is subject to the inverted-U principle.

- There are individual differences in arousal or arousability (susceptibility), and these interact with exposure intensity to affect arousal
  output, physiology, and relevant behavior.
- Moderate habitual exposure to arousal agents acts to reduce the
  impact of acute exposure. More extreme habitual exposure may act
  to exacerbate the impact of acute exposure.

2. *The caffeine hypotheses*
   - Acute exposure to an arousal agent, such as caffeine, raises arousal
     levels, subject to the inverted-U principle.
   - Acute exposure to an arousal agent reduces the rate of habituation
     of physiological responses, subject to the inverted-U principle.
   - Acute exposure smooths or regularizes the habituation function.
   - Average habitual exposure moderates the effect of acute exposure.
   - Exposure to arousal agents affects performance only in areas rele-
     vant to arousal, including particularly behavior requiring focused
     attention.

The advantage of this model for caffeine effects is that it places the
habitual and acute impact of the drug in the broader context of a theory that
recognizes caffeine as one of a number of arousal agents. The model postulates that it is the combination of overall habitual and acute arousal levels
deriving from the action of multiple agents that most accurately predicts
current physiology and behavior.

III. Neural and physiological mechanisms

The acute effects of caffeine are mediated primarily through the action of
this trimethylxanthine alkaloid on receptors for adenosine.[22-25] The latter is
generated in extracellular media by the degradation of ATP and released
from metabolically active cells through facilitated diffusion.[26] Adenosine acts
to modulate the activities of a number of cell types, including neutrophils,
platelets, mast cells, smooth muscle cells, and a number of neuronal populations. Its effects appear to be primarily cell-protective, particularly when
ischemic or other metabolic stress occurs. These effects are mediated through
four principal receptor subtypes: A1, A2a, A2b, and A3.

As a neuromodulator, adenosine inhibits the release of several neurotransmitters. Among other actions, it inhibits conductance in N-methyl-D-
aspartate (NMDA) receptor channels[27] and contributes to vasodilatation in
hypercapnia-induced increases in cerebral blood flow (CBF).[28] Adenosine is
probably also the metabolite most responsible for ischemic muscle pain.[29]

Caffeine is a base analog that acts to antagonize adenosine receptors,
thereby affecting a variety of cell populations by partially counteracting
many of the effects of adenosine. It acts primarily on the A2a adenosine
receptors, elevating energy metabolism in the brain[29-31] and producing a 30%

decrease in whole-brain CBF, with no regional differences.[32] Caffeine also modulates neural activity through its inhibitory effect on ionotropic GABA receptors[33] and, like other methylxanthines, acts on serotonin and noradrenaline neurons and affects local dopamine release.[31] These actions further explain its stimulating effects, which include activation of major components of the hypothalamic–pituitary–adrenal (HPA) axis response. In particular, caffeine increases adrenocorticotropin (ACTH) release at the pituitary, resulting in elevated cortisol production.[34]

Caffeine also acts to disturb calcium (Ca^{2+}) homeostasis, releasing Ca^{2+} from the neuronal sarcoplasmic/endoplasmic reticulum and hence depleting neural calcium stores.[35] This results in an inhibition of protein synthesis due to the suppression of translational initiation,[36] an effect similar to that seen in transient metabolic stress, such as that associated with ischemia and hypoglycemia. In chronic use, caffeine selectively increases opioid neuropeptide messenger RNA (mRNA) expression,[37] affecting both preproenkephalin (PPE) and preprodynorphin (PPD) mRNA.[38] The result is up-regulation of both adenosine and prostacyclin receptors.[39]

The metabolic breakdown of caffeine is extensive, with primary and secondary biotransformation yielding at least 17 metabolites.[40] The metabolites are identifiable through routine urinalysis, and most of the responsible enzymes are known. The 3-demethylation pathway is among the major routes for caffeine metabolism. It involves sequential catalysis by cytochrome P4501A2, xanthine oxidase, and N-acetyltransferase.[41]

IV. Effects of acute caffeine ingestion

It is widely recognized by both scientists and the general public that the arousal increment following the morning cup of coffee or tea is not primarily a placebo effect. In fact, caffeine is a powerful stimulant, and the arousal model presented above outlines more specific ways in which caffeine acutely affects physiological functioning and behavior.

A. Physiological effects

The neurophysiological mechanisms activated by the ingestion of caffeine mediate widespread effects of the drug on physiology. One of these effects is blood pressure elevation. We recorded pressures while each of 24 high hostile and 24 low hostile subjects performed stressful mental arithmetic and anagram tasks under double-blinded caffeine (400 mg) or placebo conditions. Results indicated that caffeine produced significant elevations in systolic and mean arterial pressures. Others have also noted the effects of caffeine on blood pressure,[42-49] and elevations based on increased systemic vascular resistance have been confirmed by nuclear ventriculography and impedance cardiography.[50] However, these effects are typically quite small[51,52] and may be altogether absent in habitual caffeine consumers.[43]

Blood pressure studies have also examined the development of tolerance to caffeine. In one recent study, subjects were given daily low or high doses of caffeine or a placebo, and complete tolerance to the hypertensive effects of the drug developed within five days.[53] Others have also shown blood pressure tolerance with regular consumption.[43,52] On the other hand, Lane and Manus[54] report incomplete tolerance to repeated caffeine doses, even in habitual drinkers, and a 9-week abstinence from caffeine did not affect blood pressure responses in another study.[55] In addition, groups at risk for hypertension may have higher resting pressures and hence be pushed into an unacceptably high range by caffeine, particularly when it is combined with situational stress.[49,56,57] The mixed results in these studies, particularly with regard to tolerance, suggest the need for further investigation of the acute effects of caffeine in high and low habitual consumers and in both normal and high-risk groups.

The effects of caffeine on the heart are considerably more complicated. Depending on the conditions existing in a particular study, heart rate may increase,[45,46,49,54,58-61] decrease,[62-64] or show no change.[43,65-70] Such mixed findings suggest that heart rate is, at best, a complex index of caffeine-induced arousal.

Electroencephalographic (EEG) changes following acute caffeine administration typically show elevated arousal and may also have implications for the effects of caffeine on attention. The most straightforward finding is relative EEG activation under caffeine,[71] with decreased alpha activity.[72] Caffeine also produces changes in auditory evoked potentials recorded during a visual vigilance task, including a significant decrease in P2-N2 amplitude.[72,73] These results suggest that caffeine not only elevates arousal, but also helps to maintain attentional focus, a result supported by vigilance performance data discussed below.

Further evidence for the arousing effects of caffeine is seen in research on electrodermal activity (EDA), which has been one major focus of work in our laboratories. In a series of studies, we have administered caffeine and examined its effects on EDA under a variety of conditions. These have included reaction time, paired associates, anagram, and vigilance tasks, as well as simple auditory stimulation and novel digit recall.[74-76] We have recorded primarily skin conductance level (SCL), skin conductance response (SCR), and EDA response frequency measures.

Skin conductance level is a reliable index of tonic arousal,[77] and results have generally supported the hypothesis that it will increase with acute caffeine administration. In fact, SCL following caffeine administration typically remains significantly elevated under a variety of conditions throughout the experiment.[13,74,75,78] An important additional finding is that caffeine also affects the habituation of arousal. As Sokolov[79] hypothesized, repetitive presentation of a simple stimulus typically produces response diminution or habituation.[80] Our results show that caffeine affects this process, slowing and stabilizing it. Over a series of trials, subjects operating under placebo conditions show a quadratic SCL function, with habituation followed by facilitation, while those given caffeine display a much more stable, nearly flat function.

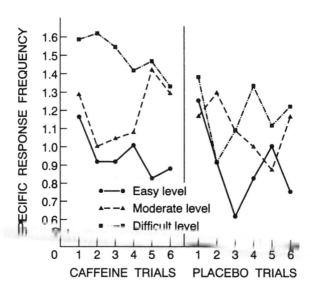

Figure 6.4 Effects of caffeine on habituation of EDA response frequency. (From Smith, B. D., Effects of acute and habitual caffeine ingestion on physiology and behavior: Tests of a biobehavioral arousal theory. *Pharmacopsychoecologia*, 151–167, 1994. With permission.)

A second electrodermal measure, response frequency, follows a similar pattern. It tends to be elevated by caffeine,[13,81] and it habituates with repeated stimulation. As with SCL, caffeine acts to somewhat reduce intertrial variability and smooth the habituation function (Figure 6.4).[12,13]

The phasic SCR provides an additional and quite different measure of the stimulant effect of caffeine, in this case on momentary changes that immediately follow a stimulus. Our studies have consistently shown that, under a variety of stimulus and task conditions, SCR amplitude is increased by caffeine, as predicted by the model.[76] Again, the less obvious predictions concerning habituation are also borne out by the data. The first effect of caffeine is to reduce the rate of SCR habituation.[12,82] The second is to smooth the SCR habituation function, such that decrementation proceeds as a relatively continuous process, rather than exhibiting the "choppy" up and down pattern often seen under placebo conditions.[12] As we will see, the slowing and smoothing of the habituation process not only supports predictions from the theoretical model, but may also provide important information relevant to our quest to determine why caffeine consumption is so widespread.

B. Interactions with other arousal agents

The predicted interactive effects of caffeine with other arousal agents have been reported in both blood pressure and electrodermal data. Some results are quite straightforward in showing that stress and caffeine demonstrate additive effects on blood pressure and other cardiovascular indicators.[44,83,84]

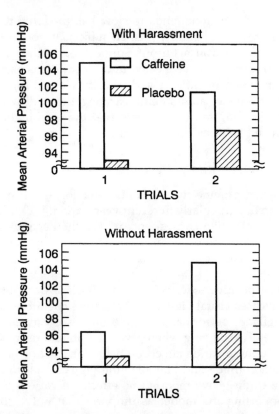

Figure 6.5 Interactive effects of caffeine and interpersonal stress (harassment) on blood pressure. (From Smith, B. D., Effects of acute and habitual caffeine ingestion on physiology and behavior: Tests of a biobehavioral arousal theory. *Pharmacopsychoecologia*, 151–167, 1994. With permission.)

In one such investigation, subjects were given caffeine (250 mg) or placebo, then subjected to stressful mental arithmetic, cold pressor, and static exercise tasks. The drug enhanced the impact of the stressors, yielding an additive effect on blood pressure.[84] A more complex result was seen in the recent blood pressure study in my laboratories described above. Subjects under caffeine or placebo performing mental arithmetic and anagram tasks received harassment (e.g., "Come on, you can do better than that") or no harassment over a series of trials. Results for SBP, DBP, and MAP were consistent in showing that caffeine without harassment produced an increase in pressure, while caffeine with harassment produced a decrease (Figure 6.5). This result supports the inverted-U postulate in that the combination of caffeine and stress pushed subjects past the peak of the function and produced a decrease in arousal output.

Similar results have been obtained with EDA data when white noise or novelty are employed as additional arousal agents. Subjects given caffeine (300 mg) displayed faster phasic SCR habituation with white noise than without.[21] Since caffeine alone tends to slow habituation,[12] the addition of the white-noise stressor apparently pushed subjects beyond the peak of the inverted-U and reduced arousal output. Support for the theory has also been reported in electrodermal data collected under conditions of novel stimulation.[13] The combination of caffeine and novelty yielded higher SCLs than either caffeine or novelty alone, demonstrating the additivity of the two arousal sources. In addition, it is notable that the combination of the two virtually eliminated habituation over trials.

C. *Effects on behavior and performance*

The effects of acute caffeine ingestion on behavior and task performance have been explored, particularly in recent years, and it is clear that there are both positive and negative consequences.[85] On the negative side, caffeine acts to disturb sleep.[ ] The drug that this has been anxiety a result seen in our recent blood pressure study, and even produce the anxiety-like syndrome *caffeinism*.[20]

On the more positive side, caffeine produces a desirable increase in alertness,[23,87] increases visual vigilance,[88] improves cognitive function,[89] and can improve vigilance performance.[90] On a more subjective level, the drug produces a sense of well-being when given to habitual users.[87]

In addition to these varied effects on behavior, caffeine has been shown to decrease reaction time[91] and liberalize the vigilance response criterion.[21] However, some studies have reported no effects on various aspects of performance,[79] supporting the model assumption that only arousal-relevant performance is affected.

D. *Caffeine and cognition*

Caffeine affects cognitive functioning in a number of ways. We will consider vigilance performance, information processing studies, memory, selective attention, and complex cognitive functioning.

1. *Vigilance*

One aspect of cognitive performance in which there is considerable consistency in the literature as to the effects of caffeine is vigilance. First, caffeine has consistently been shown to increase arousal level during prolonged vigilance tasks.[91-94] Second, caffeine improves performance on these tasks. An example is a study conducted by Lieberman and his colleagues,[91] who used the Wilkenson Vigilance Test.[91] Above a white-noise background, subjects were presented with a 400 ms tone every two seconds for 1 hour. Targets were 70 ms tones that were randomly interspersed among the longer tones, and subjects responded by pressing a key on a keyboard. Hits, misses, and false alarms were recorded. The investigators found that low (32 mg) to moderate (256 mg) doses of caffeine increased the number of correct hits without affecting the error rate. Similar results were reported in a study in which subjects received a single, timed-release dose (600 mg) of caffeine. The caffeine mimicked the benefits of sleep. Sleep-deprived subjects taking the caffeine performed the same on tests of vigilance and alertness as those who received a normal night's sleep.[95,96] Caffeine has also been shown to increase the metabolic rates of sleep-deprived subjects to their predeprived levels.[97] A number of other studies have also shown that caffeine improves vigilance performance[98-100] on both auditory[98-104] and visual[88,104] tasks. Conversely, decrements in performance have been attributed to low levels of arousal,[100] while higher levels improve performance on sustained attention tasks.[14,15,105,106]

Although some earlier work suggested negative effects on cognitive functioning in children, a meta-analysis found no supportive evidence.[107] In fact, children exhibit improved functioning on both auditory vigilance[108] and sustained attention[108] tasks when caffeine is given. Another study showed that these beneficial effects are seen not only in younger children, but also in adolescents between the ages of 11 and 15.[109] In addition, moderate doses (200 mg) decrease self-reported feelings of boredom, possibly providing a partial explanation for the positive effect of caffeine in vigilance and sustained attention tasks.[110] These are the same results reported in a study of middle-aged subjects.[111] More generally, a recent review showed that caffeine improved vigilance performance in 14 of 17 studies of adolescents.[112]

Despite the apparent overall positive effect of caffeine on vigilance, there is evidence to suggest that there may be a limitation on the size of the arousal increment (induced by caffeine, intense stimulation, or both) that will produce improvement in vigilance studies.[111,113] If that limit is exceeded, performance may approach asymptote or deteriorate. Frewer and Lader[114] found that low doses of caffeine improved vigilance performance, while high doses initially impaired it. After 2 hours, however, these high doses also improved vigilance. In another study, the senior author and colleagues preselected subjects high and low in habitual caffeine ingestion. Using double-blind controls, they were given either caffeine (4.0 mg/kg) or a placebo and exposed to 93 dB background white noise or no noise. A visual vigilance task consisted of 3 blocks of ten 1-min trials with targets appearing on trials 1, 3, 4, 7, and 10 of each block. Stimuli were angled lines appearing on a computer screen; targets were vertical lines. Results showed that caffeine decreased target response times (improved performance) when there was no white noise, but increased response times when the 93 dB noise was present.[115] In addition, subjects with histories of habitually high levels of caffeine use exhibited relative decrements in vigilance performance.

The results of these two studies and others provide further support for the theoretical inverted U-shaped arousal function. In the first case,[114] high doses of caffeine pushed subjects over the top of the curve and hence impaired vigilance performance. As time passed and the subject became more accustomed to the experimental situation, arousal returned to more optimal levels and performance improved. In the second study,[115] caffeine combined with moderately high intensity situational stimulation (white noise) to drive arousal up beyond the point of transmarginal inhibition and, again, decrease performance.

2. Information processing

Information-processing theory[116] has served as a partial basis for a number of studies of the effects of caffeine on cognitive functioning. Several studies have shown beneficial effects of the drug in rapid information processing (RIP) tasks.[117-119] Other work demonstrates that caffeine enhances problem-solving[120] and improves logical reasoning,[121,122] as well as increasing performance on mental arithmetic tasks.[123,124] Studies have also shown that caffeine counteracts the effects of sleep deprivation on spatial capacity[98] and abstract reasoning[101] tasks and partially reverses age-related deficits in cognitive functioning.[125]

As an example, Hasenfratz and Bättig[126] used a rapid information-processing task in which a subject looked for a 3-digit odd or even target on a computer screen that also presented other, nontarget digits. When the subject perceived the target, he or she pressed a response key as quickly as possible. The investigators found that a low dose (150 mg) of caffeine improved performance on the RIP task, while a higher dose (600 mg) impaired performance.

This elucidates the first potential confound in caffeine studies, namely the lack of control for dose level. Warburton[120] found that low (75 and 150 mg) doses of caffeine cause significant increases in performance on attentional, problem-solving, and memory tasks. Other studies have shown that high doses of caffeine can interfere with performance on complex tasks,[127] including the processing of ambiguous information.[128] They can also negatively affect scores on such cognitive tasks as spatial abilities in refreshed subjects.[101] In other studies, in which dose level was controlled, caffeine has not affected the search/detection domain of information processing[129] or reading comprehension at high or low dose levels.[130]

In addition to dosage variability, some of these differential cognitive results may be a product of baseline or prestimulation arousal level differences. When baseline arousal is already elevated, highly arousing tasks, such as those that are complex or involve intense stimuli, can result in overstimulation. As Hasenfratz and Bättig[126] note, it may be such excessive levels of stimulation that account for the commission errors seen in the RIP paradigm. Baseline arousal levels therefore need to be taken into account in estimating or calculating overall levels of stimulation and resultant activation.

One important influence on baseline arousal is task difficulty or complexity. Complex information-processing tasks, have been shown to increase arousal levels in otherwise unstimulated subjects.[131] The more complex the task, the higher the arousal level and the greater the chance that a given caffeine dosage would contribute to overstimulation. The additive nature of arousal sources (in this case, caffeine and complexity) means that arousal is high when caffeine is present and complexity or difficulty level is substantial. The result is that caffeine impairs performance on complex tasks.[127,128]

A second factor in prestimulation arousal level appears to be gender. When gender differences in activation have been specifically examined, female subjects have been quite consistently shown to exhibit higher levels of arousal or greater arousability than males. Physiologically, this arousal differential has been reported for both EEG[132-134] and EDA[76,135] measures. In studies involving mixed-gender groups with no analysis of the gender factor, it is possible that caffeine-induced increases in information processing in males would be counteracted by decrements in females. Depending on the gender balances of the groups, results could be considerably distorted. Stimulation that is highly arousing for males could lead to overarousal in females, whose responses would be dampened as suggested by the curvilinear (inverted U-shaped) relationship between arousal input and arousal output.

Interestingly enough, most of the studies reporting positive effects of caffeine in information-processing tasks have used primarily or exclusively male subjects.[117,120,121,123,124,136] In fact, the only study that found positive cognitive effects of caffeine on females employed fatigued women, whose baseline arousal levels may have been lower.[119] Conversely, most studies that have shown either detrimental effects[98,127] or no effects[129,130] of caffeine on information processing have used both males and females in the same study. In fact, the only study with this result that used exclusively male subjects also used moderate (250 mg) doses of caffeine with a difficult (visual Stroop) task. The combined effect of the drug and the task probably produced a degree of arousal high enough to interfere with cognitive performance. Caffeine thus contributed to an overall detrimental effect of arousal.[128] Another study, using the same Stroop task and low levels of caffeine, found that caffeine does not impair complex visual processing.[137]

A different type of study showing detrimental effects used the same caffeine dose on both tired and refreshed subjects.[101] The tired subjects were aroused to the point of showing beneficial effects from the drugs. Therefore, the same dose should be expected to overarouse the refreshed subjects and decrease their information processing abilities, as was the case.

3. Effects on memory

A major focus of cognitive psychology, memory has been studied intensively, and a number of investigations have addressed the effects of caffeine. Unfortunately, results in this important area of mental performance are quite mixed. First, a number of studies have found that caffeine enhances memory performance in several paradigms. It has been shown to improve delayed recall,[120] recognition memory,[138] semantic memory,[89] and verbal memory in general.[139] Other studies have shown significant increases in memory performance on both easy[140] and difficult[110] memory tasks. Females (but not males) in one study showed positive effects of caffeine on word list retention, though only when the words were presented at a slow rate.[141]

One study demonstrated that caffeine has cholinergic properties that can enhance cognitive functioning. In this investigation, the result was improved performance on short- and long-term memory retrieval, reading speed, and encoding efficiency.[142] In another study,[143] it was found that caffeine improves memory consolidation. Caffeine has also been shown to counteract the detrimental effects of aging on general memory performance.[125]

Complicating matters considerably, some studies have gotten opposite or equivocal results for memory tasks. Caffeine can actually decrease immediate word list recall, at least under some circumstances.[144,145] It has even been shown to amplify the detrimental effects of alcohol on memory.[146] Further clouding the picture, other studies have shown that caffeine has no effect on recall.[14,130,147] Some have found no effect on short-term memory with 64 mg of caffeine[148] or with 100,[149] and no effect on long-term memory with 64 or 128 mg.[148] Verbal learning and memory were also unaffected by doses of 125 to 500 mg.[150] In addition, no effect was seen on implicit or incidental memory,[151] as was also true for delayed recall with 3 and 6 mg/kg[152] or with 200, 400, or 600 mg.[145] This latter result is somewhat offset by a study in which delayed recall was improved by caffeine, even though immediate recall was not.[120] Finally, Linde[101] tested both fatigued and nonfatigued subjects on an auditory attention task requiring immediate recall. She found no effects of 150 or 200 mg doses of caffeine.

Given the mixed results in the literature, it is difficult to know just how caffeine does affect memory. To some extent, the differential effects may depend upon the memory assessment method (recall or recognition) and the time frame (immediate or delayed). Gender differences may also cloud the picture, as discussed above. Even when these differences are taken into account, however, unexplained discrepancies remain. One partial explanation may be that the differential effects of caffeine are a function of the subject's memory load. For example, Anderson[140] found that caffeine enhanced low load memory tasks but was detrimental in high load tasks. This could be due to the increase in arousal induced by the high load task, which, in the presence of caffeine, could produce overarousal. The drop in arousal output as the subject crossed the peak of the inverted U-shaped function could cause the memory deficits observed in some studies.

An alternative or additional explanation for the mixed results may lie in the differentially arousing effects of other factors present in the experimental situation. One study in our laboratories involved a backward recall task. Each subject received 3 blocks of 4 recall trials each. In a repetitive condition, each block involved one particular stimulus type (letters, digits, or color names). In a novel condition, each block contained one trial of each of the three stimulus types. Using a double-blind procedure, each subject was randomly assigned to receive either 300 mg caffeine or a placebo prior to performing the memory task. Half of each drug group was randomly assigned to a noise condition, involving a constant 80 dB white-noise stimulus, and half to a no-noise control condition. Results showed that both novelty and white noise improved recall performance under placebo and decreased it under caffeine.[13] It appears that the additional arousal generated by novelty and white noise served to push caffeine subjects over the top of the inverted-U curve and hence decrease their recall performance. Thus, it may be that caffeine does improve memory performance under conditions that otherwise produce low arousal. However, any condition causing over-arousal, whether or not it is a part of the memory task itself, can yield performance decrements. Confirming this hypothesis, Kaplan[153] found dose-dependent caffeine effects on memory. At lower doses (250 mg), subjects showed increased working memory ability. However, at higher doses (500 mg), subjects started to show impaired working memory performance. These results support the hypothesis that memory effects are dependent on both caffeine[154] and the level of background arousal. Studies not controlling for these factors may well have confounded results.

4. Selective attention

Attention has also been studied in paradigms other than vigilance, and results show enhancement of selective attention by caffeine.[120,155] Several studies have shown that caffeine improves performance on selective[156] and divided[105] attention tasks. For example, Lorist and Snel[156] found that 3 mg/kg of caffeine improved response time in a selective attentional task without decreasing accuracy. Similarly, another study showed that this increased attentional ability extends to self-focused processes. Subjects were better able to determine their current physical state after ingestion of 250 mg of caffeine than with placebo.[157]

Another approach to studying attention involves evoked potential responses. In particular, an EEG response occurring 300 milliseconds after presentation of a novel stimulus (P300 wave) has long been associated with visual[158-161] and auditory[162-164] attention. Larger P300 waves have been associated with more intensive attention, and smaller P300 waves with decreased attention.[159,160]

One method of presenting novel stimuli is the oddball paradigm. In this task, the subject has to attend to a certain auditory or visual stimulus. If the task is passive, the subject simply attends to the stimuli. If the task requires active attention, the subject is told to respond to any changes in the stimulus. When a novel stimulus is presented, EEG shows a consistent P300 response.[165-169] This response occurs regardless of differences in passive or active attention, although the wave is larger during the active tasks.[170,171]

Researchers have applied the oddball paradigm to an examination of the effects of caffeine on attention.[169] Kawamura[169] found that 500 mg of caffeine significantly increased the P300 wave and that its effects lasted 210 minutes. Other work has also shown that caffeine increases the P300 wave.[172] This provides further evidence that the drug can enhance attentional focus.

5. Complex cognition in work settings

Real-life settings, including work settings, often require more complex and varied cognitive functioning than is the case for many laboratory tasks. One case in point is decision-making, in which multiple types and sources of information are logically combined to yield an informed decision.[173] What effect does caffeine have on such complex reasoning processes? While little work has been done in this area, it thus far appears that the drug has both positive and negative effects. On the positive side is the reduction in time required to arrive at a decision.[174,175] On the negative side, potential resources may be neglected.[176] Streufert and colleagues[176] found that managers who drank 400 mg of caffeine or more on a daily basis often arrived at decisions much faster than other managers. However, this high-caffeine group also tended to show a decrease in their utilization of potential resources when making their decisions.[176]

Research on the effects of caffeine in the workplace is also rather sparse. However, there have been some studies, and it appears that the drug can have positive effects on mental performance. In one study, for example, managerial effectiveness was reduced when employees abstained from caffeine.[177] Another study found that moderate doses of caffeine (6 mg/kg) caused people to work harder because they underestimated how hard they were working. The workers thought they were working at the same speed they worked without caffeine, but they were actually working much faster.[178]

It is commonly assumed that the ubiquitous office coffee pot is heavily used by workers in order to increase their levels of wakefulness, alertness, and, more generally, arousal.[176] There may, however, be a number of additional perceived or actual benefits of work-related caffeine intake.[92] Headaches, for example, are often reported in work settings, and one study showed that workers sometimes consume caffeine primarily to relieve them.[179] This finding is consistent with the fact that caffeine inhibits adenosine, thereby reducing ischemia and other forms of metabolic stress,[29] and with its consequent widespread medical use in treating headache.[180,181]

A more common and salient, though more subtle, basis for on-the-job consumption may be boredom. For some people, work is perceived as boring, and boredom is, in part, a function of habituation. The habituation process takes place when there are multiple repetitions of the same stimulus complex or a continuation of that complex over time.[182] Habituation is basically a process of physiological and psychological adaptation to stimuli that cease to yield new information.[182] While it is an adaptive mechanism, in that it moves noninformative stimuli into the background and permits active attention to focus on new information, it can also have negative effects. In particular, subjects who become highly habituated or overhabituated experience psychological discomfort, fatigue, and boredom.[12] Caffeine can partially offset these detrimental effects of repeated stimulation by slowing the rate at which habituation occurs and smoothing the process.[13] Thus, workers may consume the drug not only to increase alertness, but also to slow and smooth habituation.[183]

There are also documented negative effects of the amounts of caffeine often consumed at work.[179] Most of these revolve around the stresses commonly perceived to be present in job settings.[184] Caffeine has been shown to exacerbate the effects of stress on neuroendocrine responses[185] and cardiovascular functioning, particularly blood pressure.[42,84,123,186] It is no surprise, then, that it has these same effects in work settings, predisposing the individual toward strong physiological and psychological responses to work stressors.[187] The blood pressure response to work stress appears to be particularly prone to caffeine enhancement.[52,188] In addition, a study of telemarketers showed that employees became more psychologically sensitive to job stressors after consuming caffeine.[189]

E. Caffeine and effect

If cognition is one side of the psychological coin, emotion is the other, and caffeine has been shown to have important effects on this aspect of functioning as well. Depending upon dosage level and other concurrent factors, caffeine can result in either positive or negative mood changes.

1. Caffeine and mood state

Several studies have shown that caffeine can improve mood states, increasing the frequency of positive mood self-reports [14,45,87,190-193] in both regular caffeine users and nonusers.[191] In fact, some evidence suggests that long-term use of the drug may improve overall mood.[194] Caffeine has also been shown to improve the negative moods often seen early in the morning[195,197] and after lunch[198] and can even counteract the mood deficits found after up to 48 hours of sleep deprivation.[93,97,198] Studies show that timed released of moderate amounts of caffeine can counteract the deleterious effects of sleep deprivation on mood[96] and that caffeine can be as effective as prophylactic naps.[198] In fact, the increases seen in cognitive tasks have sometimes been attributed to the positive mood effects of caffeine.[15]

In drug interaction studies, caffeine has been shown to amplify the positive mood changes induced by other stimulants.[199] In one study, for example, the drug enhanced positive mood states induced by nicotine.[200] In another, relatively low doses of caffeine enhanced mood states and increased the desire for cocaine in drug abusers.[201] At higher doses, it had a slight negative effect on mood. In addition to stimulant interactions, there have also been studies examining the interaction of caffeine with depressants. One study showed that people who have consumed moderate amounts of both alcohol (.8 g/kg) and caffeine (400 mg) report generally improved mood states.[202] Caffeine interacts with aspirin to improve mood state as well.[203]

Since caffeine consumption has resulted in improved mood, it is not surprising that caffeine abstinence in regular users can result in negative moods.[120,191] Both acute withdrawal[204] and overnight abstinence[205] appear to cause dysphoric mood states.[205,206] This has led some researchers to question whether caffeine actually improves mood or simply negates the negative mood withdrawal symptoms that result from abstinence.[207]

Whatever the mechanism, caffeine in moderate amounts does improve mood state, but the effect is dose dependent.[99] Although some work does show a positive, linear dose–response relationship,[208] most studies show mood decrements at high doses.[126] Caffeine doses of 100 mg have been shown to increase vigor[209] and feelings of well-being.[210] Even doses as high as 300 mg create positive mood states of "mental sedation" in some studies.[149] However, doses over 300 mg tend to increase tension and anxiety.[110] High caffeine doses, for example, caused inmates in one study to experience anxiety, frustration, and irritability,[211] and the drug has been shown to increase "tenseness" and "nervousness" as well.[110] Complicating matters further, there are individual differences in sensitivity to the mood state effects of varied doses of caffeine.[210] As with cognitive performance, the inverted-U function may best explain the mood effect of the drug.

Another factor that influences the effects of caffeine on mood is the expectation of the person consuming it. In one study in which subjects did not know they were getting caffeine, for example, they showed no mood effects until they reached dosage levels they could detect.[85] Similarly, another study showed that the energizing effects of cola come from the anticipation of caffeine rather than from any actual caffeine.[111] A third investigation confirmed that people who knew they were consuming caffeine reported positive mood effects.[213] On the other hand, if they unknowingly consumed the same amount of caffeine, they reported negative mood changes. The other side of this coin is that participants who anticipate caffeine but receive a placebo report significantly higher negative mood states.[214] Another study showed that subjects who chose to take caffeine felt no increase in anxiety, while those who had no choice did self-report increased anxiety with the same levels of caffeine.[215] Finally, anticipation of caffeine has been shown to increase the effect size of the actual amount of caffeine that is ingested.[47] For example, if a subject is told that he or she will be ingesting a high dosage of caffeine, a low dosage will have a greater effect on that subject.

2. The interaction of caffeine and stress

One of the major contributing factors in anxiety and the anxiety disorders is stress, and it is reasonable to hypothesize that this is an area of psychological functioning in which caffeine may be implicated.[16] Ongoing research has clearly demonstrated the destructive psychological and physiological effects of stress. One of the most serious reactions is post-traumatic stress disorder (PTSD), which was reported at least as far back as 1755 when a peasant family was trapped by an avalanche in the Italian Alps.[216] It has been widely studied in Vietnam veterans[217,218] and in veterans of World War II and the Korean War.[219] In civilians, PTSD is seen in 38% of burn victims[220] and 46% of those involved in motor vehicle accidents.[221] More generally, a study of college students revealed that any of a wide variety of prior traumatic experiences could produce the symptoms of PTSD.[222] There is now some evidence that caffeine may be a contributing factor in this disorder.[223]

Other psychological consequences of stress include its role in reducing college performance[224] and in suicide, which takes 30,000 lives in the United States each year — about one every 18 minutes.[225] Interestingly enough, however, caffeine consumption appears to have precisely the opposite effect in this regard. In a large-scale 10-year follow-up study of women, caffeine consumption was inversely related to suicide.[193] It is likely that this effect is due to the mood-enhancing properties of moderate amounts of the drug.[193]

Stress also has destructive physiological effects. Stressors significantly modify the number of NK cells and the number of T-cells, thereby reducing the immune response and becoming a contributing factor in infectious diseases and other disorders.[226,227] Stress is also a factor in cardiovascular disorders, where it contributes to the gradual development of coronary artery disease and hypertension,[191] and this effect is exacerbated when stress interacts with caffeine consumption.[226,227] Stress is also considered a precipitating factor in heart attacks[189] and can be a factor in chronic chest pain, which may precede a diagnosis of cardiac pathology.[228] In addition, stress contributes to a number of other physical disease processes, such as asthma[229] and rheumatoid arthritis.[230]

The role of caffeine in stress and stress reactions has been well documented.[231] It affects the neuroendocrine[185,232] and heart rate[233,234] responses to stress, as well as the skin conductance response to stressful anagram tasks.[115] Best documented, however, are the additive effects of stress and caffeine on blood pressure and related cardiovascular indicators.[42,123] In one investigation, subjects were exposed to stressful mental arithmetic, cold pressor, and static exercise tasks after taking caffeine (250 mg) or a placebo. The drug enhanced the impact of the stressors, yielding an additive effect on blood pressure.[186] Confirming studies show interactive effects of caffeine and stress on both resting pressures[124] and the blood pressure response.[56,67] For example, as little as 250 mg of caffeine exacerbates the cold pressor stress response.[84,126] The interaction is not only seen in adults, but also in teenagers and even prepubertal boys.[235]

One widely employed experimental paradigm is the Rapid Information Processing (RIP) task. In a recent RIP study, Lovallo and colleagues[234] had subjects respond to random light presentations of 3 ms duration. They pressed a response key with each light onset and were given 50 cents for "very rapid" responses. Using 3.3 mg/kg of caffeine (an average of 264 mg per subject), they found that the drug produced a significantly greater increase in the blood pressure response than did a placebo. Many other studies confirm the exacerbating effect of caffeine on the blood pressure response to stress.[123,185,225,236,237]

3. Anger and its concomitants

Caffeine may also be a factor in anger and its expression in aggressive behavior.[238] One of the most rampant problems in U.S. society today is violence, an extreme form of aggressive behavior often precipitated by anger. Between 1960 and 1991, mortality from all causes decreased, but deaths from murder more than doubled from 7.7 to 16.6 per 100,000 population.[239] Preliminary indications are that caffeine may well be a contributing factor. It has been clearly demonstrated that elevations in affective arousal are associated with increases in aggression.[239-242] Since caffeine contributes to emotional arousal,[243,244] it may also contribute to anger and aggression. In fact, however, very few studies have addressed this issue directly. Increases in caffeine have been shown to increase aggression in both normal subjects[211,238,245] and psychiatric patients.[246,247] In addition, studies show that caffeine increases hostility[248] and that abstinence from the drug decreases it.[249] However, these effects have not been entirely consistent, with some studies finding no effects of caffeine on aggression.[250] Considerably more research in this important area will be required.

Even our sleep is not protected from the enhancement of violent tendencies by caffeine. Some investigators have identified a population characterized by violent behavior during sleep.[251] These people experience more night terrors and hypnogogic hallucinations and have more jerks, bruxism, and REM behavior disorders than normal people. One of the most consistent findings is that these individuals consume unusually large amounts of caffeine.[251]

The inconsistencies in studies of the role of caffeine in aggression can perhaps best be explained by examining the perception of arousal. Both physiological arousal and cognitive assessment of that arousal are necessary to produce aggression.[241,242] In one study, individuated people reacted more aggressively when they were informed that their heightened arousal came from the previously unknown consumption of caffeine. Deindividuated people showed decreases in aggression levels when they were told that they had consumed caffeine.[252] The individuated person would theoretically attribute arousal to some internal factor before knowing about the caffeine. With the emergence of an external object, attention is focused on the deception, and aggression therefore increases. The deindividuated person, on the other hand, would already assume that the experimenters had caused the arousal, and specific knowledge of caffeine consumption would only confirm that suspicion. Attribution of the arousal source more generally has been shown to influence the level of resultant aggressiveness in a subject.[253] In addition, when subjects think they have no control over their stimulation, they become more aggressive,[254] and that effect may also be exacerbated by caffeine.

V. Effects of habitual caffeine exposure

Research on the effects of chronic caffeine usage has focused principally on long-term cardiovascular effects, short-term physiological effects, and impact on behavior.

A. Cardiovascular impact

A vast majority of research done on the effects of habitual caffeine consumption has focused on potential cardiovascular sequelae. Early studies suggested that coffee and other sources of caffeine significantly increase the risk of some cardiovascular problems.[18,19] However, more recent investigations report little or no elevated risk of CHD,[255,256] arrhythmia,[22,257] myocardial infarction,[258] ischemic heart disease,[259] or other disorders.[260] Caffeine does not appear to trigger episodes in patients with symptomatic idiopathic ventricular premature beats.[261] Similarly, lipid profiles now appear to be unaffected by habitual caffeine consumption.[55,255,262] In fact, decaffeinated coffee increased LDL cholesterol in one study, while caffeinated coffee did not,[263] suggesting that if coffee raises LDL levels, it is through some component other than caffeine. At the same time, Stavric[264] cautions that the questions have not yet all been answered, and there remains the possibility that boiled coffee adversely affects lipid profiles, while filtered coffee does not.[7]

B. *Physiological effects*

The effects of moderate habitual consumption on blood pressure are not yet entirely clear, though caffeine is probably not a major contributing factor in hypertension. Hofer and Bättig,[51] for example, reported no effect of habitual coffee consumption on blood pressure in 338 nonsmoking women, though elimination of caffeinated coffee in another study did reduce both systolic and diastolic pressures.[265] Other investigators report that acute caffeine intake does elevate blood pressure in chronic consumers,[54,124,191] though this effect may actually be smaller in high than in low habitual coffee drinkers.[266,267] These findings accord with the model expectation that habitual exposure to arousal agents should reduce the impact of short-term exposure on phasic physiological responses.

Similar support is found in skin conductance response (SCR) data showing that high habitual caffeine users display reduced SCR amplitudes under various kinds of stimulation.[21] Interestingly enough, tonic data also show lower overall SCLs in high than in low users.[21,80] In addition, low users, who show higher tonic levels even on the first trial of a digit recall task, demonstrate facilitation and hence arousal incrementation across multiple trials, while high users do not. Again, EDA data support the model in showing that habitual arousal exposure reduces tonic activity.

C. *Effects on behavior and performance*

There has been relatively little work concerning the effects of habitual caffeine use on task performance. While there appears to be no effect on some tasks,[137,144] including reaction time and anagram performance,[21] behaviors involving focused attention may be affected. In one investigation, we found a borderline difference ($p < .06$) in recall performance favoring high caffeine consumers.[77] In another, habitual consumption significantly facilitated paired-associate memory performance when a white-noise distractor was present but impaired it when the distractor was absent (Figure 6.6). In addition, habitual users performing a vigilance task displayed higher sensitivity, made more than twice as many commission errors, and employed a more liberal response criterion.[21] These results provide initial support for the hypothesized impact of chronic arousal on attentional tasks, but much is yet

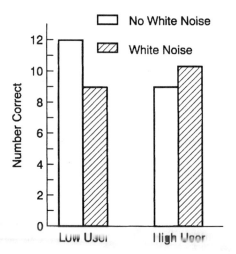

Figure 6.6 Differential effects of habitual caffeine consumption on paired-associate memory performance as a function of the presence or absence of a white-noise distractor. (From Smith, B. D., Effects of acute and habitual caffeine ingestion on physiology and behavior: Tests of a biobehavioral arousal theory. *Pharmacopsychoecologia*, 151–167, 1994. With permission.)

to be learned. Future research might further address the question of whether or not the long-term effects of habitual use are primarily on tasks involving a strong attentional, or perhaps more generally cognitive, component, as the model suggests.

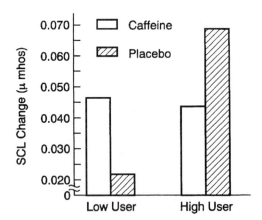

Figure 6.7 Interactive effects of habitual and acute caffeine ingestion on SCL change. (From Smith, B. D., Effects of acute and habitual caffeine ingestion on physiology and behavior: Tests of a biobehavioral arousal theory. *Pharmacopsychoecologia*, 151–167, 1994. With permission.)

VI. The interaction of habitual and acute ingestion

The theoretical model postulates that habitual consumption and acute intake should interact, and we have already seen the reduction in physiological responsivity to short-term caffeine that is induced by long-term consumption. Going beyond those results, one study found no differential effects,[187] with acute caffeine and stress showing additive effects on blood pressure, but habitual consumption making no difference. On the other hand, acute caffeine intake in another investigation virtually eliminated habituation in high habitual users exposed to auditory stimulation but had almost no effect on habituation rate in low users.[21] In addition, the change in SCL that occurs during vigilance was differentially affected by the interaction of habitual and acute caffeine intake. As can be seen in Figure 6.7, acute caffeine increased SCL change in low users and decreased it in highs.[21] In effect, the increase in arousal induced by the task was exacerbated by caffeine in low users but moderated by the drug in habitual consumers.

VII. Individual differences

Individual difference variables may affect caffeine consumption and its impact in at least three separate, though interrelated, ways. First, the model proposes individual differences in preference for and hence habitual consumption of caffeine. This habitual consumption dimension is very likely an interactive function of a genetic predisposition that we take up later and experience. A second individual difference dimension affects the impact of acute caffeine ingestion on physiology and behavior. It very likely also involves both genetic and environmental factors and helps to explain why some individuals experience a strong arousal reaction to even small amounts of caffeine, while others react minimally to much larger doses.

A final set of individual difference factors consists of those reflecting broader differences in overall arousal, arousal potential, or arousability.[77] These have typically been identified as personality dimensions that have substantial biological components and in which some form of arousal is postulated to constitute the physiological substrate for the dimension. Relevant investigations of such dimensions, including neuroticism,[268] impulsivity,[269,270] and sensation-seeking,[21,271,272] are available. However, we will focus here on extraversion, which has received the most attention. Eysenck has long hypothesized that extraversion reflects an underlying continuum of activation or arousal.[273-275] The validity of the dimension as an arousal continuum is well documented, psychophysiological studies consistently confirming that introverts are more aroused or arousable than extraverts.[276-278] As compared with extraverts, introverts show higher heart rate reactivity,[277] higher skin conductance levels,[271] larger phasic skin conductance responses,[80,272,275] and slower electrodermal habituation.[80] Since EDA has been shown to be a reliable measure of arousal,[21,76,279] these results strongly support Eysenck's hypothesis. Moreover, the extraversion dimension does differentiate high from low caffeine consumers. Specifically, extraverts, who hypothetically have low cortical arousal, consume more caffeine, presumably in order to raise their arousal to an optimal midpoint.

VIII. Caffeine and psychopathology

Caffeine has been implicated as a possible exacerbating factor in some forms of psychopathology. In limited instances, it has even been found to be a primary causal agent. Even short of actual pathology, the drug has long been known to be associated with the experience of anxiety.

A. Anxiety

Anxiety is associated with elevated physiological and behavioral arousal,[244,280] especially levels of affective arousal.[244,281] Similarly, caffeine has long been shown to increase arousal.[72,84,128,245] This parallel may suggest both underlying causal relationships and interactions.

A number of studies document an increase in anxiety with caffeine ingestion,[153,282,283] a phenomenon seen in both users and nonusers.[215] In addition, caffeine has been found to increase anxiety levels in smokers,[127,284] alcoholics,[285] and ECT[286] and anxiety[287] patients. Elevations of anxiety have also been observed in children as young as age 8,[109] and males appear to be more sensitive to this anxiety-producing effect of caffeine.[101,215] Taking this finding a step further, some investigators have reported a dose–response relationship: Higher doses of caffeine produce greater elevations in anxiety.[126,250] Doses of 400 mg increase anxiety more than doses of 200 mg,[107,288] and 600 mg elevates anxiety more than 400 mg.[145] It is not clear, however, whether or not this relationship is linear. Although some work does suggest linearity,[145] other studies have shown that any dose over 300 mg may push anxiety to asymptotic levels, with no further increase at higher doses.[250,288,289]

To the extent that caffeine increases anxiety and tracks its effects, it is worth considering the possibility of some underlying physiological relationship between the drug and the emotion. Some work suggests that feelings of anxiety and such anxiety symptoms as sweating, palpitations, nausea, irritability, loss of appetite, tension, and lack of concentration[288] may be associated with adenosine receptors in the brain.[289-291] The fact that caffeine acts as an antagonist for these same adenosine receptors,[292] may suggest a neural association between the emotion and the drug. It may also be no coincidence that both habitual levels of anxiety and the preference for and consumption of caffeine have been shown to be subject to genetic predispositions.

Increasing evidence in recent years suggests the presence of a substantial genetic factor in anxiety.[293] This predisposition may be expressed in the serotonin system of the brain[294-296] and possibly also the norepinephrine system.[297] Evidence indicates that levels of habitual caffeine consumption and the intensity of responses to single doses may also be subject to a hereditary predisposition.[298-300] It appears to involve the bimodally distributed ability to acetylate molecules possessing an amino functional group. The genetic predisposition very likely has a direct effect on the extent to which caffeine acts as an adenosine receptor antagonist in a given individual and may also influence caffeine-relevant taste preferences. Depending on the percentage of variance accounted for by the genetic predisposition, early exposure to and experience with caffeinated substances interacts with heredity to shape both the preference for caffeine and the reaction to ingestion. The potential for a genetically based neurophysiological relationship between anxiety and caffeine is clear, but a considerable amount of research will be required to further confirm it.

Despite the well-documented association of caffeine with anxiety and the possibility that there is a neural, and perhaps genetic, association between the two, the literature has not been entirely consistent. In fact, a survey of over 9,000 people in England found no relationship between caffeine and anxiety.[300] In addition, even though panic disorder patients are hypersensitive to anxiety-producing stimuli and situations,[301-303] doses of caffeine under 100 mg do not increase their anxiety levels.[302]

B. The anxiety disorders

Recent studies have begun to confirm the widely held hypothesis that caffeine can be a contributing factor in the maintenance, and perhaps even genesis, of some anxiety disorders. Included are post-traumatic stress disorder (PTSD),[223] phobia,[304,305] obsessive-compulsive disorder,[306] and panic disorder.[307-309] One study showed, for example, that excessive caffeine consumption is a common factor in the PTSD reactions seen in combat troops. Based on their results, the investigators recommended decaffeinated beverages for all troops entering combat situations.[223]

There is also an emerging body of research that implicates caffeine in the genesis of panic attacks,[309,310] and the drug has long been used as a panicogenic agent to elicit anxious reactions for clinical purposes.[307,308] One study recently showed that caffeine can cause the dysregulation of multiple neuronal systems that results in panic attacks.[308] Moreover, this panic response to excessive amounts of caffeine was found to have a strong genetic component. Another investigation showed that caffeine can induce panic attacks in healthy controls as well as panic disordered patients.[310] A patient in one case study had a panic attack when caffeine was infused during sleep.[309]

C. Caffeinism

Caffeine is only one of a number of factors that enter into the development and expression of symptomatology in most anxiety disorders, but can it be a direct, primary causal factor? It would appear that the answer is "yes." The first published report of caffeinism — essentially an anxiety disorder based on chronic high caffeine consumption — appeared in 1967 and described the case of a woman thought to have an anxiety disorder until it was determined that she was consuming 15 to 18 cups of brewed coffee per day. She showed rapid improvement when her caffeine intake was drastically reduced.[311]

In 1974, John Greden[312] alerted clinicians to this newly identified disorder, which can produce symptoms indistinguishable from those of the anxiety disorders. He highlighted the diagnostic dilemma this situation can present and noted the need for care in making a differential diagnosis. Most of the research on caffeinism followed his article and was conducted primarily in the 1970s and 1980s. While there has been little recent research, the problem remains a potentially important one that may require further empirical attention.

Wilfrid Pilette[313] defined caffeinism as "...a pharmacological state of acute or chronic toxicity that results from the ingestion of high doses of caffeine...." As Greden[312] had suggested, the symptoms are highly correlated with those of the anxiety disorders. Commonly included among observed psychological manifestations are excessive anxiety, sleep disturbances,[94] irritability, and agitation.[314] Accompanying physical symptoms can include tremulousness, muscle twitches, diuresis, arrhythmias, flushing, tachypnea, palpitations, gastrointestinal disturbances,[315] psychophysiological complaints, sensory disturbances,[314] tachycardia,[315] and respiratory distress.[316] At extreme doses, the drug can be severely toxic or even fatal. In one case, a 27-year-old male patient experienced epileptoid convulsions, shallow respiration, and unconsciousness, as well as hyperpyrexia, tachycardia, and hypertension. He had ingested 500 g of ground coffee to obtain a "high."[317] In another, a 20-year-old bulimic woman consumed 20 g of caffeine in a suicide attempt. In addition to severe manifestations of the more common symptoms of caffeine intoxication, she suffered a subendocardial infarction.[318]

The existence of caffeinism and its major symptom patterns were repeatedly confirmed in the early studies,[316] and it may be fairly common. Behar,[319] for example, diagnosed it in 16% of veterans referred to an outpatient clinic for PTSD. However, the more general prevalence of the disorder is unknown; definitive differential diagnostic signs have not been identified; and contributing causal factors are not well understood. Since many people who consume large amounts of caffeine apparently do not develop caffeinism,[5,6] while others manifest the symptoms on lower (though substantial) doses,[313] there are clearly factors other than dosage.

Gender and tobacco use both appear to be among the factors that contribute to caffeinism liability. Women and nonsmokers are at higher risk.[300] Body weight, pattern of consumption, and exposure to other psychostimulants are other obvious possibilities. Somewhat more subtle is the likelihood that a predisposing factor may be present, and the genetic factor discussed above may provide this missing link. It may affect the adenosine receptor antagonism potency of caffeine[298] and may involve the serotonin[294] or norepinephrine[297] system or both. Supporting this possibility is an interesting study in which patients with generalized anxiety disorder (GAD) were given 250 or 500 mg of caffeine prior to the recording of psychophysiological measures. They showed greater increases in EEG, skin conductance level, and blood pressure indicators of arousal than did normal controls.[289] The authors concluded that GAD patients appear to be abnormally sensitive to caffeine, an observation that both supports the likelihood of a genetic factor and helps to further understand the relationship between caffeine and GAD. Supportive studies show a similar caffeine hypersensitivity in patients with panic disorder.[149,320,321]

D. Depression

Caffeine is widely used to counteract depressive moods,[322] such as those seen in dysthymia[323] and seasonal affective disorder (SAD).[324] One study demonstrated that a combination of caffeine and light exposure effectively counteracts the depressive effects of sleep deprivation.[325] In addition, clinically depressed patients consume more caffeine than the average person,[326,327] and some researchers have suggested that caffeine should be used clinically.[328]

There are also at least three contraindications to using the drug as a therapeutic agent in the depressive disorders.[327-330] First, caffeine inhibits the effectiveness of some antidepressant medications.[327] This appears to be due to the increases in drug metabolism that result from caffeine-induced elevations in metabolic rate.[329] Second, depressed patients often substitute caffeine for their antidepressant medication, potentially causing a deepening of the depression.[327] The third problem is sleep loss. Sleep is disturbed in 90% of depressed patients, and this problem is exacerbated by caffeine consumption, worsening depressive symptoms.[331] Even more troubling is the fact that depressed patients seem to metabolize caffeine at a slower rate that nondepressed controls during sleep.[332]

A final contraindication is that chronic caffeine exposure in depressed patients can lead to tolerance and thus loss of its initial effectiveness.[330] This tolerance usually results in episodic, rebound headaches, which are a primary symptom of caffeine withdrawal.[333] Cacciatore[334] found that these headaches increase negative affect and depression. Thus, prolonged caffeine use can actually lead to a magnification of depressive symptoms.

E. Schizophrenia

Psychiatric patients consume elevated amounts of caffeine compared to controls, and schizophrenics top the list.[335,336] In fact, they may average more than 750 mg of caffeine each day.[337] It appears that they consume the caffeine in order to obtain certain of its adenosine antagonist effects,[338] such as increased motor activity.[337] Schizophrenics may also consume caffeine in order to partially counteract the negative effects of their neuroleptic medications.[339] In fact, some schizophrenics appear to actually substitute caffeine for their medication.[340] The problem is that caffeine can increase subjective distress in schizophrenic patients[340] and increase psychotic symptomatology.[337] Recent studies are showing, however, that the newer neuroleptic medications, such as clozapine, are associated with lower levels of caffeine abuse.[341]

IX. Conclusion

Caffeine is clearly a powerful and consistent arousal agent. For the average person, consuming typical amounts of caffeine, the drug appears to be essentially harmless. Indeed, it often has positive effects, in that it raises arousal to more optimal levels, decreases boredom, improves cognitive functioning, and increases work productivity. These findings and others are supportive of some of the theoretical principles and hypotheses that we articulated at the outset.

Acute caffeine intake elevates blood pressure and increases EEG, SCL, SCR, and EDA response frequency indicators of arousal. The drug also affects the process of habituation, reducing its rate and smoothing the function. The arousing effect of caffeine interacts with the activating properties of such other agents as stress and novelty in an additive fashion to further increase blood pressure and EDA indicators, and there is support for the postulated inverted-U input/output function. For example, caffeine without stressful harassment increases blood pressure, but the addition of the stressor pushes subjects over the top of the hypothetical function and decreases pressures. Similarly, caffeine alone decreases EDA habituation rate, but the addition of stressful white noise reverses that result and increases the rate.

Findings concerned with habitual caffeine use also support theoretical principles. Habitual intake decreases the impact of short-term exposure to the drug, reducing the sizes of increments in blood pressure, SCR, and SCL and differentially affecting habituation rate. Moreover, arousal induced by tasks is exacerbated by acute caffeine ingestion in low habitual users and nonusers but moderated by caffeine in high habitual users.

While there is substantially less work to date on the behavioral effects of caffeine, relevant findings are generally supportive of the theory. Basically, the drug appears to have greater effects on tasks involving concentration and, perhaps more generally, substantial cognitive components. Thus, acute caffeine ingestion improves aspects of vigilance performance, and habitual use facilitates paired associates performance under some conditions, increases vigilance sensitivity, reduces vigilance accuracy, and may increase recall performance. However, much additional work is needed to further assess the differential effects of both acute and habitual caffeine on performance indicators.

This brings us back to the question of why caffeine is such a popular, widely consumed arousal agent. Clearly, the drug does have a useful alerting effect that appears to develop relatively little tolerance over time, but just as clearly this is not the only reason for its popularity. Caffeine also serves to improve performance on a variety of tasks (particularly those involving logical reasoning, semantic memory,[89] and sustained attention[21]) elevates mood states,[91] may partially offset the effects of alcohol,[146] and increases the individual's sense of well-being,[87] at least in habitual users. A more subtle but perhaps equally important effect is that on habituation. A major, centrally mediated process, habituation in the face of repetitive stimulation reduces arousal to what is likely to be an uncomfortably low level,[76] subjectively described as boredom. Moreover, habituation is often very "choppy," and caffeine smooths the habituation function, making the decremental process more even and perhaps more comfortable for the individual.[21] Thus, in both slowing and smoothing habituation, caffeine may make a potentially aversive process less aversive or even pleasant. Finally, caffeine may be popular with many individuals for a much more basic reason — genetics. As we have seen, there appears to be a hereditary predisposition that combines with early experience to influence preference for the drug. While this possibility requires considerable additional research, it may well be a powerful factor.

Support for the current arousal theory is by no means uniform. However, there is a substantial body of literature that is consistent with the stated principles. In particular, the inverted U-shaped input/output function, the additive impact of arousal agents, the interactive influences of habitual and acute intake, and the individual difference dimensions postulated by the theory have received considerable support. However, much additional research is needed, both on the effects of caffeine and on arousal functions and agents more generally. As that research is completed, the theory will undoubtedly have to be updated and modified.

References

1. Edwards, R., Drug use among eighth grade students is increasing, *Internat. J. Addict.*, 28, 1621, 1993.
2. National Institute on Drug Abuse, *National Survey Results on Drug Use*, Washington, D.C., U.S. Government Printing Office, 1995.
3. Substance Abuse and Mental Health Administration, *National Household Survey on Drug Abuse Population Estimates*, Washington, D.C., U.S. Government Printing Office, 1995.
4. Barone, J. and Roberts, H., Caffeine consumption, *Food Chem. Toxicol.*, 34, 119, 1996.
5. Gilbert, R. M., Caffeine as a drug of abuse, in *Research Advances in Alcohol and Drug Problems*, Volume 3, Gibbons, R. J., Israel, Y., and Kalant, H., Eds., John Wiley & Sons, New York, 1976, 49.
6. Wells, S. J., Caffeine: Implications of recent research for clinical practice, *Am. J. Orthopsychiat.*, 54, 375, 1984.
7. Pirich, C., O-Grady, J., and Sininger, H., Coffee, lipoprotiens and cardiovascular disease, *Wien Klinical Wochenschr*, 105, 3, 1993.

8. Hirsh, K., Central nervous system pharmacology of the dietary methylxanthines, in *The Methylxanthine Beverages and Foods: Chemistry, Consumption, and Health Effects*, Spiller, G.A., Ed., Allan R. Liss, Inc., New York, 1984, 235.

9. Gilbert, R., Caffeine consumption, in *The Methylxanthine Beverages and Foods: Chemistry, Consumptions and Health Effects*, Spiller, G. A., Ed., Allan R. Liss, Inc., New York, 1984, 185.

10. Murray, J., Physiological aspects of caffeine consumption, *Psychol. Rep.*, 62, 575, 1988.

11. Benowitz, N. L., Clinical pharmacology of caffeine, *Annu. Rev. Med.*, 41, 277, 1990.

12. Davidson, R. and Smith, B., Arousal and habituation: Differential effects of caffeine, sensation seeking and task difficulty, *Person. Individ. Diff.*, 10, 111, 1989.

13. Davidson, R. and Smith, B., Caffeine and novelty: Effects on electrodermal activity and performance, *Physiol. Behav.*, 49, 1169, 1991.

14. Smith, A., Caffeine, performance, mood and states of reduced alertness, *Pharmacopsychoecologia*, 7, 75, 1994.

15. Rusted, J., Caffeine and cognitive performance: Effects on mood or mental processing? *Pharmacopsychoecologia*, 7, 49, 1994.

16. Smith, B., Effects of acute and habitual caffeine ingestion in physiology and behavior: Tests of a biobehavioral arousal theory, *Pharmacopsychoecologia*, 7, 151, 1994.

17. La Vecchia, C., Gentile, A., Negri, E., Parezzini, F., and Franceschi, S., Coffee consumption and myocardial infarction in women, *Am. J. Epidemiol.*, 130, 481, 1989.

18. Darragh, A., Kenny, M., Lambe, R., and O'Kelly, D., Adverse effects of caffeine, *Irish J. Med. Sci.*, 150, 47, 1981.

19. Dobmeyer, D., Stino, R., Leier, C., Greenberg, R. and Schaal, S., The arrythmogenic effects of caffeine in human beings, *N. Engl. J. Med.*, 308, 814, 1983.

20. Mackay, D. C. and Rollins, J. W., Caffeine and caffeinism, *J. Royal Naval Med. Serv.*, 75, 65, 1989.

21. Smith, B. D., Davidson, R. A., Perlstein, W., and Gonzalez, F., Sensation-seeking: Electrodermal and behavioral effects of stimulus content and intensity, *Internat. J. Psychophysiol.*, 9, 179, 1991.

22. Chou, T., Wake up and smell the coffee: Caffeine, coffee, and the medical consequences, *West. J. Med.*, 157, 544, 1992.

23. Smith, A. P., Brockman, P., Flynn, R., Maben, A., and Thomas, M., Investigation of the effects of coffee on alertness and performance during the day and night, *Neuropsychobiology*, 27, 217, 1993.

24. Rainnie, D. G., Granze, H. C., McCarley, R. W., and Greene, R. W., Adenosine inhibition of mesopontine cholinergic neurons: implications for EEG arousal, *Science*, 265, 16, 1994.

25. Myers, D. E., Zulli, T. G., and Shaikh, Z., Hypoalgesic effect of caffeine in experimental ischemic muscle contraction pain, *Headache*, 37, 654, 1997.

26. Ledent, C., Parmentier, M., Vassart, G., Heath, J. K., Costentin, J., Vanderhaeghen, J. J., El Yacoubi, M., Pedrazzini, T., Schiffmann, S.N., and Vaugeois, J.M., Aggressiveness, hypoalgesia and high blood pressure in mice lacking the adenosine A2a receptor, *Nature*, 388, 674, 1997.

27. Norenberg, W., Illes, P., and Wirkner, K., Effect of adenosine and some of its structural analogues on the conductance of NMDA receptor channels in a subset of rat neostriatal neurones, *Br. J. Pharmacol.*, 122, 71, 1997.
28. Estevez, A.Y. and Phillis, J.W., Hypercapnia-induced increases in cerebral blood flow: roles of adenosine, nitric oxide and cortical arousal, *Brain Res.*, 758, 1, 1997.
29. Ghelardini, C., Bartolini, A., and Galeotti, N., Caffeine induces central cholinergic analgesia, *Arch. Pharmacol.*, 356, 590, 1997.
30. Neuhauser-Berthold, Luhrmann, P.M., Verwied, S.C., and Beine, S., Coffee consumption and total body water homeostasis as measured by fluid balance and bioelectrical impedance analysis, *Ann. Nutr. Metab.*, 41, 29, 1997.
31. Nehlig, A., Daval, J. L., and Debry, G., Caffeine and the central nervous system: Mechanisms of action, biochemical, metabolic and psychostimulant effects, *Brain Res. Rev.*, 17, 139, 1992.
32. Cameron, O. G., Modell, J. G., and Hariharan, M., Caffeine and human cerebral blood flow: A positron emission tomography study, *Life Sci.*, 47, 1141, 1990.
33. Aoshima, H. and Tenpaku, Y., Modulation of GABA receptors expressed in Xenopus oocytes by 13-L-hydroxylinoleic acid and food additives, *Biosci. Biotechnol. Biochem.*, 61, 2051, 1997.
34. Lovallo, W. R., Wilson, M.F., Whitsett, T.L., Blick, K., and Al'Absi, M., Stress-like adrenocorticotropin responses to caffeine in young healthy men, *Pharmacol. Biochem. Behav.*, 55, 365, 1996.
35. Moore, R.A., Allen, D., Pessah, N., Glaceran, R., and Nguyen, D. A transgenic myogenic cell line lacking ryanodine receptor protein for homologous expression studies: reconstitution of Ry1R protein and function, *J. Cell. Biol.*, 140, 843, 1998.
36. Reilly, B.A., Brostrom, C.O., and Brostrom, M.A., Regulation of protein synthesis in ventricular myocytes by vasopressin: The role of sarcoplasmic/endoplasmic reticulum Ca^{2+} stores, *J. Biol. Chem.*, 273, 3747, 1998.
37. Svenningsson, P., Fredholm, B.B., and Johansson, B., Caffeine-induced expression of c-fos mRNA and NGFI-A mRNA in caudate putamen and in nucleus accumbens are differentially affected by the N-methyl-D-aspartate receptor antagonist MK-801, *Brain Res. Mol. Brain Res.*, 35, 183, 1996.
38. Datta, U., Angulo, J.A., Zhang, Y., Kraft, M., and Noailles, P.A., Concurrent elevation of the levels of expression of striatal preproenkephalin and preprodynorphin and mRNA in the rat brain by chronic treatment with caffeine, *Neurosci. Lett.*, 231, 29, 1997.
39. Paul, S., Kurunwune B., and Biaggioni, I., Caffeine withdraw: Apparent heterologous sensitization to adenosine and prostacyclin actions in human platelets, *J. Pharmacol. Exp. Therap.*, 267, 838, 1993.
40. Miners, J.O. and Birkett, D.J., The use of caffeine as a metabolic probe for human drug metabolizing enzymes, *Gen. Pharmacol.*, 27, 245, 1996.
41. Rasmussen, B.B. and Brosen, K., Determination of urinary metabolites of caffeine for the assessment of cytochrome P4501A2, xanthine oxidase, and N-acetyltransferase activity in humans, *Ther. Drug Monit.*, 18, 254, 1996.
42. Greenberg, W. and Shapiro, D., The effects of caffeine and stress on blood pressure in individuals with and without a family history of hypertension, *Psychophysiology*, 24, 151, 1987.

43. Casiglia, E., Paleari, C. D., Petucco, S., Bongiovi, S., Colangeli, G., Baccilieri, M. S., Pavan, L., Pernice, M., and Pessina, A. C., Haemodynamic effects of coffee and purified caffeine in normal volunteers: A placebo-controlled clinical study, *J. Hum. Hyperten.*, 6, 95, 1992.

44. Bättig, K., Acute and chronic cardiovascular and behavioral effects of caffeine, aspirin and ephedrine, *Intern. J. Obes. Rel. Metabol. Disord.*, 17, Suppl 1, S61, 1993.

45. Green, J.J. and Suls, J., The effects of caffeine on ambulatory blood pressure, heart rate, and mood in coffee drinkers, *J. Behav. Med.*, 19, 111, 1996.

46. DelRio, G., Menozzi, R., Zizzo, G., Avogaro, A., Marrama, P., and Velardo, A., Increased cardiovascular response to caffeine in perimenopausal women before and during estrogen therapy, *Eur. J. Endocrinol.*, 135, 598, 1996.

47. Lotshaw, S.C., Bradley, J.R., and Brooks, L.R., Illustrating caffeine's pharmacological and expectancy effects utilizing a balanced placebo design, *J. Drug Educ.*, 26, 13, 1996.

48. Van Soeren, M., Mohr, T., Iguer, M., and Graham, T.L., Acute effects of caffeine ingestion at rest in humans with impaired epinephrine responses, *J. Appl. Physiol.*, 80, 999, 1996.

49. Sung, B.H., Wilson, M.F., Whitsett, T., and Lovallo, W.R., Caffeine elevates blood pressure response to exercise in mild hypertensive men, *Am. J. Hyperten.*, 8, 1184, 1995.

50. Pincomb, G.A., Sung, B.H., Lovallo, W.R., and Wilson, M.F., Consistency of cardiovascular response pattern to caffeine across multiple studies using impedance and nuclear cardiography, *Biol. Psychol.*, 36, 131, 1993.

51. Hofer, I. and Bättig, K., Coffee consumption, blood pressure tonus and reactivity to physical challenge in 338 women, *Pharmacol. Biochem. Behav.*, 44(3), 573, 1993.

52. Myers, M. G. and Reeves, R. A., The effect of caffeine on daytime ambulatory blood pressure, *Am. J. Hyp.*, 4, 427, 1991.

53. Denaro, C. P., Brown, C. R., Jacob, P., and Benowitz, N. L., Effects of caffeine with repeated dosing, *Eur. J. Clin. Pharmacol.*, 40, 273, 1991.

54. Lane, J. D. and Manus, D. C., Persistent cardiovascular effects with repeated caffeine administration, *Psychosom. Med.*, 51, 373, 1989.

55. Bak, A. A. and Grobbee, D. E., Caffeine, blood pressure, and serum lipids, *Am. J. Clin. Nutr.*, 53, 971, 1991.

56. Lovallo, W. R., Pincomb, G. A., Sung, B. H., Everson, S. A., Passey, R. B., and Wilson, M. F., Hypertension risk and caffeine's effect on cardiovascular activity during mental stress in young men, *Health Psychol.*, 10, 236, 1991.

57. Lovallo, W. R., Wilson, M. F., Passey, R. B., Sung, B. H., Everson, S. A., Pincomb, G. A., and al'Absi, M., Caffeine and behavioral stress effects on blood pressure in borderline hypertensive Caucasian men, *Health Psychol.*, 15, 11, 1996.

58. MacDougall, J. M., Musante, L., Castillo, S., and Acevedo, M. C., Smoking, caffeine, and stress: effects on blood pressure and heart rate in male and female college students, *Health Psychol.*, 7, 461, 1988.

59. Nussberger, J., Mooser, V., Maridor, G., Juilerat, L., Waeber, B., and Brunner, H. R., Caffeine-induced diuresis and atrial natriuretic peptides, *J. Cardiovasc. Pharmacol.*, 15, 685, 1990.

60. Bender, A.M., Goldberg, S.J., Zhu, D., Samson, R.A., and Donnerstein, R.L., Hemodynamic effects of acute caffeine ingestion in young adults, *Am. J. Cardiol.*, 79, 696, 1997.

61. Van Soeren, M., Graham, T.E., Kjaer, M., and Mohr, T., Acute effects of caffeine ingestion at rest in humans with impaired epinephrine responses, *J. Appl. Physiol.*, 80, 999, 1996.
62. Pincomb, G., Lovallo, W., Passey, R., Brackett, D., Wilson, M., Caffeine enhances the physiological response to occupational stress in medical students, *Health Psychol.*, 6, 101, 1987.
63. Smits, P., Schooten, J., and Thien, T., Cardiovascular effects of two xanthines and the relation to adenosine antagonism, *Clin. Pharmacol. Ther.*, 45, 593, 1989.
64. Strickland, T. L., Myers, H. F., and Lahey, B. B., Cardiovascular reactivity with caffeine and stress in black and white normotensive females, *Psychosom. Med.*, 51, 381, 1989.
65. Greenstadt, L., Yang, L., Shapiro, D., Caffeine, mental stress, and risk for hypertension: a cross-cultural replication, *Psychosom. Med.*, 50, 15, 1988.
66. Hirsch. A. T., Gervino, E. V., Nakao, S., Come, P. C. Silverman, K. J., and Grossman, W., The effect of caffeine on exercise tolerance and left ventricular function in patients with coronary artery disease, *Ann. Intern. Med.*, 110, 593, 1989.
67. Myers, H. F., Shapiro, D., McClure, F., and Daims, R., Impact of caffeine and psychological stress on blood pressure in black and white men, *Health Psychol.*, 8, 597, 1989.
68. Pennickx, F., Vuysteke, P., and Kerremans, R., Recurrences after highly selective vagotomy in refractory and non-refractory duodenal ulcer disease, *Acta Chirurgica Belgica*, 90, 41, 1990.
69. James, J. E. and Richardson, M., Pressor effects of caffeine and cigarette smoking, *Br. J. Clin. Psychol.*, 30, 276, 1991.
70. Rakic, V., Beilin, L.J., and Burke, V., Effects of coffee and tea drinking on postprandial hypotension in older men and women, *Clin. Exp. Pharmacol. Physiol.*, 23, 559, 1996.
71. Sawyer, D. A., Julia, H. L., Turin, A. C., Caffeine and human behavior: arousal, anxiety and performance effects, *J. Behav. Med.*, 5, 415, 1982.
72. Dimpfel, W., Schober, F., and Spuler, M., The influence of caffeine on human EEG under resting conditions and during mental loads, *Clin. Invest.*, 71, 197, 1993.
73. Tharion, W. J., Kobrick, J. L., Lieberman, H. R., and Fine, B. J., Effects of caffeine and diphenhydramine on auditory evoked cortical potentials, *Percept. Mot. Skills*, 76, 707, 1993.
74. Smith, B. D., Rypma, C. B., and Wilson, R. J., Dishabituation and Spontaneous recovery of the electrodermal orienting response: Effects of extraversion, impulsivity, sociability and caffeine, *J. Res. Person.*, 15, 475, 1981.
75. Smith B. D., Wilson, R. D., Davidson, R. A., Electrodermal activity and extraversion: caffeine, preparatory signal and stimulus intensity effects, *Person. Individ. Diff.*, 5, 59, 1984.
76. Smith, B. D., Davidson, R. A., and Green, R. L., Effects of caffeine and gender on physiology and performance: further tests of a biobehavioral model, *Physiol. Behav.*, 54, 415, 1993.
77. Smith, B. D., Extraversion and electrodermal activity: arousability and the inverted-U, *Person. Individ. Diff.*, 4, 411, 1983.
78. Bruce, M., Scott, N., Lader, M., Marks, V., The psychopharmacological and electrophysiological effects of single doses of caffeine in healthy human subjects, *Br. J. Clin. Pharmacol.*, 22, 81, 1986.
79. Sokolov, E., *Perception and the Conditioned Reflex*, Oxford: Pergamon Press, 1963.

80. Smith, B. D., Kline, R., and Meyers, M., The differential hemispheric processing of emotion: a comparative analysis in strongly-lateralized sinistrals and dextrals, *Intern. J. Neurosci.*, 50, 59, 1990.

81. Zahn, T. P., Rapoport, J. L., Acute autonomic nervous system effects of caffeine in prepubertal boys, *Psychopharmacology*, 91, 40, 1987.

82. Lader, M., Comparison of amphetamine sulphate and caffeine citrate in man, *Psychopharmacologia*, 14, 83, 1969.

83. Bottcher, M., Schelbert, H., Phelps, M., Sun, K., and Czernin, J., Effect of caffeine on myocardial blood flow at rest and during pharmacological vasodilation, *J. Nucl. Med.*, 36, 2016, 1995.

84. France, C. and Ditto, B., Cardiovascular responses to the combination of caffeine and mental arithmetic, cold pressor, and static exercise stressors, *Psychophysiology*, 29, 272, 1992.

85. Silverman, K., Evans, S. M., Strain, E. C., and Griffiths, R. R., Withdrawal syndrome after the double blind cessation of caffeine consumption, *N. Engl. J. Med.*, 041, 1101, 1101

86. Bonnet, M. H. and Arand, D. L., Caffeine use as a model of acute and chronic insomnia, *Sleep*, 15, 526, 1992.

87. Smith, A. P., Kendrick, A. M., and Maben, A. L., Effects of breakfast and caffeine on performance and mood in the late morning and after lunch, *Neuropsychobiology*, 26, 198, 1992.

88. Fagan, D., Swift, C. G., and Tiplady, B., Effects of caffeine on vigilance and other performance tests in normal subjects, *J. Psychopharmacol.*, 2, 19, 1988.

89. Stein, M.A., Bender, G.B., Phillips, W., Leventhal, B.L., and Krasowski, M., Behavioral and cognitive effects of methylxanthines. A meta-analysis of theophylline and caffeine, *Arch. Pediatr. Adolesc. Med.*, 150, 284, 1996.

90. Loke, W. H. and Meliska, C. J., Effects of caffeine use and ingestion on a protracted visual vigilance task, *Psychopharmacology*, 84, 54, 1984.

91. Lieberman, H. R., Wurtman, R. J., Emde, G. G., and Roberts, C., The effects of low doses of caffeine on human performance and mood, *Psychopharmacology*, 92, 308, 1987.

92. Akerstedt, T. and Ficca G., Alertness-enhancing drugs as a countermeasure to fatigue in irregular work hours, *Chronobiol. Int.*, 14, 145, 1997.

93. Giam, G. C., Effects of sleep deprivation with reference to military operations, *Ann. Acad. Med. Singapore*, 14, 145, 1997.

94. Kelly, T.L., Bonnet, M.H., and Mitler, M.M., Sleep latency measures of caffeine effects during sleep deprivation, *Electroencephalogr. Clin. Neurophysiol.*, 102, 397, 1997.

95. Wilkinson, R., Methods for research on sleep deprivation and sleep function, in *Sleep and Dreaming*, Hartmann, E., Ed., Little/Brown, Boston, 1970, 369.

96. Sicard, B.A., Tachon, P., Vandel, B., Chauffard, F., Enslen, M., and Perault, M.C., The effects of 600 mg of slow release caffeine on mood and alertness, *Aviat. Space Environ. Med.*, 67, 859, 1996.

97. Bonnet, M.H. and Arand, D.L., Metabolic rate and the restorative function of sleep, *Physiol. Behav.*, 59, 777, 1996.

98. Linde, L., An auditory attention task: A note on the processing of verbal information, *Percept. Mot. Skills*, 78, 563, 1994.

99. Loke, W. and Meliska, C., Effects of caffeine use and ingestion on a protracted vigilance test, *Psychopharmacology*, 87, 344, 1984.

100. Lin, A.S., McCann, U.D., Slate, S.O., and Uhde, T.W., Effects of intravenous caffeine administered to healthy males during sleep, *Depress. Anxiety,* 5, 21, 1997.
101. Linde, L., Mental effects of caffeine in fatigued and non-fatigued female and male subjects, *Ergonomics,* 38, 864, 1995.
102. Zwyghuizen-Doorenbos, A., Roehrs, T., Lipshutz, L., Timms, V., and Roth, T., Effects of caffeine on alertness, *Psychopharmacology,* 100, 36, 1990.
103. Pons, L., Trenque, T., Bielcki, M. and Moulin, M., Attentional deficits of caffeine in man: Comparison with drugs acting on performance, *Psychiat. Res.,* 23, 329, 1988.
104. Fine, B. J., Kobrick, J. L., Lieberman, H. R., Marlowe, B., et al., Effects of caffeine or diphenhydramine on visual vigilance, *Psychopharmacology,* 114, 233, 1994.
105. Lorist, M., Snel, J., Kok, A., and Madler, G., Acute effects of selective attention and visual search processes, *Psychophysiology,* 33, 354, 1996.
106. Munte, T. F., Heinze, H. J., Kunkel, H., and Scholz, M., Personality traits influence the effects of diazepam and caffeine on CNV magnitude, *Neuropsychobiology,* 12, 60, 1984.
107. Stein, M., Krasowski, M., Leventhal, B., Phillips, W., and Bender, B., Behavioral and cognitive effects of methylxanthines: A meta-analysis of theophylline and caffeine, *Arch. Pediatr. Adolesc. Med.,* 150, 284, 1996.
108. Kupietz, S. and Winsberg, B., Caffeine and inattentiveness in reading-disabled children, *Percept. Mot. Skills,* 44, 1238, 1977.
109. Tynjala, J., Levalahti, E., and Kannas, L., Perceived tiredness among adolescents and its association with sleep habits and use of psychoactive substances, *J. Sleep Res.,* 6, 189, 1997.
110. Loke, W. H., Effects of caffeine on mood and memory, *Physiol. Behav.,* 44, 367, 1988.
111. Landolt, H.P., Borbely, A.A., Dijk, D.J., and Roth, C., Late-afternoon ethanol intake affects nocturnal sleep and the sleep EEG in middle-aged men, *J. Clin. Psychopharmacol.,* 16, 428, 1996.
112. Koelega, H. S., Stimulant drugs and vigilance performance: A review, *Psychopharmacology,* 111, 1, 1993.
113. Brown, S.L., Harris, T.B., Wallace, R.B., Langlois, J.A., Corti, M.C., Foley, D.J., Pahor, M., and Salive, M.E., Occult caffeine as a source of sleep problems in an older population, *J. Am. Geriatr. Soc.,* 43, 860, 1995.
114. Frewer, L. and Lader, M., The effects of caffeine on two computerized tests of attention and vigilance, *Hum. Psychopharmacol. Clin. Experimental,* 6, 119, 1991.
115. Smith, B., Rafferty, J., Lindgren, K., Smith, D. and Nespor, A., Chronic and acute effects of caffeine: Testing a biobehavioral model, *Physiol. Behav.,* 51, 131, 1991.
116. Stratta, P., Mancini, F., Mattei, P., and Casacchia, M., Information processing strategy to remediate Wisconsin Cart Sorting test performance in schizophrenia: A pilot study, *Am. J. Psychiat.,* 151, 915, 1994.
117. Hasenfratz, M., Jaquet, F., Aeschbach, D., and Bättig, K., Interactions of smoking and lunch with the effects of caffeine on cardiovascular functions and information processing, *Hum. Psychopharmacol. Clin. Exp.,* 6, 277, 1991.
118. Hasenfratz, M., Bunge, A., Dal-Pra, G., and Bättig, K., Antagonistic effects of caffeine and alcohol on mental performance parameters, *Pharmacol. Biochem. Behav.,* 46, 463, 1993.

119. Bättig, K. and Buzzi, R., Effect of coffee on the speed of subject paced information processing, *Neuropsychobiology*, 16, 12, 1986.
120. Warburton, D. M., Effects of caffeine on cognition and mood without caffeine abstinence, *Psychopharmacology*, 119, 66, 1995.
121. Bonnet, M. and Arand, D., The use of prophylactic naps and caffeine to maintain performance during a continuous operation, *Ergonomics*, 37, 1009, 1994.
122. Massaro, D. and Ferguson, E., Cognitive style and perception: The relationship between category width and speech perception, categorization and discrimination, *Am. J. Psychol.*, 106, 25, 1993.
123. France, C. and Ditto, B., Caffeine effects on several indices of cardiovascular activity at rest and during stress, *J. Behav. Med.*, 11, 473, 1988.
124. Lane, J. D. and Williams, R. B., Caffeine affects cardiovascular responses to stress, *Psychophysiology*, 22, 648, 1985.
125. Riedel, W. J. and Jolles, J., Cognition enhancers in age related cognitive decline,
126. Hasenfratz, M. and Bättig, K., Acute dose-effect relationships of caffeine and mental performance, EEG, cardiovascular and subjective parameters, *Psychopharmacology*, 114, 281, 1994.
127. Pritchard, W. S., Robinson, J. H., deBethizy, J. D., Davis, R. A., et al., Caffeine and smoking: Subjective, performance, and psychophysiological effects, *Psychophysiology*, 32, 19, 1995.
128. Foreman, N., Barraclough, S., Morre, C., Mehta, A., and Madon, M., High doses of caffeine impair performance of a numerical version of the Stroop task in men, *Pharmacol. Biochem. Behav.*, 32, 399, 1989.
129. Loke, W. H., The effects of caffeine and automaticity on a visual information processing task, *Hum. Psychopharmacol. Clin. Exp.*, 7, 379, 1992.
130. Landrum, R., Meliska, C., and Loke, W., Effects of caffeine and task experience on task performance, *Psychologia*, 31, 91, 1988.
131. de-Brabander, B., Effect of short lateralized signals on arousal vs. activation on tasks requiring visuospatial or elementary semantic visual processing, *Percept. Mot. Skills*, 67, 783, 1988.
132. Ketterer, M., Lateralized representation of affect, affect cognizance and the coronary prone personality, *Biol. Psychol.*, 15, 509, 1982.
133. Smith, B., Meyers, M., and Kline, R., Hemispheric asymmetry and emotion: Lateralized parietal processing of affect and cognition, *Biol. Psychol.*, 29, 11, 1987.
134. Smith, B., Kline, R., Lindgren, K., Ferro, M., Smith, D., and Nespor, A., The lateralized processing of affect in emotional liable extraverts and introverts: central and autonomic effects, *Biol. Psychol.*, 39, 143, 1995.
135. Roman, F., Garcia-Sanchez, F., Martinez-Selva, J., Gomez-Amor, J., and Carrillo, E., Sex differences and bilateral electrodermal activity, *Pavl. J. Biol. Sci.*, 24, 150, 1989.
136. Hasenfratz, M. and Bättig, K., Action profiles of smoking and caffeine: Stroop effect, EEG, and peripheral physiology, *Pharmacol. Biochem. Behav.*, 42, 155, 1992.
137. Edwards, S., Penri-Jones, R., Craig, C., and Brice, C., Effects of caffeine, practice, and mode of presentation on Stroop task performance, *Pharmacol. Biochem. Behav.*, 54, 309, 1997.

138. Bowyer, P., Humphreys, M. and Revelle, W., Arousal and recognition memory: The effects of impulsivity, caffeine and time on a task, *Person. Individ. Diff.*, 4, 41, 1983.

139. Jarvis, M., Does caffeine intake enhance absolute levels of cognitive performance? *Psychopharmacology*, 110, 45, 1993.

140. Anderson, K. and Revelle, W., The interactive effects of caffeine, impulsivity and task demands on visual search task, *Person. Individ. Diff.*, 4, 127, 1983.

141. Erikson, G., The effects of caffeine on memory for word lists, *Physiol. Behav.*, 35, 47, 1985.

142. Riedel, W., Jolles, J., van Praag, H., Verhey, F., Leboux, R., and Hogervorst, E., Caffeine attenuates scopolamine-induced memory impairment in humans, *Psychopharmacology*, 122, 158, 1995.

143. Cestari, V. and Castellano, C., Caffeine and cocaine interaction on memory consolidation in mice, *Arch. Int. Pharmacodyn. Ther.*, 33, 94, 1996.

144. Terry, W. and Phifer, B., Caffeine and memory performance on the AVLT, *J. Clin. Psychol.*, 42, 860, 1986.

145. Roache, J. and Griffiths, R., Interaction of diazepam and caffeine: Behavioral and subjective dose effects in humans, *Pharmacol. Biochem. Behav.*, 26, 801, 1987.

146. Oborne, D. J. and Rogers, Y., Interactions of alcohol and caffeine on human reaction time, *Aviat. Space Environ. Med.*, 54, 528, 1983.

147. Arnold, M., Petros, T., Beckwith, B., Coons, G., and Gorman, N., The effects of caffeine, impulsivity and sex on memory for word lists, *Physiol. Behav.*, 41, 25, 1987.

148. Lieberman, H. R., Beneficial effects of caffeine, in *Twelfth International Scientific Colloquium on Coffee*, Paris, ASIC, 1988.

149. Clubley, M., Bye, C. E., Henson, T. A., Peck, A. W., and Riddington, C. J., Effects of caffeine and cyclizine alone and in combination on human performance, subjective effects and EEG activity, *Br. J. Clin. Pharmacol.*, 7, 157, 1979.

150. File, S., Bond, A., and Lister, R., Interaction between effects of caffeine and lorazepam in performance and self ratings, *J. Clin. Psychopharmacol.*, 2, 102, 1982.

151. Turner, J., Incidental information processing: Effects of mood, sex and caffeine, *Intern. J. Neurosci.*, 72, 1, 1993.

152. Loke, W. H., Hinrichs, J. V., and Ghoneim, M. M., Caffeine and diazepam: Separate and combined effects on mood, memory, and psychomotor performance, *Psychopharmacology*, 87, 344, 1985.

153. Kaplan, G. B., Shader, R. I., Harmatz, J. S., Cotreau, M. M., Goddard, J. E., Ehrenberg, B. L., and Greenblatt, D. J., Dose-dependent pharmacokinetics and psychomotor effects of caffeine in humans, *J. Clin. Pharmacol.*, 37, 693, 1997.

154. Howell, L. L., Spealman, R. D., and Coffin, V. L., Behavioral and physiological effects of xanthines in nonhuman primates, *Psychopharmacology*, 129, 1, 1997.

155. Lorist, M. M., Kok, A., Mulder, G., and Snel, J., Aging, caffeine, and information processing: an event-related potential analysis, *Electroencephalogr. Clin. Neurophysiol.*, 96, 453, 1995.

156. Lorist, M. M. and Snel, J., Caffeine effects on perceptual and motor processes, *Electroencephalogr. Clin. Neurophysiol.*, 102, 401, 1997.

157. Debrah, K., Kerr, D., Murphy, J., and Sherwin, R.S., Effect of caffeine on recognition of and physiological responses to hypoglycaemia in insulin-dependent diabetes, *Lancet*, 347, 19, 1996.

158. Johnson, J. P., Muhleman, D., MacMurray, J., Gade, R., Verde, R., Ask, M., Kelley, J., and Comings, D. E., Association between the cannabinoid receptor gene (CNR1) and the P300 event-related potential, *Mol. Psychiat.*, 2, 169, 1997.
159. Wijers, A. A., Lange, J. J., Mulder, G., and Mulder, L. J., An ERP study of visual spatial attention and letter target detection for isoluminant and nonisoluminant stimuli, *Psychophysiology*, 34, 553, 1997.
160. Smid, H. G., Trumper, B. G., Pottag, G., Wagner, K., Lobmann, R., Scheich, H., Lehnert, H., and Heinze, H. J., Differentiation of hypoglycaemia induced cognitive impairments. An electrophysiological approach, *Brain*, 128, 1041, 1997.
161. Luck, S. J., Girelli, M., McDermott, M. T., and Ford, M. A., Bridging the gap between monkey neurophysiology and human perception: An ambiguity resolution theory of visual selective attention, *Cognit. Psychol.*, 33, 64, 1997.
162. Besson, M., Faita, F., Czternasty, C., and Kutas, M., What's in a pause: event-related potential analysis of temporal disruptions in written and spoken sentences, *Biol. Psychol.*, 46, 3, 1997.
163. Ehlers, C. L., Somes, C., Lopez, A., Kirby, D., and Rivier, J. E., Electrophysio logical actions of neuropeptide Y and its analogs: New measures for anxiolytic therapy? *Neuropsychopharmacology*, 17, 34, 1997.
164. Tervaniemi, M., Schroger, E., and Naatanen, R., Pre-attentive processing of spectrally complex sounds with asynchronous onsets: An event-related potential study with human subjects, *Neurosci. Lett.*, 227, 197, 1997.
165. Polich, J. and Margala, C., P300 and probability: comparison of oddball and single-stimulus paradigms, *Intern. J. Psychophysiol.*, 25, 169, 1997.
166. Ito, H., Sugiyama, Y., Mano, T., Okada, H., Matsukawa, T., and Iwase, S., Skin sympathetic nerve activity and event-related potentials during auditory oddball paradigms, *J. Auton. Nerv. Syst.*, 60, 129, 1996.
167. Houlihan, M. E., Pritchard, W. S., and Robinson, J. H., Faster P300 latency after smoking in visual but not auditory oddball tasks, *Psychopharmacology*, 123, 231, 1996.
168. Johnstone, S. J., Barry, R. J., Anderson, J. W., and Coyle, S. F., Age-related changes in child and adolescent event-related potential component morphology, amplitude and latency to standard and target stimuli in an auditory oddball task, *Inter. J. Psychophysiol.*, 24, 223, 1996.
169. Kawamura, N., Maeda, H., Nakamura, J., Morita, K., and Nakazawa, Y., Effects of caffeine on event-related potentials: Comparison of oddball with single-tone paradigms, *Psychiat. Clin. Neurosci.*, 50, 217, 1996.
170. Marshall, L., Molle, M., and Bartsch, P., Event-related gamma band activity during passive and active oddball tasks, *Neuroreport*, 7, 1517, 1996.
171. Rockstroh, B., Muller, M., Heinz, A., Wagner, M., Berg, P., and Elbert, T., Modulation of auditory responses during oddball tasks, *Biol. Psychol.*, 43, 41, 1996.
172. Kenemans, J. L. and Lorist, M. M., Caffeine and selective visual processing, *Pharmacol. Biochem. Behav.*, 52, 461, 1995.
173. Johnson, M., Thinking about strategies during, before and after making a decision, *Psychol. Aging*, 8, 231, 1993.
174. Lubit, R. and Russett-Bruce, The effects of drugs on decision-making, *J. Conflict Resolution*, 28, 85, 1984.
175. Smith, D., Tong, J., and Leigh, G., Combined effects of caffeine and tobacco on the components of choice reaction time, heart rate and hand steadiness, *Percept. Mot. Skills*, 45, 635, 1977.

176. Streufert, S., Severs, W., Roache, J., Landis, R., Gingrich, D., Pogash, R., and Satish, U., Excess coffee consumption in simulated complex work settings: detriment or facilitation of performance? *J. Appl. Psychol.*, 82, 774, 1996.

177. Streufert, S., Pogash, R., Miler, J., and Gingrich, D., Effects of caffeine deprivation on complex human functioning, *Psychopharmacology*, 118, 377, 1995.

178. Cole, K. J., Fink, W. J., Trappe, S. W., Goodpaster, B. H., Starling, R. D., and Costill, D. L., Effect of caffeine ingestion on perception of effort and subsequent work production, *Intern. J. Sport Nutr.*, 6, 14, 1996.

179. Sargent, J. and Solbach, P., Stress and headache in the workplace: The role of caffeine, *Med. Psychother.: An Intern. J.*, 1, 83, 1988.

180. Choi, A., Laurito, C., and Cunningham, F., Pharmacologic management of postdural puncture headache, *Ann. Pharmacother.*, 30, 831, 1996.

181. Ramadan, N., Headache caused by intracranial pressure and intracranial hypotension, *Curr. Opin. Neurol.*, 9, 214, 1996.

182. Smith, B., Concannon, M., Campbell, S., and Bozman, A., Regression and criterion measures of habituation: A comparative analysis in extraverts and introverts, *J. Res. Person.*, 24, 123, 1990.

183. Dekker, D., Paley, M., Popken, S. and Tepas, D., Locomotive engineers and their spouses: Coffee consumption, mood and sleep disorders, *Ergonomics*, 36, 233, 1993.

184. Luecken, L. J., Williams, R. B., Siegler, I. C., Blumenthal, J. A., Barefoot, J. C., Kuhn, C. M., and Suarez, E. C., Stress in employed women: impact of marital status and children at home on neurohormone output and home strain, *Psychosom. Med.*, 59, 352, 1997.

185. Lane, J. D., Adcock, A., Williams, R. B., and Kuhn, C. M., Caffeine effects on cardiovascular and neuroendocrine responses to acute psychosocial stress and their relationship to level of habitual caffeine consumption, *Psychosom. Med.*, 52, 320, 1990.

186. Shapiro, D. and Oakley, M., Methodological issues in the evaluation of drug behavioral interactions in the treatment of hypertension, *Psychosom. Med.*, 51, 269, 1989.

187. Lane, J., Neuroendocrine responses to caffeine in the work environment, *Psychosom. Med.*, 56, 267, 1994.

188. Jeong, D. U. and Dimsdale, J. E., The effects of caffeine on blood pressure in the work environment, *Am. J. Hyperten.*, 3, 749, 1990.

189. France, C. and Ditto, B., Cardiovascular responses to occupational stress and caffeine in telemarketing employees, *Psychosom. Med.*, 51, 145, 1989.

190. Rogers, P., Richardson, N., and Dernoncourt, C., Caffeine use: Is there a net benefit for mood and psychomotor performance? *Neuropsychobiology*, 31, 195, 1995.

191. Ratliff-Crain, J., O'Keeffe, M. K., and Baum, A., Cardiovascular reactivity, mood, and task performance in deprived and nondeprived coffee drinkers, *Health Psychol.*, 8, 427, 1989.

192. Richardson, N. J, Rogers, P. J., and Elliman, N. A., Effects of comprehensive relaxation training (CRT) on mood: A preliminary report on relaxation training plus caffeine cessation, *Pharmacol. Biochem. Behav.*, 52, 313, 1995.

193. Kawachi, I., Willett, W., Colditz, G., Stampfer, M., and Speizer, F., A prospective study of coffee drinking and suicide in women, *Arch. Intern. Med.*, 156, 521, 1996.

194. Loke, W. H., Effects of repeated caffeine administration on cognition and mood, *Hum. Psychopharmacol. Clin. Exp.*, 5, 339, 1990.

195. Johnson, L. C., Spinweber, C. L., and Gomez, S. A., Benzodiazepines and caffeine: Effect on daytime sleepiness, performance, and mood, *Psychopharmacology*, 101, 160, 1990.

196. Quinlan, P., Aspinall, L., and Lane, J., Effects of hot tea, coffee and water ingestion on physiological responses and mood: the role of caffeine, water and beverage type, *Psychopharmacology*, 134, 164, 1997.

197. Penetar, D., McCann, U., Thorne, D., and Kamimori, G., Caffeine reversal of sleep deprivation effects on alertness and mood, *Psychopharmacology*, 112, 359, 1993.

198. Bonnet, M. H., Arand, D. L., Wirth, O., and Gomez, S., The use of caffeine vs. prophylactic naps in sustained performance, *Sleep*, 18, 97, 1995.

199. Rush, C. R., Griffiths, R. R., and Sullivan, J. T., Intravenous caffeine in stimulant drug abusers: subjective reports and physiological effects, *J. Pharmacol. Exp. Therap.*, 273, 351, 1995.

200. Perkins, K. A., Sexton, J. E., Stiller, R. L., Fonte, C., et al., Subjective and cardiovascular responses to nicotine combined with caffeine during rest and casual activity, *Psychopharmacology*, 113, 438, 1994.

201. Rush, C., Sullivan, J., and Griffiths, R., Intravenous caffeine in stimulant drug abusers: Subjective reports and physiological effects, *J. Pharmacol. Exp. Therap.*, 273, 351, 1995.

202. Azcona, O., Jane, F., Torrent, J., and Barbanoj, M. J., Evaluation of the central effects of alcohol and caffeine interaction, *Br. J. Clin. Pharmacol.*, 40, 393, 1995.

203. McQuay, H. J., Juniper, R. P., Moore, R. A., Carroll, D., and Angell, K., Ibuprofen compared with ibuprofen plus caffeine after third molar surgery, *Pain*, 66, 247, 1996.

204. Cohen, C., Pickworth, W. B., Bunker, E. B., and Henningfield, J. E., Caffeine antagonizes EEG effects of tobacco withdrawal, *Pharmacol. Biochem. Behav.*, 47, 919, 1994.

205. Richardson, N. J., O'Dell, R. J., Elliman, N. A., and Rogers, P. J., Mood and performance effects of caffeine in relation to acute and chronic caffeine deprivation, *Pharmacol. Biochem. Behav.*, 52, 213, 1995.

206. Comer, S. D., Fischman, M. W., Foltin, R. W., and Haney, M., Effects of caffeine withdrawal on humans living in a residential laboratory, *Exp. Clin. Psychopharmacol.*, 5, 399, 1997.

207. Rogers, P. J., Dernoncourt, C., and Richardson, N. J., Caffeine use: is there a net benefit for mood and psychomotor performance? *Neuropsychobiology*, 31, 195, 1995.

208. Heishman, S. J. and Henningfield, J. E., Is caffeine a drug of dependence? Criteria and comparisons, *Pharmacopsychoecologia*, 7, 127, 1994.

209. Leathwood, P. and Pollet, P., Diet-induced mood changes in normal populations, *J. Psychiat. Res.*, 17, 147, 1982.

210. Griffiths, R. R., Evans, S. M., Heishman, S. J., Preston, K. L., Sannerud, C. A., Wolf, B., and Woodson, P. P., Low-dose caffeine discrimination in humans, *J. Pharmacol. Exp. Therap.*, 252, 970, 1990.

211. Hughes, G. V. and Boland, F. J., The effects of caffeine and nicotine consumption on mood and somatic variables in a penitentiary inmate population, *Addict. Behav.*, 17, 447, 1992.

212. Liguori, A., Grass, J. A., and Hughes, J. R., Absorption and subjective effects of caffeine from coffee, cola and capsules, *Pharmacol. Biochem. Behav.*, 58, 721, 1997.

213. Evans, S. M. and Griffiths, R. R., Caffeine tolerance and choice in humans, *Psychopharmacology*, 108, 51, 1992.

214. Liguori, A., Oliveto, A. H., and Hughes, J. R., Caffeine self-administration in humans: 1. Efficacy of cola vehicle, *Exp. Clin. Psychopharmacol.*, 5, 286, 1997.

215. Stern, K. N., Chait, L. D., and Johanson, C. E., Reinforcing and subjective effects of caffeine in normal human volunteers, *Psychopharmacology*, 98, 81, 1989.

216. Parry-Jones, B. and Parry-Jones, W., Post-traumatic stress disorder: Supportive evidence from an eighteenth century natural disaster, *Psychol. Med.*, 24, 15, 1994.

217. Kramer, T., Lindy, J., Green, B., Grace, M., and Leonard, A., The lombordity of post traumatic stress disorder and suicidality in Vietnam veterans, *Suicide Life Threaten. Behav.*, 24, 58, 1994.

218. Long, N., Chamberlain, K., and Vincent, C., Effect of the Gulf War on reactivation of adverse combat-related memories in Vietnam veterans, *J. Clin. Psychol.*, 50, 138, 1994.

219. Spiro, A., Schnurr, P., and Aldwin, C., Combat related post traumatic stress disorder symptoms in older men, *Psychol. Aging*, 9, 17, 1994.

220. Powers, P., Cruse, C., Daniels, S., and Stevens, B., Posttraumatic stress disorder in patients with burns, *J. Burn Care Rehab.*, 15, 147, 1994.

221. Blanchard, E., Hickling, E., Taylor, A., and Loos, W., The psychophysiology of motor vehicle accident related posttraumatic stress disorder, *Behav. Ther.*, 25, 453, 1994.

222. Vrana, S. and Lauterbuch, D., Prevalence of traumatic events and post traumatic symptoms in college students, *J. Traum. Stress*, 7, 289, 1994.

223. Iancu, I., Dolberg, O., and Zohar, J., Is caffeine involved in the pathogenesis of combat stress reaction, *Mil. Med.*, 161, 230, 1996.

224. Schreiber, E. and Schreiber, K., Using relaxation techniques and positive self esteem to improve academic achievement of college students, *Psychol. Rep.*, 76, 929, 1995.

225. Pincomb, G. A., Lovallo, W. R., Passey, R. B., and Wilson M. F., Effect of behavior state of caffeine's ability to alter blood pressure, *Am. J. Cardiol.*, 61, 798, 1988.

226. Albertsen, P. C., Urologic "nuisances": how to work up and relieve men's symptoms, *Geriatrics*, 52, 46, 1997.

227. Lovallo, W. R., al'Absi, M., Pincomb, G.A., Everson, S.A., Sung, B.H., Passey, R. B., and Wilson, M. F., Caffeine and behavioral stress effects on blood pressure in borderline hypertensive Caucasian men, *Health Psychol.*, 15, 11, 1996.

228. Stollman, N. H., Bierman, P. S., Ribeiro, A., and Rogers, A. I., CO_2 provocation of panic: symptomatic and manometric evaluation in patients with noncardiac chest pain, *Am. J. Gastroenterol.*, 92, 839, 1997.

229. Klinnert, M., Mrazek, P., and Mrazek, D., Early asthma onset: The interaction between family stressors and adaptive parenting, *Psychiat. Interperson. Biol. Processes*, 57, 51, 1994.

230. Stewart, S. and Pihl, R., Effects of alcohol administration on psychophysiological and subjective-emotional responses to aversive stimulation in anxiety sensitive women, *Psychol. Addict. Behav.*, 8, 29, 1994.

231. James, J., *Caffeine and Health*, Academic Press, London, 1991.
232. Davis, W. M., Psychopharmacologic violence associated with cocaine abuse: kindling of a limbic dyscontrol syndrome? *Prog. Neuropsychopharmacol. Biol. Psychiatry,* 20, 1273, 1996.
233. Goldstein, I. B. and Shapiro, D., The effects of stress and caffeine on hypertensives, *Psychosom. Med.,* 49, 226, 1987.
234. Lovallo, W. R., Pincomb, G. A., Sung, B. H., Passey, R. B., Sausen, K. P., and Wilson, M. F., Caffeine may potentiate adrenocortical stress responses in hypertension-prone men, *Hypertension,* 14, 170, 1989.
235. Zahn, T. and Rapoport, J., Autonomic nervous system effects of acute doses of caffeine in caffeine users and abstainers, *Intern. J. Psychophysiol.,* 5, 33, 1987.
236. James, J. E., The influence of user status and anxious disposition on hypertensive effects of caffeine, *Intern. J. Psychophysiol.,* 10, 171, 1990.
237. Lane, J. D. and Williams, R. B., Cardiovascular effects of caffeine and stress in regular coffee drinkers, *Psychophysiology,* 24, 157, 1987.
238. Branscombe, N. R. and Wann, D. L., Role of identification with a group, arousal, categorization processes, and self-esteem in sports spectator aggression, *Hum. Relations,* 45, 1013, 1992.
239. Zeichner, A., Allen, J. D., Giancola, P. R., and Lating, J. M., Alcohol and aggression: Effects of personal threat on human aggression and affective arousal, *Alcohol. Clin. Exp. Res.,* 18, 657, 1994.
240. Bond, A. J., Pharmacological manipulation of aggressiveness and impulsiveness in healthy volunteers, *Prog. Neuropsychopharmacol. Biol. Psychiat.,* 16, 1, 1992.
241. Augsburger, D., An existential approach to anger management training, *J. Psychol. Christianity,* 5, 25, 1986.
242. Anderson, C. A., Deuser, W. E., and DeNeve, K. M., Hot temperatures, hostile affect, hostile cognition, and arousal: Tests of a general model of affective aggression, *Person. Soc. Psychol. Bull.,* 21, 434, 1995.
243. Giambra, L. M., Grodsky, A., Belongie, C., and Rosenberg, E. H., Depression and thought intrusions, relating thought frequency to activation and arousal, *Imaginat. Cogn. Person.,* 14, 19, 1995.
244. Mauri, M., Sarno, N., Rossi, V. M., Armani, A., et al., Personality disorders associated with generalized anxiety, panic, and recurrent depressive disorders, *J. Person. Disord.,* 6, 162, 1992.
245. Ketterer, M. W. and Maercklein, G. H., Caffeinated beverage use among Type A male patients suspected of CAD/CHD: A mechanism for increased risk? *Stress Med.,* 7, 119, 1991.
246. Seltzer, A., Multiple personality: A psychiatric misadventure, *Can. J. Psychiatry,* 39(7), 442-445, 1994.
247. de-Freitas, B. and Schwartz, G., Effects of caffeine in chronic psychiatric patients, *Am. J. Psychiat.,* 136, 1337, 1979.
248. Carmel, H., Caffeine and aggression, *Hosp. Commun. Psychiat.,* 42, 637, 1991.
249. Zaslav, M., Psychology or comorbid posttraumatic stress disorder and substance abuse: lessons for combat veterans, *J. Psychoactive Drugs,* 26, 393, 1994.
250. Veleber, D. M. and Templer, D. I., Effects of caffeine on anxiety and depression, *J. Abnorm. Psych.,* 93, 120, 1984.
251. Ohayon, M. M., Priest, R. G., and Caulet, M., Violent behavior during sleep, *Arch. Intern. Med.,* 157, 2645, 1997.

252. Taylor, S. L., O'Neal, E. C., Langley, T., and Butcher, A. H., Anger arousal, deindividuation, and aggression, *Aggressive Behav.*, 17, 193, 1991.
253. Geen, R. G., McCown, E. J., Effects of noise and attack on aggression and physiological arousal, *Motivat. Emot.*, 8, 231, 1984.
254. Gustafson, R., Alcohol and aggression, *J. Offender Rehab.*, 21, 41, 1994.
255. Bättig, K., Cardiovascular effects of everyday coffee consumption, *Schweiz. Med. Wochenschr.*, 122, 1536, 1992.
256. Gartside, P. S. and Glueck, C. J., Relationship of dietary intake to hospital admission for coronary heart and vascular disease: The NHANES II National Probability Study, *J. Am. Coll. Nutr.*, 12, 676, 1993.
257. Myers, M. G., Caffeine and cardiac arrhythmias, *Ann. Int. Med.*, 114, 147, 1991.
258. Etherton, G. M. and Kochar, M. S., Coffee: Facts and controversies, *Arch. Fam. Med.*, 293, 317, 1993.
259. Lynn, L. A. and Kissinger, J. F., Coronary precautions: should caffeine be restricted in patients after myocardial infarction? *Heart and Lung*, 21, 365, 1992.
260. Suter, P. M. and Vetter, W., Coffee and caffeine. various selected aspects for everyday practice, *Schweiz. Rundsch. Med. Prax.*, 82, 1122, 1993.
261. Newby, D. E., Boon, N. A., Jarvie, D. R., and Neilson, J. M., Caffeine restriction has no role in the management of patients with symptomatic idiopathic ventricular premature beats, *Heart*, 76, 355, 1996.
262. Lewis, C. E., Caan, B., Funkhouser, E., Hilner, J. E., Bragg, C., Dyer, A., Raczynski, J. M., Savage, P. J., Armstrong, M. A., and Friedman, G. D., Inconsistent associations of caffeine-containing beverages with blood pressure and with lipoproteins. The CARDIA Study. Coronary Artery Risk Development in Young Adults, *Am. J. Epidemiol.*, 138, 502, 1993.
263. Superko, H. R., Bortz, W. Jr., Williams, P. T., Albers, J. J., and Wood, P. D., Caffeinated and decaffeinated coffee effects on plasma lipoprotein cholesterol, apolipoproteins, and lipase activity: A controlled, randomized trial, *Am. J. Clin. Nutr.*, 54, 599, 1991.
264. Stavric, B., An update on research with coffee/caffeine (1989-1990), *Food Chem. Toxicol.*, 30, 533, 1992.
265. Superko, H. R., Myll, J., DiRicco, C., and Williams, P. T., Effects of cessation of caffeinated-coffee consumption on ambulatory and resting blood pressure in men, *Am. J. Cardiol.*, 15, 73, 780, 1994.
266. Casiglia, E., Paleari, C. D., Daskalakis, C, Petucco, S., Bongiovi, S., and Pessina, A. C., Hemodynamic effects of "expresso" italian coffee and pure caffeine on healthy volunteers, *Cardiologia*, 35, 575, 1990.
267. Kuznicki, J. T. and Turner, L. S., The Effects of caffeine on caffeine users and nonusers, *Physiol. Behav.*, 37, 397, 1986.
268. Smith, B. D. and Wigglesworth, M., Extraversion and neuroticism in orienting reflex dishabituation, *J. Res. Person.*, 12, 284, 1978.
269. Gupta, U., Differential effects of caffeine on free recall after semantic and rhyming tasks in high and low impulsives, *Psychopharmacology*, 105, 137, 1991.
270. Gupta, U., Effects of caffeine on recognition, *Pharmacol. Biochem. Behav.*, 44, 393, 1993.
271. Smith, B. D., Rockwell-Tischer, S., and Davidson, R., Extraversion and arousal: effects of attentional conditions on electrodermal activity, *Person. Individ. Diff.*, 7, 293, 1986.

272. Smith, B. D., Davidson, R., Smith, D., Goldstein, H., and Perlstein, W., Sensation seeking and arousal: Effects of strong stimulation on electrodermal activation and memory task performance, *Person. Individ. Diff.*, 6, 671, 1989.
273. Eysenck, H. J., *The Biological Bases of Personality,* Charles C. Thomas, Springfield, IL, 1967.
274. Eysenck, H. J., A reply to Costa and McCrae: P or A and C — The role of theory, *Person. Individ. Diff.*, 13, 867, 1992.
275. Smith, B. D., Wilson, R. J., and Jones, B. E., Extraversion and multiple levels of caffeine-induced arousal: effects on overhabituation and dishabituation, *Psychophysiology,* 20, 29, 1983.
276. Stelmack, R. M., Biological bases of extraversion: psychophysiological evidence, *J. Person.*, 58, 293, 1990.
277. Pearson, G. L. and Freeman, F. G., Effects of extraversion and mental arithmetic on heart-rate reactivity, *Percept. Mot. Skills*, 72, 1239, 1991.
278. Bullock, W. A. and Gilliland, K., Eysenck's arousal theory of introversion-extraversion: A converging measures investigation, *J. Person. Soc. Psychol.*, 64, 113, 1993.
279. Davidson, R., Fedio, P., Smith, B. D., Aureille, E., and Martin, A., Lateralized mediation of arousal and habituation: differential bilateral electrodermal activity in unilateral temporal lobectomy patients, *Neuropsychologia*, 30, 1053, 1992.
280. Kenardy, J., Oei, T. P., Weir, D., and Evans, L., Phobic anxiety in panic disorder: Cognition, heart rate, and subjective anxiety, *J. Anxiety Disord.*, 7, 359, 1993.
281. Fonagy, P. and Calloway, S., The effect of emotional arousal on spontaneous swallowing rates, *J. Psychosom. Res.*, 30, 183, 1986.
282. Hughes, J. R., Higgins, S. T., and Hatsukami, D., Effects of abstinence from tobacco: A critical review, in *Research Advances in Alcohol and Drug Problems,* Vol. 10, Kozlowski, L. T., Annis, H. M., Cappell, H. D., Glaser, F. B., Goodstadt, M. S., Israel, Y., Kalant, H., Sellers, E. M., and Vingilis, E. R., Eds., Plenum Press, New York, 1990, 317.
283. Sachs, D. and Benowitz, N., The nicotine withdrawal syndrome: Nicotine absence or caffeine excess? *Proceedings of the Fiftieth Annual Meeting of the Committee on Problems of Drug Dependence: NIDA Research Monograph.* 90, p. 38. Washington, D.C., U.S. Government Printing Office, 1988.
284. Pritchard, W. S., Stiles, M. F., Davis, R.A., deBethizy, J. D., and Robinson, J. H., Caffeine and smoking: subjective, performance, and psychophysiological effects, *Psychophysiology,* 32, 19, 1995.
285. Zeiner, A. R., Stanitis, T., Spurgeon, M., and Nichols, N., Treatment of alcoholism and concomitant drugs of abuse, *Alcohol*, 2, 555, 1985.
286. Coffey, C. E., Weiner, R. D., Hinkle, P. E., Cress, M., et al., Adverse reaction to use of caffeine in ECT, *Biol. Psychiat.*, 22, 637, 1987.
287. Bruce, M., Scott, N., Shine, P., and Lader, M., Anxiogenic effects of caffeine in patients with anxiety disorders, *Arch. Gen. Psychiat.*, 49, 867, 1992.
288. Lieberman, H., Caffeine, in *Handbook of Human Performance*, Vol. 2, Smith, A.P. and Jones, D.M., Eds., London, Academic Press, 1992, 49.
289. Dubovsky, S. L., Generalized anxiety disorder: New concepts and psychopharmacologic therapies, 142nd Annual Meeting of the American Psychiatric Association (1989, San Francisco, California). *J. Clin. Psychiat.*, 51(Suppl) 3, 1990.

290. Bhattacharya, S. K., Chakrabarti, A., and Satyan, K. S., Anxiogenic action of caffeine: an experimental study in rats, *J. Psychopharmacol.*, 11, 219, 1997.

291. Jain, N., Stone, T. W., Buchanan, P., Adeyemo, O., and Kemp, N., Anxiolytic activity of adenosine receptor activation in mice, *Br. J. Pharmacol.*, 116, 2127, 1995.

292. Fredholm, B. B., On the mechanism of action of theophylline and caffeine, *Acta Med. Scand.*, 217, 149, 1985.

293. Kendler, K., Neale, M., Kessler, R., and Heath, A., Generalized anxiety disorder in women: A population based twin study, *Arch. Gen. Psychiat.*, 49, 267, 1992.

294. Rauch, S. and Jenike, M., Neurobiological models of obsessive-compulsive disorder, *Psychosomatics*, 34, 20, 1993.

295. Otto, M., Normal and abnormal information processing: A neuropsychological perspective on obsessive compulsive disorder, *Psychiat. Clinics North Am.*, 15, 825, 1992.

296. Scarone, S., Colombo, C., Livian, S., and Abbruzzese, M., Increased right caudate nucleus size in obsessive compulsive disorder: Detection with magnetic resonance imaging, *Psychiat. Res. Neuroimaging*, 45, 115, 1992.

297. Nutt, D., Glue, P., and Lawson, C., The neurochemistry of anxiety: An update, *Prog. Neuropsychopharmacol. Biol. Psychiat.*, 14, 737, 1990.

298. Hildebrand, M. and Seifert, W., Determination of acetylator phenotype in caucasians with caffeine, *J. Clin. Pharmacol.*, 37, 525, 1981.

299. Carrillo, J. and Benitez, J., Caffeine metabolism in a healthy spanish population: N-Acetylator phenotype and oxidation pathways, *Clin. Pharmacol. Therap.*, 55, 293, 1994.

300. Warburton, D. M. and Thompson, D. H., An evaluation of the effects of caffeine in terms of anxiety, depression and headache in the general population. *Pharmacopsychoecologia*, 7, 55, 1994.

301. Smith, G. A., Caffeine reduction as an adjunct to anxiety management, *Br. J. Clin. Psychol.*, 27, 265, 1988.

302. Charney, D. S., Heninger, G. R., and Jatlow, P. I., Increased anxiogenic effects of caffeine in panic disorders, *Arch. Gen. Psychiat.*, 42, 233, 1985.

303. Lee, M. A., Flegel, P., Greden, J. F., and Cameron, O. G., Anxiogenic effects of caffeine on panic and depressed patients, 41st Annual Meeting of the Society of Biological Psychiatry (1986, Washington, D.C.). *Am. J. Psychiat.*, 145, 632, 1988.

304. Holle C., Holt, C. S., Sweet, R. A., and Heimberg, R. G., Alcohol and caffeine use by social phobics: an initial inquiry into drinking patterns and behavior, *Behav. Res. Ther.*, 33, 561, 1995.

305. Potts, N. L., Davidson, J. R., and Book, S., The neurobiology of social phobia, *Intern. Clin. Psychopharmacol.*, 3, 43, 1996.

306. Perugi, G., Cassano, G. B., Ravagli, S., Lensi, P., Milanfranchi, A., Gemignani, A., Presta, S., Pfanner, C., and Akiskal, H. S., The clinical impact of bipolar and unipolar affective comorbidity on obsessive-compulsive disorder, *J. Affect. Disord.*, 46, 15, 1997.

307. Stollman, N. H., Ribiero, A., Rogers, A. I., Ribeiro, A., and Bierman, P. S., CO2 provocation of panic: Symptomatic and manometric evaluation in patients with noncardiac chest pain, *Am. J. Gastroenterol.*, 92, 1244, 1997.

308. Bourin, M., From inducers of panic attack to neurobiology of panic disorder, *Encephase*, 5, 35, 1996.

309. Krystal, J. H., Charney, D. S., and Deutsch, D. N., The biological basis of panic disorder, *J. Clin. Psychiat.*, 10, 23, 1996.
310. Bourin, M., Guitton, B., and Malinge, M., Provocative agents in panic disorder, *Therapie*, 50, 301, 1995.
311. Reimann, H., Caffeinism: A cause of long-continued, low-grade fever, *JAMA*, 202, 131, 1967.
312. Greden, J. F., Anxiety or caffeinism: A diagnostic dilemma, *Am. J. Psychiat.*, 131, 1089, 1974.
313. Pilette, W. L., Caffeine: Psychiatric grounds for concern, *J. Psycho. Nurs. Ment. Health Serv.*, 21, 19, 1983.
314. James, J. and Stirling, K., Caffeine: A survey of some of the known and suspected deleterious effects of habitual use, *Br. J. Addict.*, 78, 251, 1995.
315. Victor, B. S., Lubetsky, M., and Greden, J. F., Somatic manifestations of caffeinism, *J. Clin. Psychiat.*, 42, 185, 1981.
316. Kits van Waveren, L., Cafeinisme (Caffeinism), *Tijdschrift voor Psychiatrie*, 30, 403, 1988.
317. Wurl, P., Life threatening caffeine poisoning by using coffee as a psychoactive drug, *Wien Klin. Wochenschr.*, 106, 359, 1994.
318. Forman, J., Aizer, A., and Young, C., Myocardial infarction resulting from caffeine overdose in an anorectic woman, *Ann. Emerg. Med.*, 29, 178, 1997.
319. Behar, D., Flashbacks and posttraumatic stress symptoms in combat veterans, *Comprehen. Psychiat.*, 28, 459, 1987.
320. Newman, F. X., Stein, M. B., Trettau, J. R., Coppola, R., et al., Quantitative electroencephalographic effects of caffeine in panic disorder, *Psychiat. Res. Neuroimaging*, 45, 105, 1992.
321. Beck, J. G. and Berisford, M. A., The effects of caffeine on panic patients: Response components of anxiety, *Behav. Ther.*, 23, 405, 1992.
322. Perugi, G., Cassano, G. B., Ravagli, S., and Lensi, P., The clinical impact of bipolar and unipolar affective comorbidity on obsessive-compulsive disorder, *J. Addict. Dis.*, 16, 19, 1997.
323. Abbott, F. V. and Fraser, M. I., Use and abuse of over-the-counter analgesic agents, *J. Psychiat. Neurosci.*, 23, 13, 1998.
324. Krauchi, K., Wirz-Justice, A., and Reich, S., Eating style in seasonal affective disorder: who will gain weight in winter? *Comprehen. Psychiat.*, 38, 80, 1997.
325. Wright, K. P., Jr., Hakel, M., Plenzler, S. C., Myers, B. L., and Badia, P., Caffeine and light effects on nighttime melatonin and temperature levels in sleep-deprived humans, *Brain Res.*, 747, 78, 1997.
326. Rihs, M., Baumann, P., and Muller, C., Caffeine consumption in hospitalized psychiatric patients, *Eur. Arch. Psychiat. Clin. Neurosci.*, 246, 83, 1996.
327. Worthington, J., Rosenbaum, J. F., Pava, J. A., Nierenberg, A. A., Alpert, J., Agustin, C., and Fava, M., Consumption of alcohol, nicotine, and caffeine among depressed outpatients. Relationship with response to treatment, *Psychosomatics*, 37, 518, 1996.
328. Mainous, R. O., Research utilization: pharmacologic management of neonatal pain, *Neonatal Netw.*, 14, 71, 1995.
329. Vormfelde, S. V., Gleiter, C. H., Gundert-Remy, U. M., Meineke, I., and Bitsch, A., Non-response to maprotiline caused by ultra-rapid metabolism that is different from CY2D6? *Eur. J. Clin. Pharmacol.*, 52, 387, 1997.
330. Millwe, C. S., Chemical sensitivity: symptom, syndrome or mechanism for disease? *J. Forensic Sci.*, 40, 128, 1995.

331. Neylan, T. C., Treatment of sleep disturbances in depressed patients, *J. Clin. Psychiat.*, 56, 56, 1995.
332. Tiffin, P., Kamali, F., Marsh, R., and Ashton, H., Pharmacokinetic and pharmacodynamic responses to caffeine in poor and normal sleepers, *Psychopharmacology*, 121, 494, 1995.
333. Rapoport A., Fox, A. W., Saiers, J., Weeks, R., Markley, H., Cady, R., Gutterman, D. L., and Stang, P., Analgesic rebound headache in clinical practice: data from a physician survey, *Headache*, 36, 14, 1996.
334. Cacciatore, R., Hess, B., Jost, C., and Helbling, A., Episodic headache, diminished performance and depressive mood, *Schweiz. Rundsch. Med. Prax.*, 28, 85, 727, 1996.
335. Donnelly, C. L., Narasimhachari, N., Wilson, W.H., and McEvoy, J. P., A study of the potential confounding effects of diet, caffeine, nicotine and lorazepam on the stability of plasma and urinary homovanillic acid levels in patients with schizophrenia, *Biol. Psychiat.*, 40, 1218, 1996.
336. Van Ammers, E. C., Mulder, R. T., and Sellman, J. D., Temperament and substance abuse in schizophrenia: Is there a relationship? *J. Nerv. Ment. Disord.*, 185, 283, 1997.
337. Ferre, S., Adenosine-dopamine interactions in the ventral striatum. Implications for the treatment of schizophrenia, *Psychopharmacology*, 133, 107, 1997.
338. Erfurth, A. and Schmauss, M., Perspectives on the therapy of neuropsychiatric diseases with adenosinergic substances, *Fortschr. Neurol. Psychiatr.*, 63, 93, 1995.
339. Kruger, A., Chronic psychiatric patients' use of caffeine: pharmacological effects and mechanisms, *Psychol. Rep.*, 78, 915, 1996.
340. Hamera, E., Deviney, S., and Schneider, J. K., Alcohol, cannabis, nicotine, and caffeine use and symptom distress in schizophrenia, *J. Nerv. Ment. Disord.*, 183, 559, 1995.
341. Marcus P. and Snyder, R., Reduction of comorbid substance abuse with clozapine, *Am. J. Psychiat.*, 152, 959, 1995.

chapter seven

Is caffeine a drug of dependence? criteria and comparisons

Stephen J. Heishman and Jack E. Henningfield

Contents

I. Introduction 137
II. Drug dependence 138
III. Criteria for drug dependence 138
    A. Primary criteria for drug dependence 139
    B. Secondary criteria for drug dependence 142
IV. Conclusion 143
    A. Is caffeine a drug of dependence? 143
    B. Future issues of caffeine as a drug of dependence 144
References 145

I. Introduction

Caffeine is the most widely used psychoactive drug in the world.[1] The most notable behavioral effects after low to intermediate doses of caffeine (50 to 300 mg) are increased alertness, energy, and ability to concentrate. Higher doses can induce a state of caffeine intoxication or caffeinism, characterized by anxiety, restlessness, insomnia, and tachycardia. However, the health risks associated with moderate use of caffeine are relatively small.[2-4]

Numerous articles have discussed caffeine as a drug of dependence.[5-7] The purpose of this chapter is to review experimental and clinical data on effects of caffeine with respect to three primary and three secondary criteria for drug dependence. Within each criterion, comparative data for other drugs of dependence are presented to assess the relative dependence potential of caffeine. The chapter begins with a working definition of drug dependence and a brief discussion of the drug dependence criteria. We conclude with some thoughts concerning current and future classifications of caffeine as a drug of dependence.

II. Drug dependence

We have defined drug dependence as a pattern of behavior focused on the repetitive and compulsive seeking and taking of a psychoactive drug.[8] Thus, in this sense, drug dependence is synonymous with the term drug addiction. Drug dependence should not be confused with physical dependence. Physical dependence is determined by an abstinence or withdrawal syndrome that either occurs spontaneously when chronic drug taking ceases or is precipitated by administration of an antagonist.[9] The development of physical dependence on a drug may be one component of the drug dependence syndrome, but it is neither a necessary nor sufficient condition to establish or maintain drug dependence.[9,10]

It is also important to note that drug dependence is not simply the result of intrinsic properties of a drug, but rather results from the interaction of the drug, an individual, and their environment. As a result of this complex interaction, the likelihood that an individual will become drug dependent will vary across time and individuals. For example, the majority of people who use opiates or alcohol do not become dependent, yet these drugs are considered to be "classic" drugs of dependence.[10]

III. Criteria for drug dependence

The primary and secondary criteria for defining drug dependence used in this chapter were adapted from the 1988 Surgeon General's report, *The Health Consequences of Smoking: Nicotine Addiction.*[10] We used these criteria because they represent a consensus of many experts in the field of drug dependence and addiction.

The three primary criteria, which are necessary and sufficient to define drug dependence, are: (1) psychoactive effects, (2) highly controlled or compulsive use, and (3) drug-reinforced behavior. Psychoactive effects refer to changes in mood or feelings produced by the drug that are mediated by the central nervous system. Psychoactivity is a characteristic of all dependence-producing drugs and is used to differentiate drug dependence from other habitual behaviors, such as regular ingestion of aspirin. Highly controlled or compulsive use of a drug refers to habitual behavior that persists despite attempts to reduce or quit drug use. Such drug-seeking and drug-taking behavior typically occurs on a regular basis, and in the case of illicit drugs, usually requires much time and effort, often at the expense of other activities. Finally, it must be shown that the drug is the stimulus that functions to reinforce behavior leading to repeated drug self-administration.

The three secondary criteria to be considered in evaluating caffeine as a drug of dependence are: (1) pleasant or euphoric effects, (2) tolerance, and

(3) physical dependence. Drug-induced euphoria is typically assessed through the use of self-report measures, such as visual analog scales and standardized questionnaires. Tolerance is defined as decreased responsiveness to an effect of a drug that results from prior drug exposure. Physical dependence refers to an altered physiological state resulting from prior drug exposure, which requires the maintenance of drug administration for normal functioning and is defined by an abstinence syndrome when the drug is discontinued.[9] These three effects are often, but not always, observed after exposure to dependence-producing drugs; thus, they are neither necessary nor sufficient conditions to define drug dependence. However, each of these effects can strengthen the control of a drug on behavior, increasing the probability that initial drug use will escalate to drug dependence or that drug dependence will be maintained.

A. Primary criteria for drug dependence

1. Psychoactivity

As stated, psychoactivity is a basic criterion for drugs of dependence; however, the qualitative nature and magnitude of the psychoactive effects are also critical. The psychoactivity of a drug can be determined through studies assessing the drug's discriminative stimulus effects and subjective effects.

a. Discriminative stimulus effects. In two-choice drug discrimination studies, subjects are trained to make one response in the presence of the drug and a different response in the presence of placebo; three-choice studies add a second training drug condition. Subsequent generalization testing can identify those drugs that share discriminative stimulus effects with the training drug, producing drug-appropriate responding, and those that do not, producing placebo-appropriate responding. Drugs that can be successfully discriminated from placebo during training are defined as psychoactive and thus meet one of the essential criteria for drugs of dependence.[10] The nature of a drug's discriminative stimulus effects can be determined by whether it occasions drug- or placebo-appropriate responding. The discriminative stimulus effects of caffeine have been examined as a training drug and as a test drug during generalization testing.

Several studies have demonstrated that human volunteers with and without histories of drug abuse can readily discriminate the effects of caffeine in two-choice[11-15] and three-choice[16-18] discrimination paradigms. In these studies, the training doses of caffeine ranged from 100 to 400 mg. Using a paradigm in which subjects were trained to discriminate progressively lower doses of caffeine from placebo, two studies reported that caffeine functioned as a discriminative stimulus at doses ranging from 10 to 56 mg in low- or nontolerant individuals.[11,15] These studies clearly indicate that caffeine functions as a discriminative stimulus in humans and is therefore psychoactive. Caffeine shares this psychoactivity with all other drugs of dependence, including nicotine,[19] *d*-amphetamine,[20] cocaine,[21] ethanol,[22] and opioids.[23]

In studies in which caffeine has been used as a training drug and various doses have been examined during generalization testing, caffeine has produced dose-related increases in caffeine-appropriate responding.[13,14,16,24] Two studies have examined whether test doses of caffeine would substitute for training drugs other than caffeine. Chait and Johanson[25] trained subjects to discriminate *d*-amphetamine (10 mg) from placebo. Caffeine produced a profile of partial substitution; mean *d*-amphetamine-appropriate responding was 42% and 58% for 100 mg and 300 mg of caffeine, respectively. Heishman et al.[16] trained stimulant abusers to discriminate between *d*-amphetamine (30 mg), caffeine (400 mg), and placebo and found that lower test doses of caffeine occasioned partial *d*-amphetamine responding, whereas higher caffeine doses produced caffeine-appropriate responding. Thus, at certain doses, there is some similarity between the discriminative stimulus functions of caffeine and *d*-amphetamine. However, there are also sufficient differences that it would be inappropriate to categorize caffeine as having amphetamine-like dependence potential.

b. Subjective effects.   From the preceding section, it is evident that caffeine is psychoactive, reliably producing discriminable effects. Subjective effects data indicate that the nature and magnitude of caffeine's psychoactivity largely depends on dose. Most studies testing low to intermediate doses of caffeine (50 to 400 mg) have reported increased subjective ratings of alertness, arousal, and vigor, and decreased ratings of tiredness, fatigue, and sedation.[16,25,26] However, increases in tension, anxiety, and shaking have been reported in the range of 300 to 500 mg of caffeine.[14,27,28] The subjective effects of caffeine doses above 500 mg are usually dysphoric in nature, characterized by increased tension, anxiety, nervousness, and panic.[27,29,30]

Two subjective measures that are considered hallmarks of a drug of dependence are "liking" scales and the Morphine-Benzedrine Group (MBG) scale of the Addiction Research Center Inventory.[31,32] Low to intermediate doses of caffeine have been shown either to increase ratings of "liking"[33,34] or produce no effect,[25] whereas high doses have not increased "liking" scores.[30] On the MBG scale, caffeine has been shown either to increase scores in some or all subjects[11,27] or have no effect.[35,36] Thus, the effects of caffeine on "liking" and MBG scale scores are variable, dependent on dose, subject population, and particular experimental conditions. In the only study to directly compare caffeine and *d*-amphetamine on MBG scores, Chait and Griffiths[27] reported that the increase in MBG scores elicited by *d*-amphetamine (25 mg) was far greater than that elicited by caffeine (800 mg). These findings are in contrast to the reliable, dose-related increases in "liking" and MBG scores that have been reported for known drugs of dependence, including nicotine,[37] cocaine,[21] *d*-amphetamine,[20] and opioids.[38] Thus, under limited conditions, caffeine shares subjective effects in common with known drugs of dependence; however, these drugs typically produce such effects more reliably and under a broader range of conditions.[10]

2. Controlled or compulsive use

Information concerning the controlled or habitual use of caffeine can be found in surveys, which indicate that 92 to 98% of adults in North America consume caffeine on a regular basis.[1] Anecdotal evidence suggests that the majority of caffeine users probably consume the equivalent of one to three cups of coffee every day, which would satisfy the criterion of highly controlled or habitual use. It is estimated that 10% of caffeine users in the United States (22 to 24 million people) consume an average of about 500 mg of caffeine per day.[39] Although it is likely that some proportion of this population uses caffeine compulsively, we lack data concerning the compulsive nature of caffeine consumption.

Clinical case studies suggest that caffeine use can occur as a compulsive behavior resistant to change. Individuals who regularly consume large amounts of caffeine, usually in the form of coffee, are frequently resistant to reducing their caffeine intake on advice of their physician.[40-42] Relapse following treatment to reduce heavy caffeine consumption has been reported,[43] and Greden[44] estimated that only one third of individuals treated for caffeinism maintain moderate use of caffeine, with the majority relapsing to high rates of consumption. Although there are no data on the prevalence of individuals requiring treatment for caffeinism, it is probably a small percentage of all caffeine users. Additionally, other clinical reports[45,46] and apparent successful treatment approaches[47,48] indicate that some individuals are able to reduce or eliminate their caffeine intake. The growth of the decaffeinated beverage industry suggests that many people have readily switched from caffeinated to decaffeinated products. Thus, a tentative conclusion is that the majority of users consume caffeine in a highly controlled, habitual manner, but that a minority of individuals use caffeine compulsively, such that they find it difficult to reduce or stop their intake.

3. Drug-reinforced behavior

During the 1960s and 1970s, numerous studies demonstrated that animals would repeatedly press a lever to obtain an injection of a drug, the interpretation being that the drug reinforced the response leading to its delivery.[49] As these studies accumulated, it became clear that, in general, drugs that functioned as reinforcers in animals were likely to be abused by humans.[50] It is now evident that a drug's reinforcing effect is a primary mechanism of drug dependence by which the drug controls behavior leading to its repeated use. The experimental demonstration of the reinforcing effect of a drug is considered to be a primary measure of a drug's dependence potential.[51]

By this criterion, caffeine is a drug of limited dependence potential. Results of most animal studies have shown that caffeine does not function reliably as a reinforcer across several species and testing conditions.[52,53] Studies with humans have demonstrated some conditions under which caffeine functions as a reinforcer; however, these studies also reported large individual differences in caffeine's reinforcing effects. The majority of placebo-controlled studies in humans have shown that caffeine in the form of coffees, colas, and capsules was self-administered by some or all subjects.[28,33,54-62] However, other studies have reported minimal[63,64] or no evidence[34] for the reinforcing effects of caffeine. Thus, caffeine's potential to function as a reinforcer in humans has been demonstrated, but, in general, it is self-administered by less than half of subjects tested. In contrast, nicotine,[65,66] cocaine,[67,68] d-amphetamine,[69,70] ethanol,[67,71] and opioids[72,73] have been shown to maintain reliable self-administration behavior in animals and humans.

B. Secondary criteria for drug dependence

As discussed previously, euphoria, tolerance, and physical dependence are effects often produced by psychoactive drugs. Although neither necessary nor sufficient to define drug dependence, these states may influence drug self-administration by increasing the control of a drug on behavior.[10]

1. Pleasant or euphoric effects

Drug-induced euphoria has been postulated as a predictor of a drug's reinforcing effects and its dependence potential.[74,75] As discussed, caffeine typically produces positive mood changes at low to intermediate doses (50 to 300 mg), whereas doses in the 300 to 500 mg range can produce positive or negative subjective effects. The ability of caffeine to produce euphoria and dysphoria in the same dosage range may function to limit caffeine's dependence potential. Some studies have shown that caffeine at low to intermediate doses increased ratings of drug "liking" and scores on the MBG scale, two subjective measures of drugs of dependence.[11,33,34] However, these positive mood effects of caffeine are typically less reliably observed and of lesser magnitude than those reported for known drugs of dependence, such as d-amphetamine.

2. Tolerance

There are at least two mechanisms by which tolerance could increase the control of a drug on behavior. Tolerance to the euphoric or reinforcing effects of a drug could result in increased drug taking to maintain the desired effect, and thus the drug dependency may escalate. On the other hand, if tolerance develops to the aversive effects of high drug doses, the drug's reinforcing effects may be strengthened, resulting in increased drug taking. Tolerance develops to many of the pharmacological effects of caffeine in humans.[76,77] Several studies have reported minimal or no evidence for tolerance to caffeine-induced alertness and wakefulness.[78-80] However, Evans and Griffiths[12] clearly demonstrated tolerance to several subjective effects of caffeine, including tension, anxiety, nervousness, and magnitude of drug effect, but tolerance to the reinforcing effects of caffeine was not observed.

3. Physical dependence

If physical dependence on a drug develops, the occurrence of an aversive abstinence syndrome may precipitate drug taking to alleviate the discomfort, thereby maintaining the drug dependency. An abstinence syndrome following the cessation of caffeine use has been documented by clinical reports, surveys, and experimental studies,[7,81] and withdrawal symptoms have predicted prospectively caffeine self-administration.[82] The symptoms most frequently associated with caffeine abstinence are headache, fatigue, and lethargy.[81,83-85] Abstinence symptoms typically begin in 12 to 24 hours and reach maximum intensity 20 to 48 hours after cessation of caffeine intake. The intensity of the withdrawal syndrome can vary from mild to severe and, in some instances, can dramatically impair normal functioning. Studies have reported large individual differences in terms of the occurrence, intensity, and duration of the caffeine abstinence syndrome, which is consistent with caffeine's other effects in humans.

IV. Conclusion

A. Is caffeine a drug of dependence?

Caffeine is a drug that maintains an enigmatic existence somewhere in the middle or lower end of the continuum of dependence-producing drugs. It is not readily categorized with known drugs of dependence, such as cocaine, nicotine, heroin, and ethanol. On the other hand, caffeine does elicit some effects typical of dependence-producing drugs.

We conclude that caffeine partially meets the primary criteria of drug dependence and thus has limited dependence potential: (1) caffeine is psychoactive; (2) the use of caffeine by most individuals could be described as regular, but it appears that only a minority of individuals have significant problems due to caffeine and are unable to abstain; and (3) caffeine does not function reliably as a reinforcer in animals, and its reinforcing effects in humans are limited to certain conditions. In terms of secondary criteria for dependence, caffeine can produce pleasant effects, which, under limited conditions, could be considered euphoric; however, these effects are qualitatively and quantitatively distinct from known drugs of dependence, such as *d*-amphetamine. Tolerance and physical dependence have been shown to develop after exposure to caffeine.

The conclusion that caffeine has a limited dependence potential and a lower relative dependence potential compared with known drugs of dependence is consistent with at least two authoritative sources on drug dependence: (1) the Diagnostic and Statistical Manual of Mental Disorders[86] (DSM-IV) includes caffeine intoxication as a Substance-Induced Disorder, but does not recognize caffeine withdrawal as a Substance-Induced Disorder or caffeine abuse and dependence as diagnostic categories of Substance Use Disorders; and (2) Goldstein and Kalant[87] concluded that caffeine's relative risk of addiction was the lowest of seven drugs or drug classes. However, the International Classification of Diseases[88] (ICD-10) includes a category of "Mental and behavioral disorders due to use of other stimulants, including caffeine" that can be used to diagnose caffeine intoxication, harmful use, dependence, and withdrawal. This is a new development because the previous edition, ICD-9, did not include caffeine in either Drug Dependence or Nondependent Abuse categories.

B. Future issues of caffeine as a drug of dependence

It is likely that future editions of DSM will include a diagnostic classification of caffeine withdrawal and perhaps dependence, given the ICD-10 classification and the scientific and clinical data collected during the past few years.[7,81,89] The wide availability of over the-counter pills and mail-order capsules that contain intermediate to high doses of caffeine and the ease with which large doses can be ingested rapidly would support the possibility of diagnostic categories of caffeine abuse or caffeine dependence. Concern about the public health effects of caffeine use would increase if tolerant individuals using large doses began having numerous accidents while stimulated by caffeine or during caffeine withdrawal. However, reports of caffeine-induced accidents, toxicity, and overdose are rare.[90,91] Presumably, the dysphoria produced as caffeine dose increases is an important factor limiting the dependence potential of caffeine.

Another possible factor mitigating the dependence potential of caffeine is the delivery form in which it is most commonly used. The gradual intake of caffeine-containing beverages via the oral route, although allowing complete absorption, results in a delayed onset of subjective effects at 30 to 60 minutes after administration. It is known that when a drug is absorbed quickly via the smoked or intravenous routes and thus produces a rapid onset of effect, the potential for abuse and dependence is greater than for the same drug formulated to be absorbed more slowly. A good example of this is nicotine, which when smoked in the form of tobacco cigarettes is highly addictive, but when delivered buccally as a gum or transdermally as a patch, is of minimal dependence potential.[92,93] The dependence potential of smoked or intravenous caffeine might be greater than that associated with oral delivery forms. For example, when stimulant abusers were administered caffeine intravenously, dose-related increases in "good effects" and "liking" were observed, and most subjects reported that they had received cocaine.[94] However, there have been no reports of abuse of rapidly delivered forms of caffeine, possibly because of the widespread availability of addictive delivery forms of other drugs, such as nicotine and cocaine.

A final factor arguing against the classification of caffeine as a drug of dependence is that given the widespread use of caffeine throughout the world, there are few reports in the literature of individuals who find it difficult to decrease or stop their caffeine intake.[40] However, as Hughes et al.[89]

indicated, there are no prospective clinical studies on attempts to quit caffeine or on success rates of caffeine cessation programs. We think the ability to reduce or quit drug use is a critically important criterion of drug dependence and was described in this article as compulsive drug use. In the absence of data to the contrary, we conclude that only a small proportion of individuals who use caffeine do so compulsively, such that they find it difficult to reduce their intake. If future studies indicate that a greater proportion of caffeine users can be categorized as compulsive, then the probability of classifying caffeine as a drug of dependence would be increased.

References

1. Gilbert, R. M., Caffeine consumption, in *The Methylxanthine Beverages and Foods: Chemistry, Consumption, and Health Effects,* Spiller, G. A., Ed., Alan R. Liss, New York, 1984, 185.
2. Curatolo, P. W. and Robertson, D., The health consequences of caffeine, *Ann. Intern. Med.,* 98, 641, 1983.
3. James, J. E. and Stirling, K. P., Caffeine: A survey of some of the known and suspected deleterious effects of habitual use, *Br. J. Addict.,* 78, 251, 1983.
4. Benowitz, N. L., Clinical pharmacology of caffeine, *Annu. Rev. Med.,* 41, 277, 1990.
5. Gilbert, R. M., Caffeine as a drug of abuse, in *Research Advances in Alcohol and Drug Problems,* Vol. 3, Gibbins, R. J., Israel, Y., Kalant, H., Popham, R. E., Schmidt, W., and Smart, R. G., Eds., John Wiley & Sons, New York, 1976, 49.
6. Gilliland, K. and Bullock, W., Caffeine: A potential drug of abuse, *Adv. Alcoh. Subs. Abuse,* 3, 53, 1984.
7. Griffiths, R. R. and Woodson, P. P., Caffeine physical dependence: A review of human and laboratory animal studies, *Psychopharmacology,* 94, 437, 1988.
8. Heishman, S. J. and Henningfield, J. E., Stimulus functions of caffeine in humans: Relation to dependence potential, *Neurosci. Biobehav. Rev.,* 16, 273, 1992.
9. Jaffe, J. H., Drug addiction and drug abuse, in *The Pharmacological Basis of Therapeutics,* Gilman, A. G., Rall, T. W., Nies, A. S., and Taylor, P., Eds., Pergamon, New York, 1990, 522.
10. U.S. Department of Health and Human Services, *The Health Consequences of Smoking: Nicotine Addiction. A Report of the Surgeon General,* U.S. Department of Health and Human Services, Rockville, MD, 1988.
11. Griffiths, R. R., Evans, S. M., Heishman, S. J., Preston, K. L., Sannerud, C. A., Wolf, B., and Woodson, P. P., Low-dose caffeine discrimination in humans, *J. Pharmacol. Exp. Therap.,* 252, 970, 1990.
12. Evans, S. M. and Griffiths, R. R., Caffeine tolerance and choice in humans, *Psychopharmacology,* 108, 51, 1992.
13. Oliveto, A. H., Bickel, W. K., Hughes, J. R., Shea, P. J., Higgins, S. T., and Fenwick, J. W., Caffeine drug discrimination in humans: Acquisition, specificity and correlation with self-reports, *J. Pharmacol. Exp. Therap.,* 261, 885, 1992.
14. Oliveto, A. H., Bickel, W. K., Hughes, J. R., Terry, S. Y., Higgins, S. T., and Badger, G. J., Pharmacological specificity of the caffeine discriminative stimulus in humans: Effects of theophylline, methylphenidate and buspirone, *Behav. Pharmacol.,* 4, 237, 1993.

15. Silverman, K. and Griffiths, R. R., Low-dose caffeine discrimination and self-reported mood effects in normal volunteers, *J. Exp. Anal. Behav.*, 57, 91, 1992.
16. Heishman, S. J., Taylor, R. C., Goodman, M. L., Evans, S. M., and Henningfield, J. E., Discriminative stimulus effects of *d*-amphetamine, caffeine, and mazindol in humans. Paper presented at American Psychological Association, Toronto, 1993.
17. Schuh, L. M., Heishman, S. J., Lewis, B., and Henningfield, J. E., Discriminative stimulus effects of ephedrine and phenylpropanolamine, in *Problems of Drug Dependence 1994*, Harris, L. S., Ed., National Institute on Drug Abuse, Rockville, MD, 1995, 216.
18. Schuh, L. M., Heishman, S. J., Lewis, B. D., and Henningfield, J. E., Discriminative stimulus effects of caffeine, ephedrine, and phenylpropanolamine, in *Problems of Drug Dependence 1995*, Harris, L. S., Ed., National Institute on Drug Abuse, Rockville, MD, 1996, 347.
19. Henningfield, J. E., and Jasinski, D. R., Human pharmacology of nicotine, *Psychopharmacol. Bull.*, 19, 414, 1983.
20. Heishman S. J. and Henningfield, J. E., Discriminative stimulus effects of *d*-amphetamine, methylphenidate, and diazepam in humans, *Psychopharmacology*, 103, 436, 1991.
21. Fischman, M. W., Schuster, C. R., Resnekov, L., Shick, J. F. E., Krasnegor, N. A., Fennell, W., and Freedman, D. X., Cardiovascular and subjective effects of intravenous cocaine administration in humans, *Arch. Gen. Psychiat.*, 33, 983, 1976.
22. Carpenter, J., Effects of alcohol on some psychological processes, *Quart. J. Stud. Alcoh.*, 23, 274, 1962.
23. Preston, K. L., Bigelow, G. E., Bickel, W. K., and Liebson, I. A., Drug discrimination in human post-addicts: Agonist-antagonist opioids, *J. Pharmacol. Exp. Therap.*, 250, 184, 1989.
24. Evans, S. M. and Griffiths, R. R., Dose-related caffeine discrimination in normal volunteers: Individual differences in subjective effects and self-reported cues, *Behav. Pharmacol.*, 2, 345, 1991.
25. Chait, L. D. and Johanson, C. E., Discriminative stimulus effects of caffeine and benzphetamine in amphetamine-trained volunteers, *Psychopharmacology*, 96, 302, 1988.
26. Warburton, D. M., Effects of caffeine on cognition and mood without caffeine abstinence, *Psychopharmacology*, 119, 66, 1995.
27. Chait, L. D. and Griffiths, R.R., Effects of caffeine on cigarette smoking and subjective response, *Clin. Pharmacol. Therap.*, 34, 612, 1983.
28. Hughes, J. R., Higgins, S. T., Bickel, W. K., Hunt, W. K., Fenwick, J. W., Gulliver, S. B., and Mireault, G. C., Caffeine self-administration, withdrawal and adverse effects among coffee drinkers, *Arch. Gen. Psychiat.*, 48, 611, 1991.
29. Charney, D. S., Galloway, M. P., and Heninger, G. R., The effects of caffeine on plasma MHPG, subjective anxiety, autonomic symptoms and blood pressure in healthy humans, *Life Sci.*, 35, 135, 1984.
30. Roache, J. D. and Griffiths, R. R., Interactions of diazepam and caffeine: Behavioral and subjective dose effects in humans, *Pharmacol. Biochem. Behav.*, 26, 801, 1987.
31. Jasinski, D. R., Johnson, R. E., Henningfield, J. E., Abuse liability assessment in human subjects, *Trends Pharmacol. Sci.*, 5, 196, 1984.

32. Haertzen, C. A. and Hickey, J. E., Addiction Research Center Inventory (AR-CI): Measurement of euphoria and other drug effects, in *Methods of Assessing the Reinforcing Properties of Abused Drugs*, Bozarth, M. A., Ed., Springer-Verlag, New York, 1987, 489.
33. Griffiths, R. R., Bigelow, G. E., and Liebson, I. A., Reinforcing effects of caffeine in coffee and capsules, *J. Exp. Anal. Behav.*, 52, 127, 1989.
34. Stern, K. N., Chait, L. D., and Johanson, C. E., Reinforcing and subjective effects of caffeine in normal human volunteers, *Psychopharmacology*, 98, 81, 1989.
35. Cole, J. O., Pope, H. G., LaBrie, R., and Ionescu-Pioggia, M., Assessing the subjective effects of stimulants in casual users, *Clin. Pharmacol. Therap.*, 24, 243, 1978.
36. Noble, R., A controlled clinical trial of the cardiovascular and psychological effects of phenylpropanolamine and caffeine, *Drug Intell. Clin. Pharm.*, 22, 296, 1988.
37. Henningfield, J. E., Miyasato, K., and Jasinski, D. R., Abuse liability and pharmacodynamic characteristics of intravenous and inhaled nicotine, *J. Pharmacol. Exp. Therap.*, 234, 1, 1985.
38. Martin, W. R. and Fraser, H. F., A comparative study of physiological and subjective effects of heroin and morphine administered intravenously in postaddicts, *J. Pharmacol. Exp. Therap.*, 133, 388, 1961.
39. Barone, J. J. and Roberts, H., Human consumption of caffeine, in *Caffeine: Perspectives from Recent Research*, Dews, P. B., Ed., Springer-Verlag, Berlin, 1984, 59.
40. Greden, J. F., Anxiety or caffeinism: A diagnostic dilemma, *Am. J. Psychiat.*, 131, 1089, 1974.
41. Sours, J. A., Case reports of anorexia nervosa and caffeinism, *Am. J. Psychiat.*, 140, 235, 1983.
42. Wilkin, J. K., The caffeine withdrawal flush: Report of a case of "weekend flushing." *Mil. Med.*, 151, 123, 1986.
43. James, J. E., Paull, I., Cameron-Traub, E., Minors, J., Lelo, A., and Birkett, D. J., Biochemical validation of self-reported caffeine consumption during caffeine fading, *J. Behav. Med.*, 11, 15, 1988.
44. Greden, J. F., Caffeine and tobacco dependence, in *Comprehensive Textbook of Psychiatry*, Kaplan, H. I., Freedman, A. M., and Sadock, B. J., Eds., Williams & Wilkens, Baltimore, 1980, 1645.
45. Reimann, H. A., Caffeinism, *JAMA*, 202, 131, 1967.
46. Russ, N. W., Sturgis, E. T., Malcolm, R. J., and Williams, L., Abuse of caffeine in substance abusers, *J. Clin. Psychiat.*, 49, 457, 1988.
47. Foxx, R. M. and Rubinoff, A., Behavioral treatment of caffeinism: Reducing excessive coffee drinking, *J. Appl. Behav. Anal.*, 12, 335, 1979.
48. James, J. E., Stirling, K. P., and Hampton, B. A. M., Caffeine fading: Behavioral treatment of caffeine abuse, *Behav. Ther.*, 16, 15, 1985.
49. Thompson, T. and Pickens, R., Eds., *Stimulus Properties of Drugs*, Appleton-Century-Crofts, New York, 1971.
50. Griffiths, R. R., Bigelow, G. E., and Henningfield, J. E., Similarities in animal and human drug-taking behavior, in *Advances in Substance Abuse*, Vol. 1, Mello, N. K., Ed., JAI Press, Greenwich, CT, 1980, 1.

51. Johanson, C. E., Woolverton, W. L., and Schuster, C. R., Evaluating laboratory models of drug dependence, in *Psychopharmacology: The Third Generation of Progress*, Meltzer, H. Y., Ed., Raven, New York, 1987, 1617.
52. Heppner, C. C., Kemble, E. D., and Cox, W. M., Effects of food deprivation on caffeine consumption in male and female rats, *Pharmacol. Biochem. Behav.*, 24, 1555, 1986.
53. Schuster, C. R., Woods, J. H., Seevers, M. H., Self-administration of central stimulants by the monkey, in *Abuse of Central Stimulants*, Sjoqvist, F. and Tottie, M., Eds., Raven, New York, 1969, 339.
54. Griffiths, R. R., Bigelow, G. E., and Liebson, I. A., Human coffee drinking: Reinforcing and physical dependence producing effects of caffeine, *J. Pharmacol. Exp. Therap.*, 239, 416, 1986.
55. Griffiths, R. R. and Woodson, P. P., Reinforcing effects of caffeine in humans, *J. Pharmacol. Exp. Therap.*, 246, 21, 1988.
56. Silverman, K., Mumford, G. K., and Griffiths, R. R., Enhancing caffeine reinforcement by behavioral requirements following drug ingestion, *Psychopharmacology*, 114, 424, 1994.
57. Evans, S. M., Critchfield, T. S., and Griffiths, R. R., Caffeine reinforcement demonstrated in a majority of moderate caffeine users, *Behav. Pharmacol.*, 5, 231, 1994.
58. Hughes, J. R., Hunt, W. K., Higgins, S. T., Bickel, W. K., Fenwick, J. W., and Pepper, S. L., Effect of dose on the ability of caffeine to serve as a reinforcer in humans, *Behav. Pharmacol.*, 3, 211, 1992.
59. Hughes, J. R., Oliveto, A. H., Bickel, W. K., Higgins, S. T., and Badger, G. J., The ability of low doses of caffeine to serve as reinforcers in humans: A replication, *Exp. Clin. Psychopharmacol.*, 3, 358, 1995.
60. Hale, K. L., Hughes, J. R., Oliveto, A. H., and Higgins, S. T., Caffeine self-administration and subjective effects in adolescents, *Exp. Clin. Psychopharmacol.*, 3, 364, 1995.
61. Liguori, A., Hughes, J. R., and Oliveto, A. H., Caffeine self-administration in humans: 1. Efficacy of cola vehicle, *Exp. Clin. Psychopharmacol.*, 5, 286, 1997.
62. Liguori, A. and Hughes, J. R., Caffeine self-administration in humans: 2. A within-subjects comparison of coffee and cola vehicles, *Exp. Clin. Psychopharmacol.*, 5, 295, 1997.
63. Griffiths, R. R., Bigelow, G. E., Liebson, I. A., O'Keeffe, M., O'Leary, D., and Russ, N., Human coffee drinking: Manipulation of concentration and caffeine dose, *J. Exp. Anal. Behav.*, 45, 133, 1986.
64. Oliveto, A. H., Hughes, J. R., Higgins, S. T., Bickel, W. K., Pepper, S. L., Shea, P. J., and Fenwick, J. W., Forced-choice vs. free-choice procedures: Caffeine self-administration in humans, *Psychopharmacology*, 109, 85, 1992.
65. Goldberg, S. R., Spealman, R. D., and Goldberg, D. M., Persistent behavior at high rates maintained by intravenous self-administration of nicotine, *Science*, 214, 573, 1981.
66. Henningfield, J. E., Miyasato, K., Jasinski, D. R., Cigarette smokers self-administer intravenous nicotine, *Pharmacol. Biochem. Behav.*, 19, 887, 1983.
67. Deneau, G., Yanagita, T., and Seevers, M. H., Self-administration of psychoactive substances by the monkey: A measure of psychological dependence, *Psychopharmacologia*, 16, 30, 1969.
68. Fischman, M. W. and Schuster, C. R., Cocaine self-administration in humans. *Fed. Proc.*, 41, 241, 1982.

69. Griffiths, R. R., Brady, J.V., and Bradford, L. D., Predicting the abuse liability of drugs with animal drug self-administration procedures: Psychomotor stimulants and hallucinogens, in *Advances in Behavioral Pharmacology*, Vol. 2, Thompson, T. and Dews, P. B., Eds., Academic, New York, 1979, 163.
70. Johanson, C. E. and Uhlenhuth, E. H., Drug preference and mood in humans: *d*-amphetamine, *Psychopharmacology*, 71, 275, 1980.
71. Bigelow, G. E., Griffiths, R. R., Liebson, I. A., Experimental models for the modification of human drug self-administration: Methodological developments in the study of ethanol self-administration by alcoholics, *Fed. Proc.*, 34, 1785, 1975.
72. Thompson, T. and Schuster, C. R., Morphine self-administration, food-reinforced, and avoidance behaviors in rhesus monkeys, *Psychopharmacologia*, 5, 87, 1964.
73. Mello, N. K., Mendelson, J. H., Kuehnle, J. C., and Sellers, M. S., Operant analysis of human heroin self-administration and the effects of naloxone, *J. Pharmacol. Exp. Therap.*, 216, 45, 1981.
74. Schuster, C. R., Fischman, M. W., and Johanson, C. E., Internal stimulus control and subjective effects of drugs, in *Behavioral Pharmacology of Human Drug Dependence*, Thompson, T. and Johanson, C. E., Eds., National Institute on Drug Abuse, Rockville, MD, 1981, 116.
75. Fischman, M. W., Relationship between self-reported drug effects and their reinforcing effects: Studies with stimulant drugs, in *Testing for Abuse Liability in Humans*, Fischman, M. W. and Mello, N. K., Eds., National Institute on Drug Abuse, Rockville, MD, 1989, 211.
76. Robertson, D., Wade, D., Workman, R., Woosley, R. L., and Oates, J. A., Tolerance to the humoral and hemodynamic effects of caffeine in man, *J. Clin. Invest.*, 67, 1111, 1981.
77. Denaro, C. P., Brown, C. R., Jacob, P., and Benowitz, N. L., Effects of caffeine with repeated dosing, *Eur. J. Clin. Pharmacol.*, 40, 273, 1991.
78. Goldstein, A., Warren, R., and Kaizer, S., Psychotropic effects of caffeine in man. I. Individual differences in sensitivity to caffeine-induced wakefulness, *J. Pharmacol. Exp. Therap.*, 149, 156, 1965.
79. Coltin, T., Gosselin, R. E., and Smith, R. P., The tolerance of coffee drinkers to caffeine, *Clin. Pharmacol. Therap.*, 9, 31, 1968.
80. Kosman, M. E. and Unna, K. R., Effects of chronic administration of the amphetamines and other stimulants on behavior, *Clin. Pharmacol. Therap.*, 9, 240, 1968.
81. Strain, E. C., Mumford, G. K., Silverman, K., and Griffiths, R. R., Caffeine dependence syndrome: Evidence from case histories and experimental evaluations, *JAMA*, 272, 1043, 1994.
82. Hughes, J. R., Oliveto, A. H., Bickel, W. K., Higgins, S. T., and Badger, G. J., Caffeine self-administration and withdrawal: Incidence, individual differences and interrelationships, *Drug Alcoh. Depend.*, 32, 239, 1993.
83. Greden, J. F., Victor, B. S., Fontaine, P., Lubetsky, M., Caffeine-withdrawal headache: A clinical profile, *Psychosomatics*, 21, 411, 1980.
84. Griffiths, R. R., Evans, S. M., Heishman, S. J., Preston, K. L., Sannerud, C. A., Wolf, B., and Woodson, P. P., Low-dose caffeine physical dependence in humans, *J. Pharmacol. Exp. Therap.*, 255, 1123, 1990.

85. Silverman, K., Evans, S. M., Strain, E. C., and Griffiths, R. R., Withdrawal syndrome after the double-blind cessation of caffeine consumption, *N. Engl. J. Med.*, 327, 1109, 1992.
86. American Psychiatric Association, *Diagnostic and Statistical Manual of Mental Disorders*, 4th ed., American Psychiatric Association, Washington, D.C., 1994.
87. Goldstein, A. and Kalant, H., Drug policy: Striking the right balance, *Science*, 249, 1513, 1990.
88. World Health Organization. *International Statistical Classification of Diseases and Related Health Problems*, 10th revision. World Health Organization, Geneva, 1992.
89. Hughes, J. R., Oliveto, A. H., Helzer, J. E., Higgins, S. T., and Bickel, W. K., Should caffeine abuse, dependence, or withdrawal be added to DSM-IV and ICD-10?, *Am. J. Psychiat.*, 149, 33, 1992.
90. Garriott, J. C., Simmons, L. M., Poklis, A., and Mackel, M. A., Five cases of fatal overdose from caffeine containing "look-alike" drugs, *J. Analyt. Toxicol.*, 9, 141, 1985.
91. Morrow, P. L., Caffeine toxicity: A case of child abuse by drug ingestion, *J. Foren. Sci.*, 32, 1801, 1987.
92. Nemeth-Coslett, R., Henningfield, J. E., O'Keeffe, M. K., and Griffiths, R. R., Nicotine gum: Dose-related effects on cigarette smoking and subjective ratings, *Psychopharmacology*, 92, 424, 1987.
93. Pickworth, W. B., Bunker, E. B., and Henningfield, J. E., Transdermal nicotine: Reduction of smoking with minimal abuse liability, *Psychopharmacology*, 115, 9, 1994.
94. Rush, C. R., Sullivan, J., and Griffiths, R. R., Intravenous caffeine in stimulant drug abusers: Subjective reports and physiological effects, *J. Pharmacol. Exp. Therap.*, 273, 351, 1995.

chapter eight

Caffeine withdrawal

Malcolm H. Lader

Contents

I. Introduction 151
II. Early studies 152
III. More recent studies 152
IV. Special situations...155
V. Special tests..155
VI. Conclusion..156
References..156

Key words: Caffeine, withdrawal, abstinence

I. Introduction

Caffeine is the most widely-used of all psychotropic drugs. The acute effects have been extensively evaluated but the effects on chronic administration have received much less attention. Some work has focused on tolerance to caffeine's effects: for example, the diuretic effect of caffeine wanes on repeated dosing,[1,2] as does the stimulation of salivation.[3] Some discrepancies remain. For example, one observer claimed tolerance to occur to caffeine's effects on sleep,[4] whereas another group stated that habitual consumers failed to achieve tolerance.[5] Similarly, Landis[6] believed caffeine to lose its stimulant effects on repeated use, whereas Dews[7] did not. However, as was seen with the benzodiazepines, tolerance can occur to one function, but not to another. Tolerance also depends on the level of prior exposure to caffeine and the pattern of previous use.

While the question of tolerance remains unresolved, a more direct approach concerns caffeine withdrawal phenomena, the elicitation of which

0-8493-1166-7/99/$0.00+$.50

provide evidence for a state of dependence in the organism. The purpose of this brief review is to examine various studies in this area, including our own.

II. Early studies

Kingdon[8] noted mental confusion if the habitual cup of morning tea was omitted, other symptoms being a sense of weakness and emotional dysphoria. Severe headache was listed as a cardinal symptom by Bridge.[9] Deliberate withdrawal was engineered by Horst and his colleagues,[10] who found psychomotor performance to be impaired during the first week after discontinuation.

The first controlled caffeine withdrawal study was reported by Dreisbach and Pfeiffer.[11] For the first 7 days, caffeine intake was increased up to 850 mg/day. Placebo capsules were then substituted. The caffeine itself was associated with feelings of stimulation with lessened fatigue, slight muscle tension, and mild insomnia. On withdrawal, most subjects reported headache, often severe: the headache was throbbing in type, first central or occipital in location but then becoming more diffuse. Other symptoms included nausea, vomiting, drowsiness, disinclination to work, mental depression, and yawning. The headache was not helped by analgesics but responded to the reinstitution of caffeine.

III. More recent studies

A questionnaire survey was conducted on the effects of stopping habitual caffeine intake in 239 young housewives.[12] Among both moderate (3 to 4 cups of coffee per day) and heavy (5 to 10 cups) drinkers, common withdrawal symptoms were headache, restlessness, lethargy, and inability to work. However, a wide range of other symptoms were reported on a sporadic basis. A further study[13] compared users and nonusers with respect to withdrawal. Nonusers complained of jitteriness, nervousness, and upset stomach when given caffeine but noted no change when switched to placebo. By contrast, users noted increased alertness, decreased irritability, and feelings of contentment when taking caffeine, but they became sleepy and dysphoric when the caffeine was withdrawn. Heavy caffeine users also developed headache.

A fairly detailed description of caffeine withdrawal was provided by Roller,[14] during the course of a study on theophylline kinetics which necessitated abstention from all methylxanthines. Headache came on about 6 hours after the last caffeine intake, followed by tiredness and lassitude, nasal stuffiness and leg pains. Ten hours later, the subject complained of "flu-like" feelings and general muscle pain. Caffeine relieved all these symptoms. A longer time-scale was discerned in the very careful studies of Griffiths' group.[15] The onset-latency of the caffeine withdrawal headache averaged 19 hours. The symptoms in general peaked over the first 24 to 36 hours and then waned over the subsequent 5 to 6 days. Observer ratings tallied well with self-ratings and comprised increases in fatigue and decreases in vigor and friendliness. However, no increases in anxiety, irritability, or gastrointestinal symptoms were detected. All 9 volunteers were male smokers, and cigarette smoking is known to induce the metabolism of caffeine. Smoking also decreased during abstinence, so complex metabolic changes could have ensued. A second study from this group studied placebo substitution in low-dose caffeine users.[16] In particular, the replicability of the withdrawal syndrome was estimated. During placebo substitution, ratings of alertness, motivation, sociability, concentration, well-being, contentedness, and energy levels decreased in parallel with increases in ratings of headache, tiredness, depression, muscle pains, feelings of fullness in the head, flu-like feelings, and craving for caffeine. As usual, headache supervened 1 to 2 days into withdrawal. Griffiths and Woodson,[17] reviewing the data, concluded that a definable withdrawal syndrome was demonstrable and was indicative of a state of preexisting caffeine dependence.

In a choice paradigm, subjects were deprived of caffeine and then asked what amount of money they were prepared to forfeit in order to obtain caffeine or to avoid placebo administration in the future.[18] The results suggested that choice of caffeine is more closely related to avoidance of withdrawal effects than positive effects of caffeine. However, caffeine self-administration may not necessarily be related to deprivation conditions alone, but might be a more complex behavior.[19] It was estimated to be demonstrable in one quarter to one third of subjects, caffeine withdrawal symptoms in up to one half.[20]

A double-blind study concentrated on awareness of caffeine withdrawal.[21] Subjects were allocated randomly to either five cups/day of regular coffee (84 mg caffeine/cup) or decaffeinated coffee, over six weeks. The 45 subjects were moderate coffee drinkers but nonsmokers. Most of the subjects failed to detect the switch from caffeinated to decaffeinated coffee, but one half reported some increase in headaches over the first week of withdrawal.

One detailed and careful study concentrated on low to moderate users of caffeine.[22] Sixty-two normal adults were selected whose daily intake of caffeine was low to moderate, with a mean intake of 235 mg. Various symptom questionnaires were completed, and tests of mood and performance were carried out when subjects consumed their normal diets and at the end of each of two two-day periods during which they ate caffeine-free diets and, under controlled, double-blind conditions, took capsules containing caffeine equivalent to their daily caffeine consumption or placebo.

During the placebo period, more subjects had abnormally high Beck Depression Inventories than when on caffeine (11% vs. 3%). On placebo, in caffeine withdrawal, more had high anxiety scores (8 vs. 2), high fatigue scores (8 vs. 0), and most strikingly, over half (52%) complained of headache, compared with only 6% on caffeine replacement. Performance of a tapping task was slowed during the placebo period as compared with baseline and caffeine replacement.

In another study, even short-term caffeine withdrawal was associated with reports of decreased vigor, increased fatigue, sleepiness, and yawning.[23] Blood pressure was reduced by 5 mm Hg but psychomotor performance was unaffected. Similar results were found in a 17-day study in 12 normal volunteers: subjective effects were induced by caffeine withdrawal but no psychological performance deficits.[24] Thus, even short times of caffeine deprivation, equivalent to missing regular morning coffee, can produce unpleasant effects.[25]

Thus, the existence of a caffeine withdrawal syndrome was confirmed and shown to be not uncommon even in low to moderate users of caffeine, headache being the cardinal feature. The clinical implication is that physicians should bear in mind the possibility of caffeine withdrawal in patients who have headaches, depression, fatigue, and drowsiness.[26]

However, the issues may be quite complicated. The mood and performance effects of short-term (90 min), overnight, and one week's caffeine deprivation and ingestion (70 and 250 mg) were assessed in moderate consumers (N = 49) or nonusers (N = 18).[27] Dysphoric symptoms developed after overnight deprivation and persisted. Acute caffeine intake affected withdrawn consumers, nonwithdrawn consumers, and nonconsumers similarly. The authors conclude, as dysphoric symptoms follow both the use and deprivation of caffeine, usage has to be precisely titrated to pursue the middle course.

A study from my laboratory contrasted withdrawal symptoms in subjects who had abstained for 24 hours with those who had done so for 7 days.[28] On test days, subjects were given either 250 or 500 mg of caffeine (as the base equivalent) or placebo, and underwent a series of physiological and psychological tests. Headache was the most common complaint in the 24-hour withdrawal group but not in the 7-day withdrawal group. Tiredness was the most common indicator of withdrawal, and this was counteracted by the administered caffeine.

Forty normal subjects (mean age 36) had their caffeine intake estimated by keeping a diary (n = 40) and also by analyzing samples of tea and coffee (n = 28). A test dose of caffeine (500 mg) was given and a series of salivary samples analyzed to estimate pharmacokinetic measures of the rate of caffeine metabolism. They then underwent 48 h of placebo substitution using double-blind procedures. A wide range of physiological, psychological, and subjective measures was taken on successive days during withdrawal and resumption of caffeine. On withdrawal, 27 subjects reported tiredness, and 18 developed headache. Electroencephalograph, skin conductance, and blood pressure changes were apparent. Sleep improved on withdrawal, but subjects reported feeling less alert and more tired. The higher the usual caffeine intake, the greater the unpleasant feelings on withdrawal and the more marked the reversal of feelings on resumption. The faster the metabolism of caffeine, the less the drop in anxiety during withdrawal and the less its return on resumption. These correlations were, however, rather weak and sporadic.

If caffeine discontinuation is followed by withdrawal symptoms, subjects should compensate by drinking more coffee when free access is allowed again. Hughes et al.[29] gave their subjects such free access after abstinence. Headache, drowsiness, and fatigue were reported on caffeine-free days. Nine of the 22 subjects drank more regular coffee for a time when access to it was reinstituted.

IV. Special situations

Psychiatric patients are notoriously heavy coffee and tea drinkers. As caffeine in excess may be related to disturbed behavior, the possible beneficial effects of restricting or banning caffeine in hospital wards need exploration. Edelstein and his colleagues[30] evaluated the effects of caffeine withdrawal on bodily complaints in psychiatric patients, and in a later study[31] on nocturnal enuresis, insomnia, and episodes of disturbed behavior. Caffeine abstinence was followed by drops in heart-rate and blood-pressure and in gastrointestinal complaints. Incidents of disturbed behavior were reduced, together with enuretic episodes and insomnia. The wider applicability of these interesting data would bear close scrutiny.

In hospital, a general anesthetic is often followed by complaints of headache. It has been suggested that headache is actually due to caffeine withdrawal ensuing upon hospital admission and preparation for surgery. In one study, caffeine abstainers had no postoperative headaches, whereas the incidence of headaches in the remainder increased in direct relationship to habitual caffeine intake.[32] Thus, of light users (10 to 100 mg/day), 13% suffered headaches, but three quarters of heavy users (over 1000 mg/day) developed this syndrome. Unfortunately, these clear-cut results have not been replicated.[33] Nevertheless, administration of caffeine substitution in a placebo-controlled study proved effective in preventing postoperative headache.[34]

Caffeine passes readily across the placenta into the fetus and is present in breast milk. The possibility of the newborn suffering from caffeine has been raised — "the jittery baby syndrome."[35,36] The toxicity might then be followed by caffeine withdrawal if the baby is not breast fed, although as caffeine is very slowly metabolized in the neonate, this withdrawal would be protracted. Some evidence in support of these concerns has been adduced.

The clinical uses of caffeine lie outside the scope of this chapter and have been reviewed by Sawynok.[37]

V. Special tests

As well as clinical and subjective ratings, some studies have used objective biochemical, physiological, or psychological measures to monitor possible withdrawal effects. The noradrenaline metabolite MHPG (3-methoxy,4-hydroxyphenylene ethylglycol) was markedly increased in the urine of a heavy coffee drinker (10 to 15 cups/day) on withdrawal.[38] Levels tripled over 3 days. The subject also experienced headache and anxiety, but it is unlikely that the very high MHPG levels attained were a nonspecific response to the discomfort.

Caffeine withdrawal is associated with significant increases in alpha and theta power in the EEG.[39] These changes reverted when caffeine was reintroduced.

Cerebral blood-flow was measured in another study.[40] High and low caffeine users were compared with respect to cerebral blood flow under normal and abstinence conditions. In the high caffeine users, abstinence was associated with marked frontal lobe flow increases, whereas the administration of caffeine reversed this. Thus, caffeine causes cerebral vasoconstriction, and the vasodilation on withdrawal probably underlies the throbbing vascular-type headache.[41]

Rizzo et al.[42] measured the effects of caffeine status on reaction time. Users and nonusers were compared, defined as consumers of five-or-more and one-or-less caffeine-containing beverages per day. Subjects were tested twice per week. Users maintained normal intake for 5 days then 2 days abstinence before testing; nonusers continued unchanged throughout. There were no performance differences between the groups while on caffeine, but users abstaining from caffeine had slowed reaction times. In the Bruce et al.[28] study, tapping rate was impaired in the subjects 24 hours into withdrawal, but not 7 days after withdrawal. Caffeine restored the slowed performance. White and associates[43] obtained similar results for reaction time. Caffeine improved performance, but the period of abstinence (3 hours) was too short for withdrawal to have set in.

VI. Conclusion

The evidence reviewed in this chapter suggests that caffeine withdrawal is followed by clear physiological changes, with subjective complaints, the most obvious of which is headache. The severity of the withdrawal reaction is dependent to a large extent on previous patterns of usage, particularly dosage. Decrements in performance can also be detected and can be reversed by restarting caffeine usage. Thus, caffeine meets the criteria for a typical psychoactive substance of dependence.[44] The possibility of such dependence as manifested by both the typical withdrawal syndrome and by drug-seeking behavior should be borne in mind.

References

1. Myers, H. B., Renal tolerance of caffeine, *J. Pharmacol. Exp. Therap.*, 23, 465, 1924.
2. Eddy, N. B. and Downs, A. W., Tolerance and cross tolerance in the human subject to the diuretic effects of caffeine, theobromine and theophylline, *J. Pharmacol. Exp. Therap.*, 33, 167, 1928.
3. Winsor, A. L. and Strongin, E. I., A study of the development of tolerance for caffeinated beverages, *J. Exp. Psychol.*, 16, 725, 1933.

4. Colton, T., Gosselin, R., and Smith, R., The tolerance of coffee drinkers to caffeine, *Clin. Pharmacol. Therap.*, 9, 31, 1968.
5. Goldstein, A., Warren, R., and Kaizer, S., Psychotropic effects of caffeine in man. I. Individual differences in sensitivity to caffeine induced wakefulness, *J. Pharmacol. Exp. Therap.*, 149, 156, 1965.
6. Landis, C., Physiological and psychological effects of the use of coffee, in *Problems of Addiction and Habituation*, Hoch, P. and Zubin, J., Eds., Grune & Stratton, New York, 1958, 37.
7. Dews, P. B., Caffeine, *Annu. Rev. Nutr.*, 2, 323, 1982.
8. Kingdon, W., Effects of tea and coffee drinking, *Lancet*, ii, 47, 1833.
9. Bridge, N., Coffee drinking as a frequent cause of disease, *Trans. Assoc. Am. Physicians*, 8, 281, 1893.
10. Horst, K., Buxton, R. E., and Robinson, W.D., The effect of the habitual use of coffee or decaffeinated coffee upon blood pressure and certain motor reactions of normal young men, *J. Pharmacol. Exp. Therap.*, 52, 322, 1934.
11. Dreisbach, R. H. and Pfeiffer, C., Caffeine withdrawal headache, *J. Lab. Clin. Med.*, 28, 1212, 1943.
12. Goldstein, A. and Kaizer, S., Psychotropic effects of caffeine in man. III. A questionnaire survey of coffee drinking and effects on a group of housewives, *Clin. Pharmacol. Therap.*, 10, 477, 1969.
13. Goldstein, A., Kaizer, S., and Whitby, O., Psychotropic effects of caffeine in man. IV. Quantitative and qualitative differences associated with habituation to coffee, *Clin. Pharmacol. Therap.*, 246, 489, 1969.
14. Roller, L., Caffeinism: subjective quantitative aspect of withdrawal syndrome, *Med. J. Aust.*, 1, 146, 1981.
15. Griffiths, R. R., Bigelow, G. E., and Liebson, I. A., Human coffee drinking: reinforcing and physical dependence producing effects of caffeine, *J. Pharmacol. Exp. Therap.*, 239, 416, 1986.
16. Griffiths, R. R., Evans, S. M., Heishman, S. J., Preston, K. L., Sannerud, A., Wolf, B., and Woodson, P. P., Low dose physical dependence in humans, *J. Pharmacol. Exp. Therap.*, 255, 1123, 1990.
17. Griffiths, R. R. and Woodson, P. P., Caffeine physical dependence: a review of human and animal studies, *Psychopharmacology*, 94, 437, 1988.
18. Schuh, K. J. and Griffiths, R. R., Caffeine reinforcement: the role of withdrawal, *Psychopharmacology*, 130, 320, 1997.
19. Mitchell, S. H., de Wit, H., and Zacny, J. P., Caffeine withdrawal symptoms and self-administration following caffeine deprivation, *Pharmacol. Biochem. Behav.*, 51, 941, 1995.
20. Hughes, J. R., Oliveto, A. H., Bickel, W. K., Higgins, S. T., and Badger, G. J., Caffeine self-administration and withdrawal: incidence, individual differences and interrelationships, *Drug Alcoh. Depend.*, 32, 239, 1993.
21. Van Dusseldorp, M. and Katan, M. B., Headache caused by caffeine withdrawal among moderate coffee drinkers switched from ordinary to decaffeinated coffee — a 12 week double-blind trial, *Br. Med. J.*, 300, 1558, 1990.
22. Silverman, K., Evans, S. M., Strain, E. C., and Griffiths, R. R., Withdrawal syndrome after the double-blind cessation of caffeine consumption, *N. Engl. J. Med.*, 327, 1111, 1992.
23. Phillips-Bute, B. G. and Lane, J. D., Caffeine withdrawal symptoms following brief caffeine deprivation, *Physiol. Behav.*, 63, 35, 1997.

24. Comer, S. D., Haney, M., Foltin, R. W., and Fischman, M. W., Effects of caffeine withdrawal on humans living in a residential laboratory, *Exp. Clin. Psychopharmacol.*, 5, 399, 1997.

25. Lane, J. D., Effects of brief caffeinated-beverage deprivation on mood, symptoms and psychomotor performance, *Pharmacol. Biochem. Behav.*, 58, 203, 1997.

26. Hughes, J. R., Clinical importance of caffeine withdrawal, *N. Engl. J. Med.*, 327, 1160, 1992.

27. Richardson, N. J., Rogers, P. J., Elliman, N. A., and O'Dell, R. J., Mood and performance effects of caffeine in relation to acute and chronic caffeine deprivation, *Pharmacol. Biochem. Behav.*, 52, 313, 1995.

28. Bruce, M. S., Scott, N. R., Shine, P., and Lader, M. H., Caffeine withdrawal: a contrast of withdrawal symptoms in normal subjects who have abstained from caffeine for 24 hours and for 7 days, *J. Psychopharmacol.*, 5, 129, 1991.

29. Hughes, J. R., Higgins, S. T., Bickel, W. K., Hunt, W. K., Fenwick, H. W., Gullivor, S. R. and Mireault, G. C., Caffeine self administration, withdrawal and adverse effects among coffee drinkers, *Arch. Gen. Psychiat.*, 48, 611 1991,

30. Edelstein, B. A., Keaton-Brastead, C., and Burg, M. M., Effects of caffeine on cardiovascular and gastrointestinal responses, *Health Psychol.*, 2, 343, 1979.

31. Edelstein, B. A., Keaton-Brastead, C., and Burg, M. M., Effects of caffeine withdrawal on nocturnal enuresis, insomnia and behaviour restraints, *J. Consult. Clin. Psychol.*, 52, 857, 1984.

32. Galletely, D., Fennelly, M., and Whitwam, J. G., Does caffeine withdrawal contribute to post anaesthetic morbidity? (letter), *Lancet*, i, 1335, 1989.

33. Verhoeff, F. H. and Millar, J. M., Does caffeine contribute to postoperative morbidity?, *Lancet*, ii, 632, 1990.

34. Hampl, K. F., Schneider, M. C., Ruttimann, U., Ummenhofer, W., and Drewe, J., Perioperative administration of caffeine tablets for prevention of postoperative headaches, *Can. J. Anesthesiol.*, 42, 789, 1995.

35. Kulkarni, P. B. and Dorand, R. D., Caffeine toxicity in a neonate, *Pediatrics*, 64, 254, 1979.

36. Banner, W. and Czajka, P. A., Acute caffeine overdose in the neonate, *Am. J. Dis. Child.*, 134, 495, 1980.

37. Sawynok, J., Pharmacological rationale for the clinical use of caffeine, *Drugs*, 49, 37, 1995.

38. Gibson, C. J., Caffeine withdrawal elevates urinary MHPG excretion, *N. Engl. J. Med.*, 304, 363, 1981.

39. Reeves, R. R., Struve, F. A., Patrick, G., and Bullen, J. A., Topographic quantitative EEG measures of alpha and theta power changes during caffeine withdrawal: preliminary findings from normal subjects, *Clin. Electroencephalogr.*, 26, 154, 1995.

40. Mathew, R.J. and Wilson, H., Caffeine consumption, withdrawal and cerebral blood flow, *Headache*, 25, 305, 1985.

41. Couturier, E. G., Laman, D. M., van Duijn, M. A., and Duijn, H., Influence of caffeine and caffeine withdrawal on headache and cerebral blood flow velocities, *Cephalalgia*, 17, 188, 1997.

42. Rizzo, A. A., Stamps, L. E., and Fehr, L. A., Effects of caffeine withdrawal on motor performance and heart rate changes, *Intern. J. Psychophysiol.*, 6, 9, 1988.

43. White, B. C., Lincoln, C. A., Pearce, N. W., Reeb, R., and Vaida, C., Anxiety and muscle tension as consequences of caffeine withdrawal, *Science*, 209, 1547, 1980.
44. Strain, E. C., Mumford, G. K., Silverman, K., and Griffiths, R. R., Caffeine dependence syndrome. Evidence from case histories and experimental evaluations, *JAMA*, 272, 1043, 1994.

chapter nine

Caffeine, caffeine withdrawal, and performance efficiency

Andrew P. Smith

Contents

Abstract...161
   I. Introduction 162
  II. Methods 163
     A. Study design..163
     B. Subjects 163
     C. Procedure 165
     D. Regular caffeine consumption...166
 III. Results...166
     A. First acute challenge...166
     B. Control of caffeine consumption over a week..............................171
     C. Acute effects of caffeine after 7 days of caffeine withdrawal.....174
  IV. Discussion 175
Acknowledgment..178
References..178

Abstract

It has recently been claimed that the beneficial effects of caffeine on performance might represent deleterious effects of caffeine deprivation rather than an actual net benefit due to caffeine use. This view was tested here by examining the effects of two doses of caffeine (1.5 mg/kg and 3 mg/kg) and then the effects of caffeine withdrawal. The results showed that administration of caffeine improved performance, whereas there was no evidence of impairment when subjects were deprived of caffeine. It has been argued that subjects should be deprived of caffeine for about a week to remove negative effects of deprivation before studies of the acute effects of caffeine are carried out. This was done here, and beneficial effects of caffeine were still observed. These results support the view that caffeine consumption leads to beneficial effects on performance, whereas caffeine withdrawal produces no impairments.

I. Introduction

A large number of studies have examined the effects of caffeine on mental performance and many of these have demonstrated beneficial effects (see Reference 1 for a review). It is also well established that caffeine is especially beneficial when alertness is low,[2,3] and this has been examined by considering its effects at night,[4] in the early morning,[5] after lunch,[6] after prolonged work,[3] and when subjects are suffering from a cold.[7] Some of these studies have also shown that even low doses (<200 mg) can be beneficial when alertness is reduced. Furthermore, the practical importance of such efforts has been demonstrated in a study which examined caffeine and driving performance of sleepy drivers.[8] This study clearly demonstrated that 150 mg caffeine (equivalent to 2 cups of coffee) significantly reduced driving impairments, subjective sleepiness, and EEG activity indicating drowsiness.

James[9] has questioned whether the superior performance seen in caffeine conditions are due to actual enhancement or merely reflect performance being impaired in the caffeine-free conditions. Smith[10] has suggested that the evidence for negative effects of caffeine withdrawal is not strong, and this has been confirmed in a recent study of caffeine withdrawal and headaches[11] and a recent review.[12] Indeed, Rogers et al.[12] conclude that "... in a review of recent studies we find no unequivocal evidence of impaired psychomotor performance associated with caffeine withdrawal."

Another problem for the caffeine withdrawal explanation is that it cannot account for effects in naive users or animals. Indeed, Rogers et al.[12] have shown that the beneficial effects of caffeine on performance can be demonstrated in nonusers, and users who had had caffeine withdrawn for varying periods of time (1.5 h, 13 h, and 7 days). This confirms previous findings showing that caffeine has comparable effects when it is given after abstinence of 1 h[13,14] or 12 h.[5]

The most important aspect of James's article is not the emphasis on caffeine withdrawal but the identification of the need to use a range of experimental paradigms to examine effects of caffeine and its withdrawal. One of the problems in comparing studies of caffeine and caffeine withdrawal is that they have used different paradigms, have different designs, and vary in experimental power. The aim of the present study was to test James's view using the methods he recommended. The first part of the study involved a normal caffeine challenge. James argued that any positive effects of caffeine actually reflect negative effects of withdrawal in the caffeine-free condition. To directly examine withdrawal effects, half the subjects then continued using caffeinated products for a week, whereas the others were provided with caffeine-free drinks. If there are negative effects of caffeine withdrawal, these should show up at this stage of the study. James argues that after 7 days the negative effects of caffeine withdrawal should have gone. If a caffeine challenge is then repeated with these subjects, there should be no beneficial effects of caffeine because the negative effects of caffeine withdrawal have gone. An alternative view is that caffeine will improve performance, both following short-term withdrawal and 7 days of withdrawal, and that it will be difficult to demonstrate effects of caffeine withdrawal *per se*. The following experiment tested these views, and it had the following methodological features. First, two doses of caffeine were compared with placebo in a double-blind study. The fact that larger doses of caffeine produce bigger effects than the smaller doses also produces problems for the withdrawal explanation. If a person is given a dose that is equivalent to their normal intake, they have not had caffeine withdrawn. Yet the literature shows that additional caffeine may lead to beneficial effects. Second, tests which were known to be sensitive to effects of caffeine were used. Finally, caffeine abstinence is difficult to assess unless saliva samples are taken, and that was done here.

II. Methods

A. Study design

The experiment consisted of three parts. In the first and third parts, the effects of a single caffeine challenge were examined. Subjects were allocated to one of the following conditions: placebo, 1.5 mg/kg caffeine, 3 mg/kg caffeine. Subjects remained in the same condition for parts 1 and 3 of the study. Following the first challenge, study subjects were assigned to the caffeine consumption condition or the decaffeinated condition. All subjects were supplied with coffee and tea bags to last them a week. For this week, they were required to consume coffee and tea made only from these supplies and abstain from consuming any other caffeinated products. Both the acute caffeine challenges and the withdrawal part of the study were double blind.

B. Subjects

One hundred and forty four subjects were selected from the Health Psychology Research Unit Panel. There were equal numbers of males and females in each condition. Similarly, the conditions were matched in terms of age, weight, regular caffeine consumption, smoking, alcohol consumption, use of milk and sugar in drinks, and the personality dimensions of trait anxiety, impulsivity, sociability, morningness, and obsessionality. These dimensions were chosen because a recent factor analysis of a number of personality scores has shown these to be important measures (see Reference 15 for details of the personality measures and factor analysis).

Descriptive statistics for the male and female subjects are shown in Table 9.1.

Table 9.1   Descriptive Statistics For Male Subjects

Variable	Mean	Standard Error	Median	95% Confidence Lower	Upper
AGE	21.605	0.396	21	20.815	22.396
TAQ	34.986	0.819	34	33.353	36.619
EPIEXT	13.781	0.416	14	12.952	14.61
MORN	47.055	0.913	45	45.235	48.875
EPIIMP	5.068	0.205	5	4.66	5.477
EPISOC	7.849	0.266	8	7.319	8.379
OPQ	2.822	0.153	3	2.517	3.126
WEIGHT (kg)	73.233	0.943	73	71.352	75.113
SMOKE (y)	32.90%				
CIGSDAY	1.055	0.308	0	0.44	1.67
WEEKALC	14.973	1.366	14	12.25	17.695
MILKCOF	79.20%				
SWEETCOF	62.50%				
DAYCOFCF	146.027	12.917	130	120.278	171.777
TOTDAYCF	192.603	12.371	195	167.942	217.263
AGE	21	0.326	20	20.349	21.651
TAQ	38.296	0.917	36	36.467	40.125
EPIEXT	13.254	0.436	13	12.385	14.122
MORN	47.324	0.935	48	45.459	49.189
EPIIMP	4.746	0.228	4	4.291	5.202
EPISOC	7.986	0.239	8	7.508	8.464
OPQ	3.116	0.175	3	2.767	3.464
WEIGHT (kg)	60.239	0.859	61	58.525	61.954
SMOKE (y)	32.4%				
CIGSDAY	1.577	0.409	0	0.761	2.394
WEEKALC	10.225	0.851	10	8.527	11.923
MILKCOF	72.9%				
SWEETCOF	13.4%				
DAYCOFCF	133.714	12.952	130	107.876	159.553
TOTDAYCF	190.143	13.427	195	163.357	216.929

TAQ = Trait Anxiety   WEEKALC = Weekly Alcohol Consumption (units)
EPIEXT = Extroversion   MILKCOF = % Take Milk
MORN = Morningness   SWEETCOF = % Take Sweetener
EPIIMP = Impulsivity   DAYCOFCF = Daily Caffeine Intake (mg) From Coffee
EPISOC = Sociability   TOTDAYCF = Total Daily Caffeine (mg)
OPQ = Obsessional
        Personality

1. Exclusion criteria

Those who drank coffee less frequently than once a week were not eligible for the study. Similarly, those who smoked more than five cigarettes in the daytime were excluded.

2. Informed consent

All subjects were required to sign an informed consent form which outlined the study, explained they were free to withdraw at anytime, and confirmed the confidentiality of all information collected.

C. Procedure

All subjects were weighed and familiarized with the computerized performance tests in one session prior to the test day. They then completed a caffeine diary for the 24 hours prior to the start of the experiment. The evening before their test day, subjects were required to limit their alcohol consumption to a maximum of four units. On the test day, they abstained from drinking any alcohol, doing any strenuous physical exercise, and drinking caffeinated beverages for 2 hours before the test session.

Saliva samples were taken to determine caffeine levels at baseline and over the test session. During the test session, subjects completed a questionnaire about their previous night's sleep, food consumed, and alcohol consumption.

Subjects started their tests at 8 00, 11 00, 14 00, or 18 00 hours. Equal numbers of subjects in the different caffeine conditions were tested at these times. Subjects completed a baseline session, then provided a saliva sample, followed by consumption of a cup of coffee. One hour later, a second saliva sample was taken and the next performance session started. At the end of the tests, another saliva sample was taken and this part of the experiment then ended.

1. Nature of the drink

All drinks were made with one rounded teaspoon of decaffeinated coffee in a 150ml mug of boiling water. To this, either the placebo or caffeine solution was added in accordance with the condition code. The placebo solution consisted of preserved water, while the caffeine solutions contained 15% w/v caffeine (equivalent to 1.5 mg/kg) or 30% w/v caffeine (equivalent to 3 mg/kg) accordingly. Milk and sugar were added in accordance with the subject's usual preference, and this was recorded.

2. Performance tasks

These tasks were selected because previous studies have shown that they are sensitive to effects of caffeine.

a. Variable fore-period simple reaction time task.   In this task, a box was displayed on the screen and at varying intervals (from 1 to 8 secs) a square would appear in the box. Subjects were required to press a response key as soon as they detected the square. This task lasted for approximately 3 minutes.

b. Five choice serial reaction time task.   Five boxes were displayed on the screen, and a light appeared in one of the boxes. The subject pressed the corresponding key, and the light then appeared in another box and the subject was required to press the next key. This task lasted for 3 minutes.

c. Focused attention choice reaction time task.   Target letters appeared as upper case A's and B's. On each trial, three warning crosses were presented on the screen. The outside crosses were separated from the middle one by either 1.02 or 2.60 degrees. Subjects were told to respond to the letter presented in the center of the screen and ignore any distracters presented in the periphery. The crosses were on the screen for 500 msec and were then replaced by the target letter. The central letter was accompanied either by (1) nothing, (2) asterisks, (3) letters which were the same as the target, or (4) letters which differed (the two distracters were identical and the targets and accompanying letters were always A or B). The correct response to A was to press a key with the fore-finger of the left hand, while the correct response to B was to press a different key with the fore-finger of the right hand. Subjects were given 10 practice trials, followed by three blocks of 64 trials. In each block, there were equal numbers of near/far conditions, A or B responses, and equal numbers of the four distracter conditions. The nature of the previous trial was controlled.

d. Repeated digits detection task.   Subjects were shown three-digit numbers on the screen at the rate of 100 per minute. Each digit was normally different from the preceding one, but occasionally (8 times a minute) the same number was presented on successive trials. Subjects had to detect these repetitions and respond as quickly as possible. The number of hits, reaction times for hits, and false alarms were recorded. The task lasted for 3 minutes.

D. Regular caffeine consumption

Following the challenge, study subjects were given their supplies for the next week. During the course of the week, subjects kept a log of coffee and tea consumption. Saliva samples were also taken on days 1, 2, and 3 to assess whether subjects were complying with instructions. Performance was also assessed on these days, and the procedures were as in the previous part of the experiment.

III. Results

A. First acute challenge

Analyses of covariance, with the baseline data as covariates and the post-drink data as dependents, were carried out. The between factors were caffeine dose, time of day, regular caffeine consumption (categorized as high or low on a median split, median = 195 mg), and gender. The focused attention

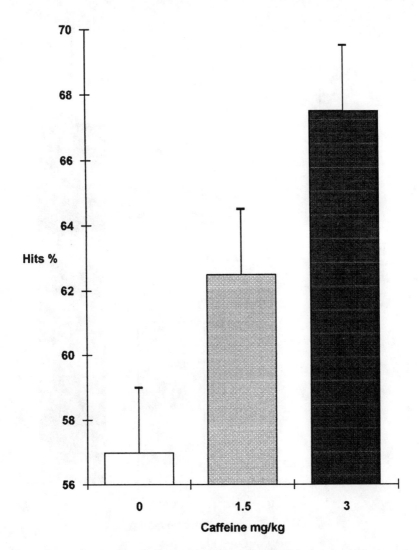

Figure 9.1 Effect of caffeine on hits in the repeated digits task. (Scores are the adjusted means from the analysis of covariance; SEs shown as bars.)

task, five-choice task, and repeated digits task all showed significant benefits of caffeine (repeated digits hits: F 2,119 = 7.66 p<0.001; five-choice task: F 2,119 = 3.28 p<0.05; focused attention task: F 2,119 = 5.37, p<0.01). These results are shown in Figures 9.1 to 9.3. The five-choice and repeated digits tasks showed a clear dose–response effect, with best performance occurring in those who were given 3 mg/kg caffeine. Comparable effects were seen in the 1.5 mg and 3 mg groups for the focused attention task. The variable fore-period simple reaction time task showed a similar numerical trend, although the effect of caffeine dose just failed to achieve significance.

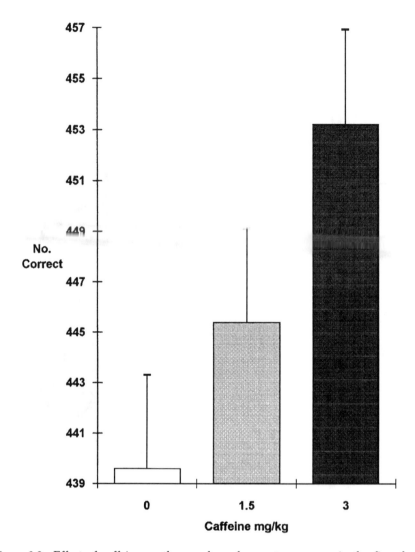

Figure 9.2 Effect of caffeine on the number of correct responses in the five-choice serial response task. (Scores are the adjusted means from the analysis of covariance; SEs shown as bars.)

Analyses were conducted to examine the effects of caffeine with time on task. When this was done, it was found that the effects of caffeine were constant over the task. In the focused attention task, it was possible to examine speed of encoding stimuli by looking at differences in reaction times to repeated stimuli and different successive stimuli (alternations). Caffeine had a beneficial effect when a new stimulus was presented, which suggests that it is enhancing the encoding of new stimuli (focused attention task — alternations — F 2,119 = 3.87, p<0.05). This is shown in Figure 9.4.

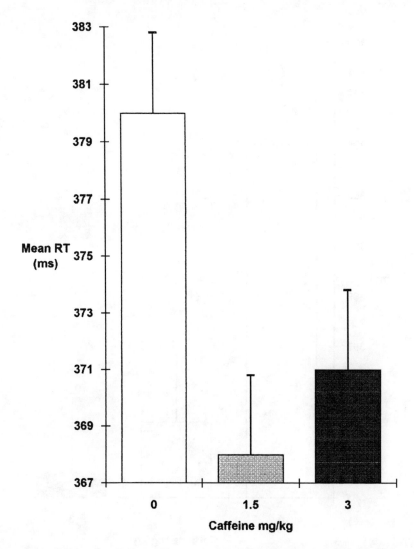

Figure 9.3 Effect of caffeine on mean reaction time in the focused attention task. (Scores are the adjusted means from the analysis of covariance; SEs shown as bars.)

The effects of caffeine dose were not modified by regular level of caffeine consumption, time of day or gender. In further analyses, personality categories were included and, again, no interactions between these variables and caffeine were observed.

1. Summary of the first acute challenge

These results demonstrate quite clearly that caffeine consumption is related to performance, usually in a dose–response fashion. None of these effects was modified by regular level of caffeine usage, time of day, or personality.

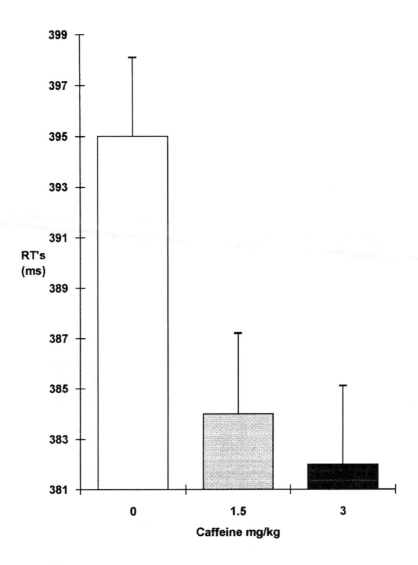

Figure 9.4 Effects of caffeine on reaction times to stimuli which differ from those presented on the previous trial. (Scores are the adjusted means from the analysis of covariance; SEs shown as bars.)

In other words, the present methodology has demonstrated significant and robust effects of caffeine. If these results reflect negative effects of caffeine withdrawal, then such effects should be apparent when subjects consume caffeine-free beverages over a longer time period. This was examined in the next part of the study.

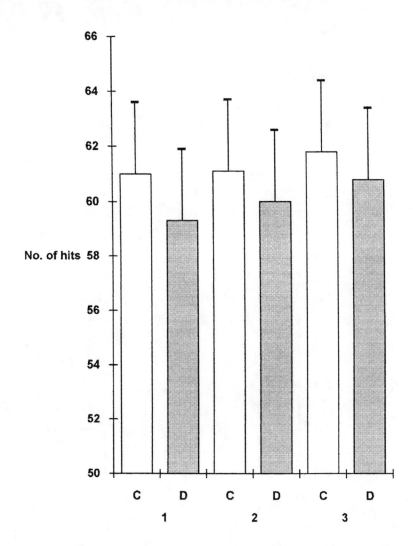

1, 2 & 3 = **Refer to testing days during regular coffee consumption**
☐ **C = Caffeinated** ▨ **D = Decaffeinated**

Figure 9.5 Effects of caffeine deprivation on hits in the repeated digits task.

B. Control of caffeine consumption over a week

Analyses of covariance revealed no effect of caffeine withdrawal on the performance tasks which were sensitive to the acute effects of the caffeine challenge. These data are shown in Figures 9.5 to 9.7.

It is possible that these negative results reflected poor compliance by the subjects. However, analyses of the saliva levels of the decaffeinated group

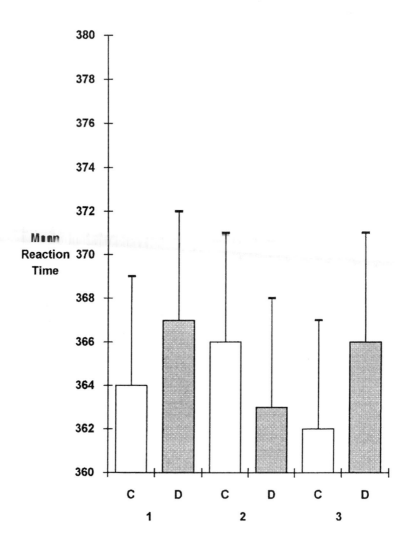

1, 2 & 3 = Refer to testing days during regular coffee consumption
☐ C = Caffeinated   ▦ D = Decaffeinated

Figure 9.6 Effects of caffeine deprivation on mean reaction time in the focused attention task.

showed that the mean level was not significantly different from zero, whereas the caffeine group had levels which showed that caffeinated beverages had been consumed. Another possibility is that caffeine withdrawal effects might depend on other factors, such as regular level of usage, time of day, or personality. This suggestion seems unlikely, given the lack of interactions in the analyses.

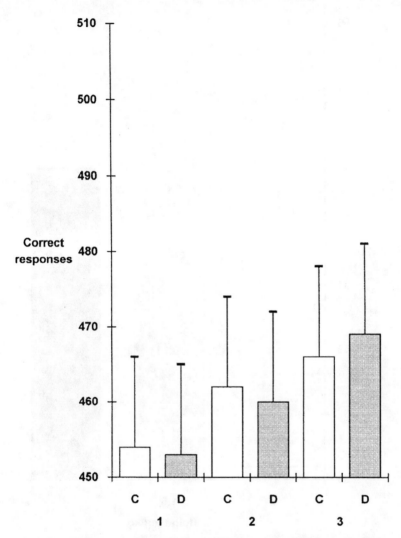

Figure 9.7 Effects of caffeine deprivation on the number of correct responses in the five-choice serial response task.

1. Summary of controlled consumption phase

This part of the experiment failed to demonstrate any significant effects of caffeine withdrawal on tests which had previously been shown to be sensitive to acute caffeine challenge. The final part of the study examined whether withdrawing caffeine for 7 days modified the effects of a caffeine challenge on performance.

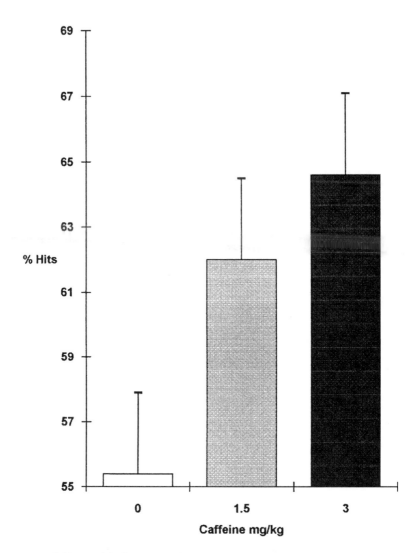

Figure 9.8 Effects of caffeine in subjects who had been deprived of caffeine for 7 days: hits in the repeated digits task.

C. *Acute effects of caffeine after 7 days of caffeine withdrawal*

Results from the five-choice and repeated digits tasks revealed an identical dose–response pattern to that seen in the first part of the study. Similarly, the effect of caffeine on alternations in the focused attention task was also significant (repeated digits hits: $F\ 2{,}67 = 3.92$, $p{<}0.05$; five-choice task: $F\ 2{,}67 = 5.48$, $p{<}0.01$; focused attention alternations: $F\ 2{,}67 = 3.56$, $p{<}0.05$). These results are shown in Figures 9.8 to 9.10.

Effects are clearly present even after withdrawal, which argues against James's view. Indeed, rather than removing the effects of caffeine, withdrawal

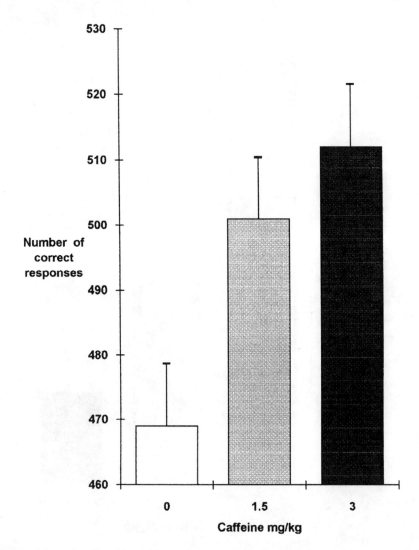

Figure 9.9   Effects of caffeine in subjects who had been deprived of caffeine for 7 days: number of correct responses in the five-choice serial response task.

appeared to make the acute effects of caffeine slightly bigger. This meant that the simple reaction task now showed a significant effect of caffeine dose (simple reaction time: F 2,67 = 4.29, p<0.05). These data are shown in Figure 9.11.

IV. Discussion

The present results confirm that performance of psychomotor tasks is improved following consumption of caffeinated coffee. The improvement

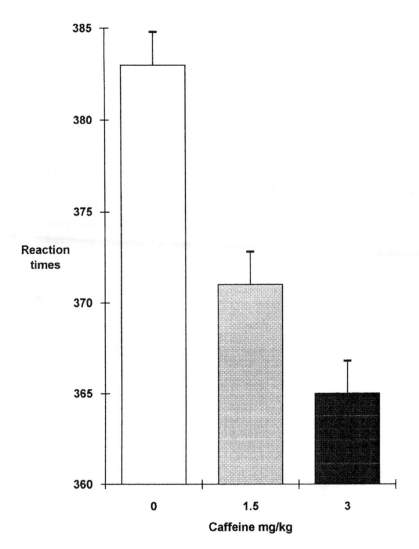

Figure 9.10 Effects of caffeine in subjects who had been deprived of caffeine for 7 days: alternation reaction times in the focused attention task.

was generally in the form of dose response, with 3 mg/kg caffeine being associated with the best performance. These results were apparent both before and after withdrawal, which suggests that the beneficial effects of caffeine cannot be accounted for by impairments in the caffeine-free condition. The view is further supported by the absence of negative effects when caffeine is withdrawn. A direct test of James's explanation of the effects of caffeine on psychomotor performance shows, therefore, that there is little support for his view.

The present findings do not argue against the occurrence of caffeine withdrawal effects in certain contexts. Indeed, Rogers et al.[13] argue that

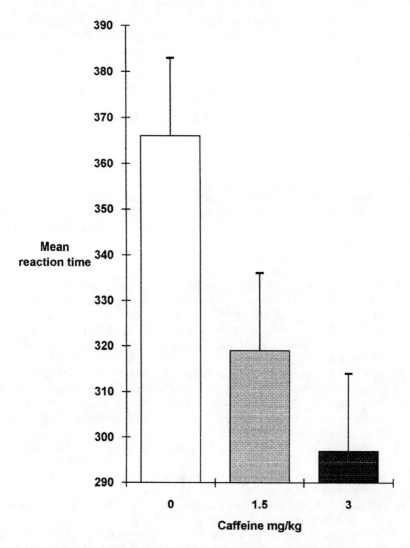

Figure 9.11 Effects of caffeine in subjects who had been deprived of caffeine for 7 days: simple reaction time task.

caffeine withdrawal influences mood but not performance. Smith[12] has suggested that effects of caffeine withdrawal on subjective reports do not necessarily reflect a pharmacological effect but may, at least in part, be due to expectancy effects. This is an issue that needs to be examined in further studies, but it now appears that in moderate caffeine users it is unlikely that caffeine withdrawal will produce impairments that show up in objective measures. On the other hand, caffeine has beneficial effects on performance, and the present results suggest that this may reflect faster encoding of new information. Further research must now clarify which of the many CNS mechanisms influenced by caffeine[16] underlie the present findings. Research should also continue to evaluate the practical benefits of caffeine consumption when operational efficiency and safety is endangered by reduced alertness.

Acknowledgment

This research was supported by a grant from the Institute for Scientific Information on Coffee.

References

1. Lieberman, H. R., Caffeine, in *Handbook of Human Performance*, Vol. 2, Smith, A. P. and Jones, D. M., Eds., London, Academic Press, 1992, 49.
2. Loriot, M. M., Snel, J., Kok, A., and Mulder, G., Influence of caffeine on selective attention in well-rested and fatigued subjects, *Psychophysiology*, 31, 525, 1994.
3. Smith, A. P., Caffeine, performance, mood and states of reduced arousal, *Pharmacopsychoecologia*, 7, 75, 1994.
4. Smith, A. P., Brockman, P., Flynn, R., Maben, A., and Thomas, M., Investigation of the effects of coffee on alertness and performance during the day and night, *Neuropsychobiology*, 27, 217, 1993.
5. Smith, A. P., Kendrick, A. M., and Maben, A. L., Effects of breakfast and caffeine on performance and mood in the late morning and after lunch, *Neuropsychobiology*, 26, 198, 1992.
6. Smith, A. P., Rusted, J. M., Eaton-Williams, P., Savory, M., and Leathwood, P., Effects of caffeine given before and after lunch on sustained attention, *Neuropsychobiology*, 23, 160, 1990.
7. Smith, A., Thomas, M., Perry, K., and Whitney, H., Caffeine and the common cold, *J. Psychopharmacol.*, 11, 319, 1997.
8. Horne, J. A. and Reyner, L. A., Counteracting driver sleepiness: Effects of napping, caffeine and placebo, *Psychophysiology*, 33, 306, 1996.
9. James, J. E., Does caffeine enhance or merely restore degraded psychomotor performance? *Neuropsychobiology*, 30, 124, 1994.
10. Smith, A. P., Caffeine, caffeine withdrawal and psychomotor performance: A reply to James, *Neuropsychobiology*, 31, 200, 1995.
11. Smith, A. P., Caffeine dependence: an alternative view, *Nat. Med.*, 2, 494, 1996.
12. Rogers, P. J., Richardson, N. J., and Dernoncourt, C., Caffeine use: Is there a net benefit for mood and psychomotor performance, *Neuropsychobiology*, 31, 195, 1995.
13. Smith, A. P., Maben, A., and Brockman, P., Effects of evening meals and caffeine on cognitive performance, mood and cardiovascular functioning, *Appetite*, 22, 57, 1994.
14. Warburton, D. M., Effects of caffeine on cognition and mood without caffeine abstinence, *Psychopharmacology*, 119, 66, 1995.
15. Smith, A. P., Chappelow, J., and Belyavin, A., Cognitive failures, focused attention and categoric search, *Appl. Cogn. Psychol.*, 9, 115, 1995.
16. Nehlig, A., Daval, J. L., and Debry, G., Caffeine and the central nervous system: mechanisms of action, biochemical, metabolic and psychostimulant effects, *Brain Res. Rev.*, 17, 139, 1992.

chapter ten

The association of anxiety, depression, and headache with caffeine use

David M. Warburton

Contents

Abstract 179
   I. Introduction 180
  II. Participants and methods...180
 III. Results..181
     A. Depression...182
     B. Anxiety 184
     C. Headache..185
 IV. Discussion 187
     A. Depression...188
     B. Anxiety 188
     C. Headache..188
References..189

Abstract

The objective of this study was to evaluate the association of caffeine consumption and anxiety, depression, and headache in a sample from a 9003-person survey of the general population of Great Britain. An analysis of the database gave no evidence of any association between caffeine consumption and depression in the general population. Similarly, no evidence was found of any association between caffeine consumption and anxiety in the general population or any association between caffeine avoidance and anxiety in the

0-8493-1166-7/99/$0.00+$.50
© 1999 by CRC Press LLC

general population. With respect to headache, drinking three to four cups of coffee reduced headache occurrence to below that of nonusers. Given the positive association between caffeine use and beverage alcohol consumption, a possible explanation is that caffeine may be used as a form of self-medication for headache, resulting from beverage alcohol consumption.

I. Introduction

The notion that people may inadvertently suffer from their lifestyle is one of the tenets of health psychology. An alternative view of this hypothesis is that individuals may adopt a lifestyle as a form of control of their psychological state. These views have been applied to caffeine use in the form of the following hypotheses

First, depression is a positive predictor of caffeine use, because it is mood elevating.[1] Second, caffeine use is a positive predictor of anxiety levels, if caffeine is anxiogenic.[2,3] Third, anxiety level is a negative predictor of caffeine use, because some individuals will avoid its anxiogenic effects.[4-9] Fourth, caffeine use is a predictor of headaches during abstinence.[10-14]

These hypotheses, based on small experimental studies and large population surveys, have been influential in shaping attitudes to caffeine use. Access to a large health and lifestyle database[15] has enabled us to test these hypotheses with respect to caffeine consumption.

II. Participants and methods

The Health and Lifestyle Survey[15] consists of data from 9003 individuals aged 18 and over and living in Great Britain. The respondents were chosen at random from representative constituencies in England, Scotland, and Wales, and the information was collected during 1984–85.

The survey asked about health knowledge, attitudes to health, lifestyles, and health status, and focused on the four habits or behaviors most often implicated in studies of ill health: physical exercise, smoking, beverage alcohol consumption, and diet, including tea and coffee consumption. We have analyzed some of these factors with respect to smoking status, and these results have been published elsewhere.[16,17]

Our analyses were concerned with the questions of caffeine consumption, mental health, and headache. The respondents were asked how many cups of coffee they usually drank in a day (none, 1 or 2, 3 or 4, 5 or 6, more than 6). The same question was asked about tea consumption. The number of cups in each block was averaged so that 1 or 2 cups was scored as 1.5 and so on, while more than 6 was scored as 6.5.

From these data, it was also possible to estimate "total caffeine" consumption based on the assumption that a cup of tea was half the caffeine strength of coffee, giving a scale ranging from 0 to more than 8.[18] If the amount of caffeine in a cup is 80 mg, then the daily use range is from 0 to over 640 mg. This measure can be criticized when used for analysis of acute

Table 10.1a Age Differences in Coffee Consumption

	18–29	30–39	40–49	50–59	60–69	70+
Percent of men	70	77	72	65	63	61
Percent of women	69	79	76	71	66	60

Table 10.1b Age Differences in Tea Consumption

	18–29	30–39	40–49	50–59	60–69	70+
Percent of men	81	82	88	94	96	98
Percent of women	79	83	86	91	95	98

measures like performance or blood pressure, because it assumes that the individuals had consumed their full amount at the time of measurement. However, this criticism would not apply to our study, in which we were examining questions related to a person's state over a period of time.

Two of these questions were included in the 30-item version of the General Health Questionnaire,[19] which is concerned with a person's mental state and asked whether the person had recently experienced the symptoms of depression and anxiety. In addition, in Section 23 of the Survey, a number of questions were asked about health in the last month. For headache, the question was: "Within the last month have you suffered from any problems with headache?"

III. Results

The data for age differences in coffee and tea consumption for men and women are shown in Tables 10.1a and 10.1b. The data for coffee, tea consumption and total caffeine consumption for men and women are shown in Tables 10.2a and 10.2b. It can be seen that there are age differences in consumption, with coffee drinking being an activity of the younger generation, while the older generation drink tea. However, there is no evidence of gender differences in coffee and tea consumption, and so it is possible to calculate a combined caffeine consumption which is expressed in coffee cup equivalents for the combined population, and this is shown in Table 10.2c.

Although there is no evidence of gender differences in coffee and tea consumption, the half-time of elimination for ovulating women is 20 to 39% shorter than for men, but the half-time of elimination for women taking contraceptive steroids is twice that of ovulating women.[20] Consequently, controls were made for age and sex, as well as social class and beverage alcohol consumption, when appropriate.

The data for anxiety, depression, and headache were analyzed for "total caffeine" and coffee consumption using chi-square and likelihood ratio chi-square tests of association. Because the chi-square and likelihood ratio chi-square tests of association gave exactly the same results, only the results of analysis by the chi-square test will be presented.

Table 10.2a Data for Coffee Consumption
for Men and Women

Cups/day	0	1–2	3–4	5–6	6+
Percent of Men	31	36	18	8	8
Percent of Women	29	36	19	8	8

Table 10.2b Data for Tea Consumption
for Men and Women

Cups/day	0	1–2	3–4	5–6	6+
Percent of Men	12	16	26	20	27
Percent of Women	13	18	29	19	22

Table 10.2c Total Caffeine Consumption
in Coffee Cup Equivalents

Cups/day	0	1–2	3–4	5–6	6–7	8+
Approx. %	4	9.62	35.5	31.5	14.3	5

We used the Mantel–Haensel chi-square test in order to examine whether there were significant trends in the data, in our case level of consumption. In the cases of anxiety and depression, analyses were also done to determine whether levels of anxiety or depression were predictors of caffeine consumption.

It should be noted that the numbers in the Tables of the Results section, and so the number in each analysis, vary depending on the number of participants who provided information on the relevant variables.

A. Depression

The first hypothesis is that depression is a positive predictor of caffeine use, if it is mood elevating. The relevant data are shown in Table 10.3. If depressed individuals were using caffeine as a form of self-medication, then it would be anticipated that consumption would increase with severity of depression. It can be seen that there is no evidence for this hypothesis, and this is confirmed by the statistical analysis ($X^2 = 37.74$, df = 30; p = 0.16). There were no significant trends in the data, i.e., with level of consumption; the Mantel–Haensel chi-square was 1.20 with df = 1 (p = 0.27).

It is possible that any association might be obscured by the use of the caffeine consumption index, and so an analysis was done for coffee drinkers alone, and the reduced sample was divided into nondepressed (0–2) and depressed (3–6), and the data are shown in Table 10.4. Once again, inspection reveals that there is no evidence for this hypothesis of increasing with severity of depression, and this is confirmed by the statistical analysis ($X^2 = 3.65$, df = 4; p = 0.46). There were no significant trends in the data, i.e., with level of consumption; the Mantel–Haensel chi-square was 1.434 with df = 1 (p = 0.27).

Table 10.3 Caffeine Intake as Function of Depression

Severity	0	1	2	3	4	5	6	Total
Consumption (cups/day)								
0	1.61	0.98	2.10	2.53	2.58	5.26	0.00	
1–2	9.36	11.81	10.88	10.55	13.55	3.51	14.04	
3–4	37.61	35.53	37.60	35.02	34.84	50.88	40.35	
5–6	33.76	33.66	29.96	32.07	30.32	29.32	26.32	
7–8	14.94	15.26	14.89	16.46	15.48	7.02	15.79	
8+	2.71	2.76	4.58	3.38	3.32	3.51	3.51	
n =	3354	1016	524	237	153	58	57	5399

Table 10.4 Coffee Intake as Function of Depression

Severity	Nondepressed	Depressed	
Consumption (cups/day)			
0	12.27	19.12	
1–2	10.47	7.35	
3–4	23.24	26.47	
5–6	22.75	19.12	
6+	31.26	27.94	
n =	611	68	679

Table 10.5 Percentage Depressed by Caffeine Use (Cups in Coffee Equivalents)

Severity	Depression score							n
	0	1	2	3	4	5	6	
Consumption (cups/day)								
0	61.36	11.36	12.50	6.82	4.55	3.41	0	88
1–2	57.40	21.94	10.42	4.57	3.84	0.37	1.46	547
3–4	62.80	17.98	9.81	4.13	2.69	1.44	1.15	2008
5–6	63.38	19.15	8.79	4.26	2.63	0.95	0.84	1786
7–8	61.85	19.14	9.63	4.81	2.96	0.49	1.11	810
8+	56.88	17.50	15.00	5.00	3.13	1.25	1.25	160
Total								5399

Although there has been no suggestion in the literature that caffeine may induce depression, the data were recast in order to examine this possibility. The data are shown in Table 10.5. If this hypothesis was correct, we would expect that an association between depression and caffeine intake would be reflected in an increase in severity of depression as consumption increased. Inspection shows no evidence of this sort of trend, and no statistical analysis was carried out.

Table 10.6 Percentage of Anxiety by Caffeine Use (Cups in Coffee Equivalents)

	Anxiety Score							
Severity	0	1	2	3	4	5	6	n
Consumption (cups/day)								
0	51.14	18.18	11.36	6.82	4.55	4.55	3.41	88
1–2	55.76	18.10	7.31	7.86	5.30	3.66	2.01	547
3–4	59.66	15.85	8.00	5.46	5.37	4.02	1.64	2013
5–6	60.03	17.78	8.16	4.42	4.36	3.80	1.45	1789
7–8	57.95	16.40	7.77	6.66	5.55	3.70	1.97	811
8+	54.38	17.50	9.38	5.63	4.38	7.50	1.25	160
Total								5408

In summary, there was no evidence for the hypothesis that depressed individuals had increased caffeine or coffee consumption, as one might expect if individuals were using caffeine as a mood-elevating compound. The alternative possibility that caffeine would result in more depressive episodes was examined, and no evidence for this potentiality was seen.

B. *Anxiety*

The second hypothesis is that caffeine use is a positive predictor of anxiety levels, if caffeine is anxiogenic. From Table 10.6, there is no obvious increase in anxiety with increased consumption, and this was confirmed by the statistical analysis ($X^2 = 30.75$, df = 30; $p = 0.43$). There were no significant trends in the data, i.e., with level of consumption; the Mantel–Haensel chi-square was 0.272 with df = 1 ($p = 0.60$).

We also examined the possibility that an association might be hidden by the use of the caffeine consumption index. Consequently, the analysis was restricted to individuals who drink coffee but not tea and dividing the reduced sample into nonanxious (0–2) and anxious (3–6). The data are shown in Table 10.7. Once again, there is no evidence for the hypothesis, and this was confirmed by the statistical analysis. The results were $X^2 = 2.69$, df = 4; $p = 0.61$). There were no significant trends in the data, i.e., with level of consumption; the Mantel–Haensel chi-square was 0.34 with df = 1 ($p = 0.56$).

The third hypothesis, that individuals with high anxiety level might avoid caffeine, was examined in Table 10.8. The scores show no evidence of a shift from the median consumption of three to four cups with increasing anxiety levels. The statistical analysis confirmed that there was no significant effect ($X^2 = 28.32$, df = 30; $p = 0.55$). There were no significant trends in the data, i.e., with level of consumption; the Mantel–Haensel chi-square was 0.461 with df = 1 ($p = 0.53$).

In summary, there was no evidence that caffeine or coffee was anxiogenic. In addition, the data give no indication that individuals with high levels of anxiety avoid caffeine.

Table 10.7   Coffee Intake as Function of Anxiety

Severity	Non-anxious	Anxious	
Consumption (cups/day)			
0	12.66	14.41	
1–2	10.70	7.63	
3–4	22.82	27.12	
5–6	22.10	23.73	
6+	31.73	27.12	
n =	561	118	679

Table 10.8   Percentage Caffeine Use by Anxiety (Cups in Coffee Equivalents)

	Anxiety Score							
Severity	0	1	2	3	4	5	6	Total
Consumption (cups/day)								
0	1.41	1.75	2.30	1.99	1.48	1.86	3.30	
1–2	9.59	10.84	9.20	14.29	5.30	3.66	2.01	
3–4	37.74	34.94	37.01	36.54	39.85	37.67	36.26	
5–6	33.74	34.83	33.56	26.25	28.78	31.63	28.57	
7–8	14.77	14.57	14.48	17.94	16.61	13.95	17.58	
8+	2.73	3.07	3.45	2.99	2.58	5.58	2.20	
n =	3182	913	435	301	271	215	91	5408

Table 10.9   Prevalence of Headache in Last Month

Age (years)	18–24	25–34	35–44	45–54	55–64	65–74	75+
Men	25%	24%	26%	19%	14%	13%	13%
Women	40%	38%	38%	37%	28%	25%	21%

C. Headache

The fourth hypothesis is that caffeine use is a predictor of headaches. As a preliminary to the analysis, the prevalence of headache in the survey population was examined for age and sex differences. The specific question was: "Within the last month have you suffered from any problems with headache?" The frequency of scores is shown in Table 10.9. First, it is clear that there is an age-related decline in prevalence, which was marked for both women and men. Second, the levels are higher for women than men. Accordingly, age and sex, as well as socioeconomic status, were controlled for in the analysis of caffeine use as a predictor of headache.

Since the question in the survey was: "Within the last month have you suffered from any problems with headache?," it would be anticipated that

Table 10.10   Prevalence of Headache by Caffeine Use (Cups in Coffee Equivalents)

	No	Yes	n	
Consumption (cups)				
0	213	78	291	4.04%
	73.20	26.80		
1–2	525	168	693	9.63%
	75.76	24.24		
3–4	1998	554	2552	35.44%
	78.29	21.71		
5–6	1672	584	2268	31.50%
	74.25	25.75		
7–8	761	267	1028	14.28%
	74.03	25.97		
8+	273	95	368	5.11%
	74.19	25.81		
n =	5442	1746	7200	100%

if caffeine abstinence was associated with headache, then regular consumers would have experienced some period of involuntary abstinence during the previous month, with the consequent occurrence of a headache.

The data for headache as a function of caffeine use are shown in Table 10.10. It can be seen that there is no consistent trend in the data for increased incidence of headache with increased consumption. In fact, headache is reduced at intermediate levels of consumption and the statistical analysis showed that this effect was significant. ($X^2 = 18.43$, df = 5; p = 0.005). There were no significant trends with level of consumption (Mantel–Haensel chi-square was 0.683, df = 1; p = 0.41).

The reduced effect was examined for coffee drinkers alone, and the data are shown in Table 10.11. Once again there was a significantly reduced incidence of headache at the intermediate levels of consumption, three to four cups of coffee ($X^2 = 13.17$, df = 4; p = 0.010). There were no significant trends in the data, i.e., with level of consumption; the Mantel–Haensel chi-square was 0.285 with df = 1 (p = 0.59).

In summary, the consumption of three to four cups of coffee or its caffeine equivalent significantly reduced headache. There was no evidence that either less or more caffeine intake increased headache over the level of no caffeine use.

In order to examine the headache data further, we performed a set of analyses using beverage alcohol consumption as the independent variable. The prevalence data are shown in Table 10.12. There are significant differences in the occurrence of headache with different levels of beverage alcohol consumption. The significant Mantel–Haensel chi-square shows that there was a significant difference in the incidence of headache with different amounts of beverage alcohol consumption ($X^2 = 15.01$, df = 4; p = 0.005).

Table 10.11 Percentage of Headache
by Coffee Use

	No	Yes	n
Consumption (cups)			
0	1200	430	1630
	73.62%	26.38%	
1–2	1657	525	2182
	75.94%	24.06%	
3–4	1030	278	1308
	78.75%	21.25%	
5–6	406	149	555
	73.15	26.85	
6+	394	139	533
	73.92	26.08	
Total	4687	1521	6208

Table 10.12 Percentage of Headache in Last Month
by Beverage Alcohol Use

Consumption	No	Yes	n	
Non-drinker	826	251	1077	12%
	76.70%	23.30%		
Very occasional drinker	1607	459	2066	23%
	77.78%	22.22%		
Regular, but none last week	1010	247	1257	14%
	80.35%	19.65%		
Light drinker	1681	653	2334	26%
	71.03%	27.97%		
Moderate/heavy	1369	876	2245	25%
	60.98%	39.02%		
	6493	2486	8979	100%
	72.31%	27.68%		

Men: Light 1–10 units; Moderate 11–50 units; Heavy >50 units

Women: Light 1–5 units; Moderate 6–35 units; Heavy >35 units

There was a significant increasing trend in headache with the amount of beverage alcohol consumption; the Mantel–Haensel chi-square was 0.456 with df = 1 (p = 0.041).

IV. Discussion

The following hypotheses were tested using the Health and Lifestyle Database. First, depression is a positive predictor of caffeine use. Second, caffeine use is a positive predictor of anxiety levels. Third, anxiety level is a negative predictor of caffeine use. Fourth, caffeine use is a predictor of headaches.

A. Depression

The hypothesis of Neil et al.,[1] that depression might be a cause of consumption, was tested by examining caffeine consumption as a function of depression level. There was no evidence that level of depression was a predictor of caffeine use, which is in accord with a study of 143,000 men and women, which showed there was a high association between high coffee consumption and depression in women but not in men. The association disappeared when there was control for smoking status.[21] Of course, this control does not point to the influence of smoking on depression but may be due to a common association, such as emotional problems, as the authors acknowledge.

B. Anxiety

The analyses showed no association between total caffeine intake index and anxiety. In addition, there was no association between anxiety and coffee consumption. The hypothesis that individuals with high anxiety levels might avoid caffeine was examined by caffeine consumption as a function of anxiety level. There was no evidence that level of anxiety was a negative predictor of caffeine use.

This finding is in accord with the only other general population study (N = 3854), which found no association between anxiety and caffeine intake as measured by the number of cups of coffee and tea drunk.[22] However, no distinction was made between whether the beverage was tea or coffee. In our study, we made an analysis for both coffee alone and for combined coffee and tea consumption as an index of total caffeine consumption. No relationship was seen with either analysis.

C. Headache

Evidence has been presented in support of the assertion that caffeine use is associated with an increased risk of headache during abstinence.[14,23] In the survey, the question was asked. "Within the last month have you suffered from any problems with headache?"

Our analysis of the survey responses showed no greater incidence of headache in the past month as a function of caffeine intake. In fact, it was found that individuals were **less** likely to have headaches if they drank three to four cups of coffee, or its caffeine equivalent, per day than if they did not use coffee. One interpretation is that moderate use of coffee can reduce headache, because of its vasoactive properties.[24]

When we examined the general hypotheses of predictors of headache, it was found that beverage alcohol consumption was a predictor of headache when the factors of age and social class were controlled. Thus, it could be argued that caffeine may be used as a form of self-medication for headache, e.g., resulting from beverage alcohol consumption. Certainly, it is clear from the analyses of the Health and Lifestyle Survey that there is positive association between caffeine use and beverage alcohol consumption.[15,18]

References

1. Neil, J. F., Himmelhoch, J. M., Mallinger, A. G., Mallinger, J., and Hanin, I., Caffeinism complicating hypersonic depressive syndromes, *Comprehen. Psychiat.*, 19, 377, 1978.
2. Greden, J. F., Fontaine, P., Lubetsky, M., and Chamberlin, K., Anxiety and depression associated with caffeinism among psychiatric inpatients, *Am. J. Psychiat.*, 135, 963, 1978.
3. James, J. E. and Crosbie, J., Somatic and psychological health implications of heavy caffeine use, *Br. J. Addict.*, 82, 503, 1987.
4. Lynn, R., National differences in anxiety and the consumption of caffeine, *Br. J. Soc. Clin. Psychol.*, 12, 92, 1973.
5. Hire, J. N., Anxiety and caffeine, *Psychological Reports*, 42, 833, 1978.
6. Boulenger, J. P., Uhde, T. W., Wolff, E. A., and Post, R. M., Increased sensitivity to caffeine in patients with panic disorders, *Arch. Gen. Psychiat.*, 41, 1067, 1984.
7. Uhde, T. W., Boulenger, J. P., Jimerson, D. C., and Post, R. M., Caffeine: Relationship to human anxiety, plasma MHPG, and cortisol, *Psychopharmacol. Bull.*, 20, 426, 1984.
8. Lee, M. A., Cameron, O. G., and Greden, J. F., Anxiety and caffeine consumption in people with anxiety disorders, *Psychiat. Res.*, 15, 211, 1985.
9. Lee, M. A., Flegel, P., Greden, J. F., and Cameron, O. G., Anxiogenic effects of caffeine on panic and depressed patients, *Am. J. Psychiat.*, 145, 632, 1988.
10. Goldstein, A. and Kaizer, S., Psychotropic effects of caffeine in man. III. A questionnaire survey of coffee drinking and its effects on a group of housewives, *Clin. Pharmacol. Therap.*, 10, 477, 1969.
11. Goldstein, A., Kaizer, S., and Whitby, O., Psychotropic effects of caffeine in man. IV. Quantitative and qualitative differences associated with habituation to coffee, *Clin. Pharmacol. Therap.*, 10, 489, 1969.
12. Griffiths, R. R. and Woodson, P. P., Caffeine physical dependence: a review of human and laboratory animal studies, *Psychopharmacology*, 94, 437, 1988.
13. Hughes, J. R., Higgins, S. T., Bickel, W. K., Hunt, W. K., Fenwick, J. W., Gulliver, S. B., and Mireault, G. C., Caffeine self-administration, withdrawal, and adverse effects among coffee drinkers, *Arch. Gen. Psychiat.*, 48, 611, 1991.
14. Silverman, K., Evans, S. M., Strain, E. C., and Griffiths R., Withdrawal syndrome after the double-blind cessation of caffeine consumption, *N. Engl. J. Med.*, 327, 1109, 1992.
15. Cox, B. D., Blaxter, M., Buckle, A. L. J. et al., *The Health and Lifestyle Survey*, The Health Promotion Research Trust, London, 1987.
16. Thompson, D. H. and Warburton, D. M., Lifestyle differences between smokers, ex-smokers and nonsmokers, and implications for their health, *Psychol. Health*, 7, 311, 1992.
17. Thompson, D. H. and Warburton, D. M., Dietary and mental health differences between never smokers living in smoking and nonsmoking households, *J. Smoking-Related Disord.*, 4, 203, 1993.
18. Jarvis, M. J., Does caffeine intake enhance absolute levels of cognitive performance? *Psychopharmacology*, 110, 45, 1993.
19. Goldberg, D. P., *The Detection of Psychiatric Illness by Questionnaire*, Oxford, Oxford University Press, Oxford, 1972.

20. Callahan, M. M., Robertson, R. S., Branfman, A. R., McComish, M. F., and Yesair, D. W., Comparison of caffeine metabolism in three nonsmoking populations after oral administration of radiolabelled caffeine, *Drug Metab. Dispos.*, 11, 211, 1983.
21. Jacobsen, B. K. and Hansen, V., Caffeine and health, *Br. Med. J.*, 296, 291, 1988.
22. Eaton, W. W. and McLeod, J., Consumption of coffee or tea and symptoms of anxiety, *J. Pub. Health*, 74, 66, 1984.
23. James, J. E., *Understanding Caffeine*, Sage Publications, London, 1997.
24. Mathew, R. J. and Wilson, W. H., Caffeine consumption, withdrawal and cerebral blood flow, *Headache*, 25, 305, 1985.

chapter eleven

Caffeine, impulsivity, and performance

Uma Gupta and B.S. Gupta

Contents

I. Introduction 191
   A. Caffeine, personality, and cognitive performance..........................192
   B. Caffeine, personality, and psychomotor performance..................194
II. Methods 196
   A. Subjects..196
   B. Experimental design...196
   C. Procedure 196
   D. Materials and tasks...197
III. Results..199
   A. Cognitive performance 199
   B. Psychomotor performance 200
IV. Discussion 202
References...203

I. Introduction

The behavioral effects of caffeine, a widely consumed[1-3] substance and a recognized potent stimulant,[4-6] have been studied extensively for a variety of reasons[7]: (1) it is readily available; (2) it is generally recognized as safe; (3) it has a rapid onset of action; and (4) it is active when taken orally. However, the major disadvantage is that when administered in doses generally found in foods and drinks, its effects are slight, subtle, and almost negligible.[7]

   The behavioral tasks employed for evaluating the treatment effect are of critical importance. The tasks must be sensitive enough to detect subtle variations in behavior observed under the influence of the treatment. Moreover, any variation in the manner in which a test is employed can also affect the likelihood of detecting an effect. Thus caffeine effects on human performance may not only be "task specific," i.e., the performance under the influence of caffeine is facilitated on certain tasks[6,8-16] and not on others,[17-21] but they are also "situation specific," i.e., the effects depend upon the conditions in which the drug is administered and consumed (see Reference 5 for a review). This, however, does not suggest that contradictory evidence is not available. To complicate matters further, caffeine interacts with the characteristics of the person to whom the drug is administered, particularly the trait of impulsivity.[22-27] However, studies are also available in which the observed effects of caffeine were not influenced by individual differences.[28,29] Other confounding factors[30] may be: habituation to caffeine, initial response level, personal characteristics associated with caffeine's metabolism, withdrawal symptoms (notably headache), and diurnal arousal variations.

   As the present study is devoted to investigating the interaction of caffeine and personality in their effects on cognitive performance as well as the psychomotor performance, some of the important findings of experiments in these areas will be described in the following section.

A. Caffeine, personality, and cognitive performance

1. Effects on intelligence test scores

Gilliland[31] investigated the effects of caffeine administered in doses of 0, 2, or 4 mg/kg body weight on the verbal ability of introverts and extraverts. The measure of verbal ability consisted of verbal analogies, antonyms, and sentence completions. A pre- and post-test design was used. The results revealed an interaction of caffeine and introversion–extraversion (I–E) demonstrating an improvement in the performance of introverts under the influence of a smaller dose (2 mg/kg body weight), but a decrement in performance with a larger dose (4 mg/kg body weight), and an improvement in the performance of extraverts with increasing dose levels. Revelle et al.[32] not only provided substantial support for the reliability of caffeine by I–E interaction but also demonstrated that this interaction changed as a function of the time of day the subjects were tested: the performance of high impulsives under the influence of caffeine was facilitated in the morning but was impaired in the evening; the reverse was observed for low impulsives — caffeine impaired performance in the morning but facilitated it in the evening. Revelle et al. also reported that the impulsivity subscale of the I–E scale, when used to define the subject's level of trait arousal, led to more clear interactive effects of caffeine and trait arousal, and suggested that the impulsivity subscale was a better predictor of performance than the global I–E measure.

Gupta[22] examined the effects of caffeine on the intelligence test performance of 75 high and 75 low impulsive groups of subjects. Male postgraduate students, using a double-blind procedure and a between-subjects design, were administered placebo or one of four doses of caffeine (1, 2, 3, or 4 mg/kg body weight) in a laboratory situation. All experimental sessions were arranged for 9 a.m. The dependent variable was the composite score on the culture Fair Intelligence Test, Scale 3, which comprises four subtests: series, classification, matrices, and topology. When analyzed by analysis of variance, the data revealed a highly significant ($p < 0.001$) caffeine × impulsivity interaction, demonstrating a monotonic increase in performance for high impulsives under the influence of caffeine (the increase in performance, as compared with placebo, was statistically significant only for larger doses, i.e., 3 mg/kg and 4 mg/kg body weight), and a nonsignificant effect for low impulsives. The dose trends in performance were essentially curvilinear, following an inverted-U pattern, for low impulsives, but it was predominantly linear for high impulsives. Another interesting aspect of the results was that the performance of low impulsives was much less affected by caffeine than that of high impulsives: 94.51% variation in the main effect of treatments, and the treatment × impulsivity interaction was due to the effects of caffeine for high impulsives. It was only 5.49% for low impulsives. In a way, this indicates the existence of a stronger tendency in low impulsives (who may be introverts) for a more extensive use of the central nervous system that attempts to compensate for the adverse effects of supraoptimal arousal.[33] In another study,[34] subjects were classified into four groups on the basis of having extreme scores on both extraversion (E) and neuroticism (N) dimensions of personality according to a procedure known as "zone-analysis"[35]: stable extraverts (N–E+), stable introverts (N–E–), neurotic extraverts (N+E+), and neurotic introverts (N+E–). These groups are assumed to have different positions on the scale of arousal.[36] The first (N–E+) group is low in arousal; second (N–E–) and third (N+E+) are moderate in arousal; the fourth (N+E–) is high in arousal. Male postgraduate students served as subjects. The study design and procedure, caffeine dosages, and the instruments used were similar to those of the earlier study.[22] In each cell of the experimental design there were 10 subjects. A statistically significant caffeine × personality interaction was discovered. The treatment effect was found to be statistically significant only for extraverts (i.e., N–E+ and N+E+ groups). The overall findings of this study were consistent with those of the earlier study.[22]

2. Effects on memory

Erikson et al.[28] examined the effects of two doses of caffeine, 2 and 4 mg/kg body weight, on immediate recall for words of 47 males and 60 females classified as either high or low impulsives. After an absorption period of 10 min, subjects were presented with one practice list and eight experimental lists (four lists presented at a rate of one word per second — fast rate, and four at a rate of one word every three seconds — slow rate). Each list consisted of 12 words. Immediately after each of the lists was presented, subjects orally recalled the words. Caffeine led to an impaired recall at the slow rate of presentation for females but had no effect on males' recall performance. Caffeine effects were not influenced by subjects' amount of caffeine consumption (caffeine dosage), verbal ability, or level of impulsivity. In another study[29] on short-term memory for words, 82 male and 75 female college students, classified as either high or low impulsives, were administered 2 or 4 mg/kg body weight of caffeine. After a latency period of 30 min, subjects listened to 12 word lists presented at one of four rates (two words per second, one word per second, one word every three seconds, and one word every five seconds). Each list consisted of 12 words. The authors could not find any indication of caffeine × impulsivity interaction for the male or female or mixed sample, although theoretically the possibility of discovering such an interaction could not be ruled out.

The effects of caffeine on free recall immediately after acquisition on semantic and rhyming tasks, in an incidental learning paradigm, were examined in a recent study.[23] Following a between-subjects design and a double-blind procedure, 150 high and 150 low impulsive male postgraduate students were administered placebo or one of four doses of caffeine (1, 2, 3, or 4 mg/kg body weight) in a laboratory situation. The study revealed a highly significant (p < 0.001) higher-order interaction of caffeine, impulsivity, and tasks, demonstrating that caffeine facilitated free recall in high impulsives after rhyming acquisition but hindered it after semantic acquisition. Low impulsives' free recall performance under the influence of caffeine remained uninfluenced irrespective of whether the acquisition was achieved rhythmically or semantically. Subsequently, this study was replicated using recognition as the retrieval measure.[24] Recognition and free recall differ not only in retrieval processes[37-39] but also in the kind of information that must be retrieved to lead to successful performance.[37] The study design, caffeine dosages, and the instruments used for the recognition study were the same as for the free recall study. The higher-order interaction (caffeine × impulsivity × tasks) was statistically significant and demonstrated that caffeine facilitated recognition performance after rhyming acquisition but impaired it after semantic acquisition in high impulsives; caffeine had no influence on the recognition performance of low impulsives.

B. Caffeine, personality, and psychomotor performance

Smith et al.[27] examined the effects of caffeine (3 mg/kg body weight) on a vigilance task — a visual version of the Bakan task.[40] Subjects were required to detect sequences of either three consecutive odd or three consecutive even digits. The duration of the task was 10 minutes; 20 male and 21 female subjects provided data for the study. The study showed a statistically significant (p < 0.01) caffeine × impulsivity interaction, demonstrating that high impulsives compared to low impulsives benefited more from caffeine.

Gupta and Gupta[25] reported a statistically significant caffeine × impulsivity interaction for kinesthetic after effect (KAE), a perceptual phenomenon consisting of changes in the width judgment of a standard block caused by touching a block differing in width from the standard block. Following a between-subjects design, four doses of caffeine (1, 2, 3, or 4 mg/kg body weight) and a placebo were administered to 75 high and 75 low impulsive male postgraduate students. The perceptive distortion as measured by the KAE task was diminished by all caffeine doses, compared to placebo, in the high impulsives; contrarily it was enhanced in the low impulsives but only with the largest dose.

A significant caffeine × impulsivity interaction was also discovered in a recent study on perceptual judgment.[26] The caffeine dosages and the experimental design used in this study were similar to those of the earlier study.[25] Eighty high impulsive and 80 low impulsive male subjects provided data. The results demonstrated that compared to placebo, caffeine in all doses facilitated perceptual judgment (reduced error) in high impulsives but had no effect on the performance of low impulsives. In a subsequent study,[41] the possibility of caffeine interacting with extraversion and neuroticism dimensions of personality in producing effects on perceptual judgment, was investigated. The subjects having extreme scores on both extraversion and neuroticism dimensions were administered one of four doses of caffeine (1, 2, 3, 4 mg/kg body weight) or placebo. Following a between-subjects randomized-block design, 200 subjects provided data for the study. The dependent variable was the error in perceptual judgment. Caffeine facilitated performance by diminishing error in perceptual judgment in subjects who had relatively lower levels of basal arousal, i.e., stables (having relatively lower positions on the neuroticism scale), extraverts (high scorers on the extraversion scale), and stable extraverts. The decrease in judgment error of these groups was monotonic and demonstrated predominantly a linear trend. The study also provided some evidence of the existence of a complex type of relationship between caffeine effects, and moderate and high arousal.

As the effects of caffeine on human performance are known to be "task-specific," the purpose of the present study was to investigate whether caffeine × impulsivity interaction occurs in the effects on performance across a variety of tasks. It has been suggested that this interaction is likely to be influenced by situational variables, such as the time of day[32] the drug is administered and the subject's performance is measured. It was necessary, therefore, to administer caffeine and subsequently test each subject at the same hour so that the confounding of effects produced by time of day in the caffeine × impulsivity interaction could be minimized. The present study was conducted in the morning. The evidence is not absolutely compelling; however, it does indicate that high impulsives are less aroused in the morning than low impulsives.[42] It was predicted that high impulsives would benefit more from caffeine in the morning.[42] Arousal is not a unitary concept[33] and involves multidimensional conceptualizations.[43] Hence, the arousal induced by caffeine may not be equivalent to the arousal assumed to be attributed to the trait of impulsivity.

II. Methods

The effects of three doses of caffeine citrate (1, 2, or 3 mg/kg body weight) or placebo (citric acid), dissolved in a glass of orange-flavored cold drink, were examined on cognitive and psychomotor performance of subjects differing in the trait of impulsivity.

A. Subjects

The subjects were male volunteer postgraduate students aged 19 to 24 years. They were selected on the basis of preliminary testing with the Hindi version[44] of the Eysenck Personality Inventory (EPI).[45] The 9-item impulsivity scale, containing items 1, 3, 5, 8, 10, 13, 22, 39, and 41 of the extraversion scale of the EPI provided a measure of impulsivity. On the basis of their impulsivity scores, two groups of subjects were selected; high impulsives, having a score of 6 or more on the impulsivity scale; and low impulsives, having a score of 2 or less on this scale. The criterion for selection of subjects was: high, mean + 1.5 SD; low, mean −1.5 SD (mean = 4.28; SD = 1.43). The mean and SD were based on the sample of 1604 students. Eight subjects (40 high impulsives and 40 low impulsives) were selected for participation in the study for all the dependent variables except for memory (free recall) where the number of subjects was 128, 64 high impulsives and 64 low impulsives.

Subjects who did not take coffee at all or were only casual coffee drinkers (taking not more than one cup of coffee a day for only 3 to 4 days a week) and were also nonsmokers and only casual cola and tea consumers were accepted as subjects. This was done because it is well known that the habitual level of caffeine/nicotine consumption influences response to challenge doses and consequently affects performance. In addition, only those persons qualified as subjects who did not use any kind of drug, such as opioid analgesics, CNS depressants, CNS sympathomimetics, tranquilizers, cannabinoids or psychedelics.

B. Experimental design

A between-subjects randomized-block design involving two levels of impulsivity (high and low) and four treatments (caffeine and placebo conditions) with 10 replications was used for all the dependent variables except for memory, where a $2 \times 2 \times 4$ factorial design (two impulsivity levels, two tasks, and four treatments) with 8 replications was used.

C. Procedure

Prior to the day of testing, subjects were advised to have their normal night's sleep and to abstain from caffeinated/nicotinic beverages, alcohol, and drugs as mentioned above, 10 h prior to the experiment.

The experimental session began at 9 a.m. All subjects were weighed and a saliva sample was taken to determine the caffeine level at baseline. After having given written consent for the administration of caffeine, the subject was given the orange-flavored cold drink that contained either caffeine citrate or placebo using a double-blind procedure. Subjects then completed a questionnaire about their previous night's sleep, food consumption, smoking, and alcohol or drug consumption. One hour after the drink, a second saliva sample was taken, which was followed by the performance session. Another saliva sample was taken at the end of the performance session.

D. Materials and tasks

1. Cognitive performance

a. Intelligence measures. *Culture Fair Intelligence Test (CFIT), Scale 3, Form A.*[46] This test is a time-bound measure and comprises four subtests: series, classification, matrices, and topology. There are 50 items in each of the four subtests. The subject is required to give the correct response from the given alternatives. After instructions, 12.5 minutes (as suggested in the manual) were allowed for test administration.

Raven's Standard Progressive Matrices (SPM).[47] This test comprises 60 multiple choice problems, each consisting of a design or "Matrix" from which a part has been removed. The subject is required to complete the design by one of the 6 or 8 (set C onwards) alternatives. The problems are divided into 5 sets (A, B, C, D, E) of 12 problems each. The test was used as a speed test, and the subject was given 20 minutes to work on it.

b. Memory: immediate free recall. A list containing 34 words from the Hindi language was constructed. The words were selected in such a manner that they could be equally potent as alternative sorts into the conceptual or acoustic (rhyming) classification; the subjects could, therefore, classify the words in terms of conceptual or acoustic relationships. The number of categories in which the words were to be classified conceptually or acoustically was six. All 34 words were printed in six horizontal lines on the upper part of the sorting sheet in such a manner that the words to be categorized in a category were not closeted together. The subject was required to sort words according to their conceptual or acoustical relationships and write them on the sorting sheet under the given conceptual or acoustic category names. Immediately after the acquisition task presented in an incidental learning paradigm, subjects were asked to recall words from the sorted list and write as many as they could remember and in any order that occurred to them in three minutes on a sheet provided for this purpose.

2. Psychomotor performance

a. Perceptual judgment. An electronic apparatus with automatic functions (Medicaid Systems, Chandigarh) and comprising an adjustable comparison block with a scale which was adjustable in units of 0.10 cm and a 4.0-cm-wide test block was used for the measurement of perceptual judgment. The blindfolded subject was required to hold lightly the outer edges of the comparison and test blocks with the thumb and index finger of the left and right hand, respectively, and to report when he or she felt the width of the comparison block to be identical to that of the test block. Sixteen determinations were made in four units; in each unit there were four determinations in a counterbalanced order (first, starting with a narrower width of the comparison block than that of the test block and then moving gradually in the ascending direction; second and third, starting with a wider width of the comparison block than that of the test block and then moving gradually in the descending direction; fourth, following the procedure as in the first determination). Thus, each subject provided 8 ascending and 8 descending measures. For each determination, a difference score was derived, as estimated by ascending or descending width minus the width of the test block. There was underestimation in most of the ascending determinations and overestimation in most of the descending determinations. As the direction of the difference was not particularly interesting, the absolute difference score was calculated for each subject by averaging his unsigned ascending and descending difference scores for all 16 determinations. This averaged absolute difference score, provided by each subject, was the unit of analysis.

This task employing haptic presentation was specifically chosen so that caffeine effects on perceptual judgment are not confounded with the attentional factors generally associated with visual and auditory presentations. Gupta and Gupta[41] also state that this task provides a relatively unbiased measure of perceptual judgment.

b. Figure tracking. A star-shaped 2-mm-wide figure tracking pattern, which was mounted on a metal plate and had automatic error and time recording systems, was used to measure hand-steadiness. The star-shaped pattern had six outward-projecting arms. This task involves the ability to make precise arm–hand positioning movements where strength and speed are minimized, the critical feature being the steadiness with which such movements are made. The subject is required to move a stylus at arm's length through a groove without touching the sides. Error was judged to be when either of the sides was touched. Subjects were free to track the groove from a given point in any direction, clockwise or anticlockwise, but were required to return to the starting point. Each subject was given five trials; inter-trial interval was 30 seconds. The total time spent in tracking and the total number of errors committed in five trials were recorded for each subject. The units of analysis were: (1) average tracking time, in seconds, per trial; (2) average number of errors committed per trial.

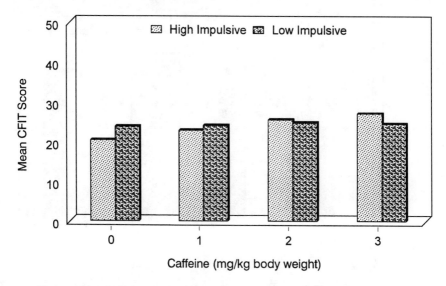

Figure 11.1   Caffeine and performance on Culture Fair Intelligence Test.

III. Results

A. Cognitive performance

1. Intelligence measures

The criterion measure for the CFIT as well as the SPM was the sum of the correct answers for the total test (i.e., composite test score). The data were treated by analysis of variance (ANOVA).

Caffeine facilitated the performance of high impulsives on the CFIT (caffeine × impulsivity; F 3, 72 = 4.56, $p < 0.01$) as well as the SPM (caffeine × impulsivity; F 3, 72 = 3.26, $p < 0.05$) but did not have a statistically significant effect on the performance of low impulsives. These results are shown in Figures 11.1 and 11.2. The dose response trends for high impulsives were essentially linear on both the CFIT and the SPM. The analysis also showed that 78.86% and 68.02% variation in the main effect of treatment and the treatment × impulsivity interaction was due to the effects of caffeine in high impulsives for the CFIT and SPM, respectively; the corresponding figures for low impulsives were 15.23 and 17.61.

2. Memory: immediate free recall

The ANOVA revealed a highly significant second-order interaction among the three variables, i.e., impulsivity, task, and caffeine (F 3, 112 = 6.23, $p < 0.001$). This interaction, as is evident from Figure 11.3, showed that caffeine, compared to placebo, facilitated free recall in high impulsives after acoustic acquisition

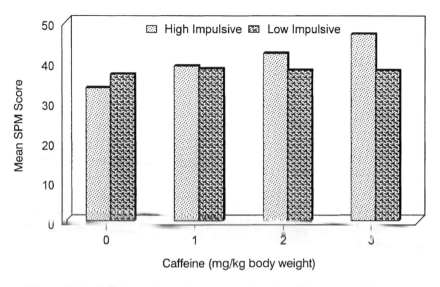

Figure 11.2 Caffeine and performance on Standard Progressive Matrices.

(p < 0.01 in each case) but inhibited it after conceptual acquisition (1 mg/kg body weight, p < 0.05; 2 mg/kg and 3 mg/kg body weight, p < 0.01), and produced no effect on the free recall of low impulsives, irrespective of whether the acquisition was done conceptually or acoustically. The differential effect of caffeine on conceptual and acoustic tasks in high impulsives who become more aroused under the influence of caffeine may be attributed to the hypothesis that higher levels of arousal facilitate utilization of conceptual cues.[48] This inference is corroborated by the observation that in the no-drug condition high impulsives (presumably less aroused persons) recall better after conceptual than acoustic acquisition (p < 0.01), while the low impulsives (presumably more aroused persons) recall better after acoustic than conceptual acquisition (p < 0.01).

B. Psychomotor performance

1. Perceptual judgment
Caffeine diminished errors in perceptual judgment in high impulsives but produced no effect in the performance of low impulsives (caffeine × impulsivity; F 3, 72 = 2.94, p < 0.05). These data are shown in Figure 11.4. It was also observed that in high impulsives 64.25% of the variation in effects on perceptual judgment was due to the effects of caffeine. The corresponding figure for low impulsives was 27.39%.

2. Figure tracking
Caffeine produced no effects on the time spent in tracking the star-shaped figure; neither the main-effect of caffeine nor the caffeine × impulsivity interaction reached an acceptable level of statistical significance. However,

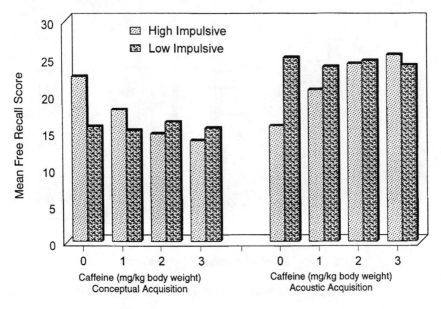

Figure 11.3 Caffeine and immediate free recall.

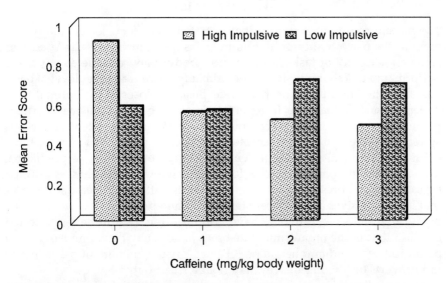

Figure 11.4 Caffeine and perceptual judgment (difference from true value = error in judgment).

caffeine did affect the other criterion variable, i.e., the average number of errors committed per trial. Caffeine diminished errors in high impulsives but produced no effect in low impulsives (caffeine × impulsivity; $F\ 3, 72 = 3.89$, $p < 0.02$). The results are shown in Figure 11.5. Caffeine accounted for more variation in the error scores of high impulsives (69.25%) than low impulsives (22.76%).

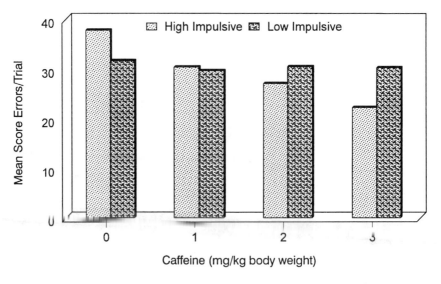

Figure 11.5   Caffeine and figure tracking (errors per trial).

IV. Discussion

The present results consistently demonstrate that caffeine leads to improvement in the performance of high impulsives (presumably less aroused persons) on a variety of tasks but does not produce any reliable effect on the performance of low impulsives, who probably possess a higher level of basal arousal in the morning[32] and have their position closer to the optimal level of performance at that time. Introverts (who may be low impulsives) perhaps possess a stronger tendency for a more extensive use of the central nervous system that attempts to compensate for the adverse effects of supraoptimal arousal.[33] The results of the present study bear out the above prediction, showing that the performance of low impulsives is relatively less affected by caffeine. The present results are also consistent with the earlier findings that low impulsives are more resistant to caffeine-provoked variations in the organismic state.[22-25] Introverts have also been reported to be less affected by variations in the organismic state brought about by the administration of barbiturates[49,50] and amphetamine.[51] Hence, greater stability of performance in introverts or low impulsives may not be ruled out.

Since caffeine has been found to produce differential effects on the performance of high and low impulsives in a variety of tasks, this personality variable needs to be examined in future research associated with the effects of caffeine on human performance. Further research may also concentrate on identifying other personality traits that might affect behavioral responsivity to caffeine.

References

1. Graham, D. M., Caffeine — Its identity, dietary sources, intake and biological effects, *Nutr. Rev.*, 36, 97, 1978.
2. Barone, J. J. and Roberts, H., Human consumption of caffeine, in *Caffeine*, Dews, P.B., Ed., Springer-Verlag, Berlin, 1984, 59.
3. Gilbert, R. M., Caffeine as a drug of abuse, in *Research Advances in Alcohol and Drug Problems*, Vol. 3, Gibbins, R.J., Israel, Y., Kalant, H., Propham, R.E., Schmidt, W. and Smart, R. G., Eds., John Wiley & Sons, New York, 1976, 49.
4. Dews, P.B., Caffeine, *Annu. Rev. Nutr.*, 2, 323, 1982.
5. Rall, T. W., Central nervous system stimulants: The xanthines, in *The Pharmacological Basis of Therapeutics*, 6th ed., Goodman, L.S., and Gilman, A., Eds., Macmillan, New York, 1980, 592.
6. Regina, E. G., Smith, G. M., Keiper, C. G., and McKelvey, R. K., Effects of caffeine on alertness in simulated driving, *J. Appl. Psychol.*, 59, 483, 1974.
7. Dews, P. B., Behavioral effects of caffeine, in *Caffeine*, Dews, P.B., Ed., Springer-Verlag, Berlin, 1984, 86.
8. Baker, W. J. and Theologus, G. C., Effects of caffeine on visual monitoring, *J. Appl. Psychol.*, 56, 422, 1972.
9. Lieberman, H. R., Wurtman, R. J., Emde, G. G., and Coviella, I. L., The effects of caffeine and aspirin on mood and performance, *J. Clin. Psychopharmacol.*, 7, 315, 1987.
10. Lieberman, H. R., Wurtman, R. J., Emde, G. G., and Coviella, I. L., The effects of low doses of caffeine on human performance and mood, *Psychopharmacology*, 92, 308, 1987.
11. Ivy, J. L., Costill, D. L., Fink, W. J., and Lower, R. W., Influence of caffeine and carbohydrate feeding on endurance performance, *Med. Sci. Sports*, 11, 6, 1979.
12. Pons, L., Trenque, T., Bielecki, M., Moulin, M., and Poitier, J. C., Attentional effects of caffeine in man: Comparison with drugs acting upon performance, *Psychiatr. Res.*, 23, 329, 1988.
13. Zwyghuien-Doorenbos, A., Roehrs, T. A., Lipschutz, L., Timms, V., and Roth, T., Effects of caffeine on alertness, *Psychopharmacology*, 100, 36, 1990.
14. Frewer, L. J. and Lader, M., The effects of caffeine on two computerized tests of attention and vigilance, *Hum. Psychopharmacol.*, 6, 119, 1991.
15. Bättig, K. and Buzzi, R., Effect of coffee on the speed of subject-paced information processing, *Neuropsychobiology*, 16, 126, 1986.
16. Hasenfratz, M., Jaquet, F., Aeschbach, D., and Bättig, K., Interactions of smoking and lunch with the effects of caffeine on cardiovascular functions and information processing, *Hum. Psychopharmacol.*, 6, 277, 1991.
17. Weiss, B. and Laties, V. G., Enhancement of performance by caffeine and the amphetamines, *Pharmacol. Rev.*, 14, 1, 1962.
18. Nash, H., Psychological effects and alcohol antagonizing properties of caffeine, *Q. J. Stud. Alcohol*, 27, 727, 1966.
19. Ghoneim, M. M., Hinrichs, J. V., Chiang, C. K., and Loke, W. H., Pharmacokinetic and pharmacodynamic interactions between caffeine and diazepam, *J. Clin Psychopharmacol.*, 6, 75, 1986.
20. Estler, C. J., Caffeine, in *Psychotropic Agents*, Part III, *Alcohol and Psychotomimetics, Psychotropic Effects of Central Acting Drugs*, Hoffmeister, F. and Stille, G., Eds., Springer-Verlag, Berlin, 1982, 369.

21. Kerr. J.S., Sherwood, N., and Hindmarch, I., Separate and combined effects of the social drugs on psychomotor performance, *Psychopharmacology*, 104, 113, 1991.

22. Gupta, U., Effects of impulsivity and caffeine on human cognitive performance, *Pharmacopsychoecologia*, 1, 33, 1988.

23. Gupta, U., Differential effects of caffeine on free recall after semantic and rhyming tasks in high and low impulsives, *Psychopharmacology*, 105, 137, 1991.

24. Gupta, U., Effects of caffeine on recognition, *Pharmacol. Biochem. Behav.*, 44, 393, 1993.

25. Gupta, U. and Gupta, B. S., Caffeine differentially affects kinesthetic aftereffect in high and low impulsives, *Psychopharmacology*, 102, 102, 1990.

26. Gupta, U., Dubey, G. P., and Gupta, B.S., Effects of caffeine on perceptual judgment, *Neuropsychobiology*, 30, 185, 1994.

27. Smith, A. P., Rusted, J. M., Savory, M., Eaton-Williams, P., and Hall, S.R., The effects of caffeine, impulsivity and time of day on performance, mood and cardiovascular function, *J. Psychopharmacol.*, 5, 120, 1991.

28. Erikson, G. C., Hager, L. B., Houseworth, C., Dungan, J., Petros, T., and Beckwith, B. E., The effects of caffeine on memory for word lists, *Physiol. Behav.*, 35, 47, 1985.

29. Arnold, M. E., Petros, T. V., Beckwith, B. E., Coons, G., and Gorman, N., The effects of caffeine, impulsivity, and sex on memory for word lists, *Physiol. Behav.*, 41, 25, 1987.

30. Lieberman, H. R., Caffeine, in *Handbook of Human Performance*, Vol. 2, Smith, A.P. and Jones, D.M., Eds., Academic Press, London, 1992, 49.

31. Gilliland, K., The interactive effect of introversion-extraversion with caffeine induced arousal on verbal performance, *J. Res. Person.*, 14, 482, 1980.

32. Revelle, W., Humphreys, M. S., Simon, L., and Gilliland, K., The interactive effects of personality, time of day, and caffeine: A test of arousal model, *J. Exp. Psychol. [Gen]*, 109, 1, 1980.

33. Eysenck, M. W., *Attention and Arousal: Cognition and Performance*, Springer-Verlag, Berlin, 1982.

34. Gupta, U., Personality, caffeine and human cognitive performance, *Pharmacopsychoecologia*, 1, 79, 1988.

35. Eysenck, H. J., *The Biological Basis of Personality*, Charles C Thomas, Springfield, Illinois, 1967.

36. Eysenck, H. J., Intelligence assessment: A theoretical and experimental approach, *Br. J. Educ. Psychol.*, 37, 81, 1967.

37. Mandler, G., Recognizing: The judgement of previous occurrence, *Psychol. Rev.*, 87, 252, 1980.

38. Mandler, G., The recognition of previous encounters, *Am. Sci.*, 69, 211, 1981.

39. Tversky, B., Encoding processes in recognition and recall, *Cogn. Psychol.*, 5, 275, 1973.

40. Smith, A. P. and Miles, C., Acute effects of meals, noise and nightwork, *Br. J. Psychol.*, 77, 377, 1986.

41. Gupta, U. and Gupta, B. S., Effects of caffeine on perceptual judgement: A dose–response study, *Pharmacopsychoecologia*, 7, 215, 1994.

42. Revelle, W., Personality and motivation: Sources of inefficiency in cognitive performance, *J. Res. Person.*, 21, 436, 1987.

43. Thayer, R. E., Toward a psychological theory of multidimensional activation (arousal), *Motiv. Emot.*, 2, 1, 1978.

44. Gupta, B. S. and Poddar, M., Personality traits among Hindi knowing Indian students, *J. Soc. Psychol.*, 107, 279, 1970.
45. Eysenck, H. J. and Eysenck, S. B. G., *Manual of the Eysenck Personality Inventory,* University of London Press, London, 1964.
46. Cattell, R. B., *Manual of the IPAT Culture Fair Intelligence Scales 1, 2 and 3* (3rd ed.), Institute of Personality and Ability Testing, Champaign, Illinois, 1965.
47. Raven, J. C., *Standard Progressive Matrices, Sets A, B, C, D, and E,* H.K. Lewis & Co., Ltd., London, 1960.
48. Schwartz, S., Individual differences in cognition: Some relationships between personality and memory, *J. Res. Person.*, 9, 217, 1975.
49. Gupta, U., Phenobarbitone and kinaesthetic after-effect, *Curr. Psychol. Res.*, 2, 171, 1982.
50. Gupta, U., Phenobarbitone and the relationship between extraversion and reinforcement in verbal operant conditioning, *Br. J. Psychol.*, 75, 499, 1984.
51. Gupta, B. S. and Gupta, U., Dextroamphetamine and individual susceptibility to reinforcement in verbal operant conditioning, *Br. J. Psychol.*, 75, 201, 1984.

chapter twelve

Behavioral effects of caffeine: the role of drug-related expectancies

Mark T. Fillmore

Contents

I. Overview 207
II. Expectancies about drug effects 208
III. Expectancy and the behavioral effects of caffeine 210
   A. Research findings 210
IV. Implications 214
   A. Individual differences 214
   B. Placebo control groups 216
   C. Caffeine consumption 216
V. Conclusion 217
Acknowledgment 217
References 217

I. Overview

This chapter reviews the concept of drug-related expectancy and describes studies that have examined the relationship between caffeine-related expectancies and behavioral responses to caffeine and to placebo. The research shows that expectancies about caffeine effects can influence behavioral responses to the drug. The findings also demonstrate the importance of using appropriate control groups for investigating the behavioral effect of caffeine and other psychoactive drugs. The chapter examines implications of this research for explaining individual differences in responses to caffeine, as well as for understanding factors that may contribute to the development and maintenance of caffeine consumption.

Caffeine is a widely used psychoactive substance. It occurs naturally in several species of plants and is found in many beverages, foods, and pharmaceutical products. Ninety percent of caffeine is consumed in the form of coffee and tea.[1] Estimates of caffeine consumption based on world production of coffee and tea suggest an average daily consumption of 50 mg per person. In Canada and the United States, daily estimates of caffeine intake per person are 238 mg and 211 mg, respectively.[2]

During recent decades, the effects of caffeine on humans have received much research attention. Caffeine acts to stimulate the central nervous system and produces a wide range of physiological, affective, and behavioral effects.[3] Behavioral effects of caffeine in humans have been examined using several measures of cognitive processing and psychomotor functioning, including choice reaction time,[4] simple reaction time,[5] simulated automobile driving,[6] arithmetic,[7] compensatory tracking,[8] tapping rate,[3,10] body sway,[11] perceptual restructuring,[12] and divided attention.[13]

Weiss and Laties[14] provided the first extensive review of studies that examined the behavioral effects of caffeine. They concluded that caffeine had equivocal effects on psychomotor and cognitive function. More recent reviews have also failed to find consistent effects of caffeine on behavior. Doses between 200 mg and 300 mg have been reported to produce no detectable behavioral effects,[3] while other reviews report inconsistent behavioral effects in this dose range.[15,16]

It is difficult to account for the equivocal findings of caffeine effects on behavior. Attempts to explain these inconsistencies have often focused on differences in the methodologies and measures used to assess behavioral effects of caffeine. The types of behavioral tasks and testing techniques are diverse and vary in their sensitivity to drug effects.[15,17] In addition, differences in caffeine doses and administration procedures may contribute to the inconsistent findings.

There has been little research attention concerning the influence of subject variables on caffeine effects. Some evidence suggests that personality characteristics, such as introversion/extroversion[18] and impulsivity,[19] can interact with caffeine effects to determine an individual's response. One characteristic that has recently received attention concerns the expectations or beliefs that individuals have about the effect of caffeine on their behavior. This chapter reviews the concept of drug-related expectancies and describes research showing that these expectancies can account for variability in behavioral responses to caffeine. The implications of these findings are also discussed with respect to understanding individual differences in response to caffeine and factors that affect caffeine consumption.

II. Expectancies about drug effects

Learning theories propose that expectancies represent information acquired about a reliable association between events.[20,21] In relation to drug effects, an expectancy about a specific behavioral effect of a drug may be acquired in a situation where the administration of a drug is reliably associated with a particular type of drug effect on a given activity. In such cases, an individual comes to expect this specific type of drug effect. For example, an individual may learn to expect that alcohol will make him or her more sociable and outgoing in a party situation. Expectancies concerning drug effects may also be acquired from observing others who consume the drug, or from reports of the drug's effect on a specific activity (e.g., alcohol impairs driving). Individuals often engage in a variety of activities in many different situations following the administration of a psychoactive drug. The behavioral effect of the drug may differ depending on the activity and the situation. In addition, individuals themselves may differ in their responses to a drug in a particular situation. Given these circumstances, expectancies about a drug effect may differ markedly among individuals and in different situations.

Much research has focused on the behavioral effect of alcohol-related expectancies. For some time it has been known that simply expecting to receive alcohol can affect social and affective responses displayed under alcohol.[22,23] In addition, expectancies concerning the consequences of displaying a particular response under alcohol have been shown to influence behavior under the drug.[24] More recent research has investigated a different expectation about alcohol: *the type of effect* that alcohol is expected to exert on behavior.[25,26]

Research has examined drinkers' expectations about the effects of alcohol on several different types of behavior.[27,28] Studies have measured expectancies about alcohol effects in a wide range of subjects, including alcoholics and adolescence, but most findings are based on college-age social drinkers. Evidence has shown that these individuals report expectations about alcohol effects on many different types of behaviors, including sociability, cognitive and motor functioning, sexual arousal, and tension reduction. Individual differences among drinkers in their expectancies have also been related to their drinking habits.[27] Social drinkers who report greater alcohol consumption tend to expect more positive effects from alcohol, such as tension reduction, enhanced sociability, and possible improvement in cognitive/motor functioning.

A number of studies have demonstrated that expectancies about alcohol effects also contribute to individual differences in the behavioral responses displayed under alcohol.[29-31] This research examined subjects' expected effect of alcohol on their psychomotor performance or cognitive performance. The results showed that subjects differed in the degree of alcohol-induced impairment they expected, and these individual differences predicted their performance of laboratory tasks under alcohol and under a placebo when alcohol was expected. Subjects who expected the most impairment from alcohol tended to perform most poorly when alcohol or a placebo was received. These findings led to the suggestion that unexplained individual differences in behavioral responses to alcohol and to placebo may be due, in part, to individual differences in the expected effect of alcohol.

III. Expectancy and the behavioral effects of caffeine

Although little research has been conducted on expectancies about drugs other than alcohol,[32] some studies have examined expectancies about the behavioral effect of caffeine. Caffeine, like alcohol, is a legal, commonly used, psychoactive drug. Most individuals likely have had experience using the drug in one form or another (e.g., coffee or tea) and in several different situations (e.g., social or work settings). Thus, like alcohol drinkers, caffeine consumers use the drug in many situations, and thus could have opportunities to acquire a variety of expectations about its effects.

A series of experiments have investigated the relationship of caffeine-related expectancies to behavioral responses under caffeine and under placebo when caffeine was expected.[30,33,34] These studies adopted the same general experimental procedure. The subjects were male university students whose ages ranged from 18 to 34 years. Their psychomotor behavior was measured on a pursuit rotor tracking task. The pursuit rotor task involves tracking a rotating target with a hand-held device. This task has been used extensively in investigations of drug effects on psychomotor functioning.[35] The next section reviews the results of these studies.

A. Research findings

Fillmore and Vogel-Sprott[30] examined individual differences in subjects' expectations about the effect of caffeine on psychomotor performance. The purpose of the experiment was to show that individuals differ in their expectancies about caffeine effects on psychomotor performance, and such individual differences predict responses to caffeine and to a placebo when caffeine is expected. It was hypothesized that when subjects expect caffeine, those who expect caffeine to produce the most impairment should perform most poorly, and those who expect the most improvement should display superior performance.

The subjects in the experiment varied in their typical daily caffeine intake. The mean daily amount of caffeine consumed was 2.24 mg/kg. For a 70-kg person, this amount would approximately equal the caffeine contained in one to two cups of brewed coffee. Prior to the experiment, all subjects abstained from caffeine for 8 hours. The subjects were tested individually during week days between 9 a.m. and 5 p.m. All subjects received 20 minutes of pretreatment practice to familiarize them with the pursuit rotor task. Following this practice, subjects used a rating scale to indicate the type of effect they expected caffeine to have on their psychomotor performance. Possible ratings ranged from −30 ("Extremely Impair") to +30 ("Extremely Enhance"), with zero indicating that "No Effect" was expected. After rating expectancies, one group of subjects (group C) received 2.93 mg/kg of caffeine in the form of coffee (approximately equal to the caffeine content of 1.5 cups of brewed coffee, for a 70-kg person). Another group (group P) received a placebo in the form of decaffeinated coffee. A

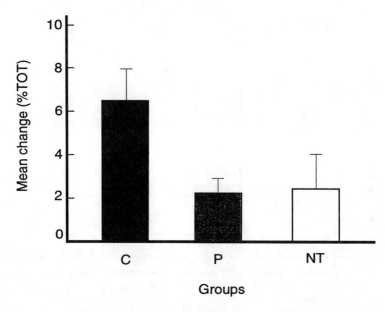

Figure 12.1 Mean change in percentage of time on target (%TOT) from pretreatment baseline for three groups (group n = 8). A positive change indicates improvement relative to pretreatment baseline %TOT. Vertical bars indicate standard error of the mean. (From Fillmore, Mark T., Investigating the behavioral effects of caffeine, *Pharmacopsychoecologia*, 1994. With permission.)

third group (group NT) served as a no-treatment control group and received no beverage. Because subjects in group NT neither expected nor received caffeine, their expectancies about the effect of caffeine were not predicted to relate to their post-treatment performance. After receiving the various treatments, subjects' post-treatment performance was measured on the task. Post-treatment performance required 20 minutes, and the total duration of the session was 1 hour and 15 minutes.

The results of this study are illustrated in Figures 12.1 and 12.2. Figure 12.1 shows the mean change in performance from pretreatment to post-treatment for each group. The ordinate plots the mean change in percentage of time on target (%TOT) for each group in response to the treatment. Higher mean change scores indicate greater time on target, and hence greater tracking accuracy. The figure shows that group C displayed superior performance compared to groups P and NT. Thus, when subjects received caffeine, their tracking performance became more accurate, as shown by the comparatively large increase in the %TOT. The figure also shows no significant difference between groups P and NT. Thus, no evidence of a placebo response was observed by comparing the mean change in performance of these two groups.

Individual differences in subjects' expected effects of caffeine were also examined in relation to their post-treatment changes in performance.

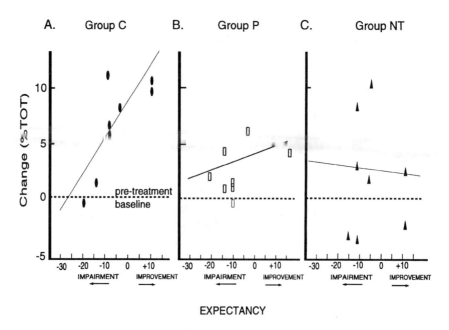

Figure 12.2 Relationship between changes in subjects' performance and their expectancy ratings in each group. The least squares regression lines are shown for each group separately. Panel A shows the regression line for group C. Panel B shows the regression line for group P. Panel C shows the regression line for the NT group. (From Fillmore, Mark T., Investigating the behavioral effects of caffeine, *Pharmacopsychoecologia*, 1994. With permission.)

Figure 12.2 plots the change in performance of the individual subjects within each group in relation to their expected effect of caffeine. Expectancy ratings, presented on the abscissa, show that subjects varied in their expectancies concerning the effect of caffeine. Some subjects expected various degrees of impairment, and others expected caffeine to improve performance. Panel A of the figure shows the data for subjects who received caffeine (group C). Change in performance, plotted on the ordinate, shows that subjects' responses to caffeine varied considerably. A few subjects displayed little or no improvement, and others displayed various degrees of improved performance under caffeine. The figure also shows that these individual differences in response to caffeine are related to subjects' expectancies about the effect of caffeine. Those who expected the most improvement displayed the best performance under the drug. Further support for this relationship was obtained by a regression of subjects' change in performance on their expectancy ratings. The analysis revealed a statistically significant positive slope relating performance to expectancy.

The same general pattern of results were obtained when subjects expected caffeine but received a placebo (group P). Panel B shows that subjects' performance under placebo was also characterized by individual differences. Moreover, these differences related to the type of effect subjects expected. Those who expected caffeine to produce the most improvement displayed the best performance under placebo. This evidence demonstrated that when subjects believed that caffeine had been administered, their expectancies about its effect predicted their performance in the absence of a drug. A regression analysis revealed that the positive slope relating performance to expectancy in group P did not differ significantly from the positive slope displayed by group C. Thus, expectancies bore similar relationships to performance under caffeine and placebo. In contrast, no relationship between expectancy and change in performance was displayed by subjects in the NT group (Panel C). Thus, when subjects did not believe that caffeine had been administered, their expectations about its effect did not predict their performance.

Demonstrating a relationship between expectancies and responses to caffeine is a novel and important finding. The results extend the previous findings with alcohol by showing that drug-related expectancies also predict individual differences in response to caffeine. The findings also provide a novel method to detect a placebo response. Evidence of a placebo effect is often assessed by comparing the mean change in performance of a group that receives a placebo to the mean change in performance of a group that receives no treatment. The present study shows no difference between the mean change in performance of the placebo group and the no-treatment control group. Thus, based on comparing group means, no placebo effect can be detected. In contrast, when the changes in performance of individual subjects are examined in relation to their expectancies, evidence of a placebo effect emerges. The positive relationship between performance and expectancies in group P shows that when subjects expected caffeine but received a placebo, their performance changed in accordance with the type of caffeine effect they expected. In contrast, no positive relationship was evident in group NT when no placebo was administered. Thus the placebo effect can be detected by comparing the expectancy–performance relationship under placebo vs. no-treatment conditions.

The findings of the study suggest that if a group of individuals all expect the same type of caffeine effect, they should display a placebo response that is observable as a mean performance change in that group. The rationale is that when subjects expect to receive caffeine, and all expect the same particular effect, they should all respond similarly in accordance with the expected effect. To explore this possibility other studies have manipulated subjects' expectations about the type of effect that caffeine would have on performance.[33,34] These studies used only caffeine placebos to obtain a "pure" measure of the expected drug effect on behavior when caffeine was expected but not administered.

In one study, four groups of subjects all practiced the pursuit rotor task.[33] Three groups of subjects were then led to expect caffeine but received a placebo instead (decaffeinated coffee). Prior to the administration of the placebo, two of these groups were given information about the effect of caffeine that aimed to manipulate their expectancies about how the drug would affect their performance. One group (P+) was led to expect improved performance. In casual conversation, the experimenter told these subjects that "research has found that caffeine improves fine motor coordination ... the purpose of the study is to determine how caffeine produces this improvement." Subjects in the other group (P–) were given information that led them to expect impairment: "Research has found that caffeine impairs fine motor coordination ... the purpose of the study is to determine how caffeine produces this impairment." No information was given to the third group (P), which served as the "traditional" placebo group. A fourth group (NT) served as a no-treatment control and received no information about caffeine and no placebo and therefore did not expect caffeine or any effect on performance. Because group P received no information that could lead all subjects to expect the same effect, the type of placebo response displayed would differ among the subjects. Thus the mean performance of this group was not predicted to differ from the no-treatment control (group NT). Subjects in groups P+, P–, and P received decaffeinated coffee, and after a brief period, all groups were retested on the pursuit rotor task.

Figure 12.3 illustrates the effect of manipulating the expectancies by showing the mean changes in performance of each group. Compared to the control group (NT), the P+ group, led to expect improvement, displayed superior performance following the placebo administration. The P– group, led to expect impairment, displayed poorer performance compared to group NT. Group P did not differ from group NT and thus, as expected, showed no evidence of a placebo response. This is consistent with the findings of the previous study that found no evidence of a placebo response based on a comparison of group means.[30]

The experimental manipulation of expectancies shows that individuals' expectations about caffeine effects can be altered to affect behavior when the drug is expected. When a group of individuals all expect the same particular type of caffeine effect, they display a placebo response that is consistent with that expectancy. Because each subject in a group expects the same type of effect, their responses are similar, and this can be detected in the mean change of the group. This evidence also suggests that expectancies could be manipulated to affect responses to an actual dose of caffeine. More research is needed to examine this possibility.

IV. Implications

A. Individual differences

Individual differences in response to caffeine are commonly reported.[16,36] Researchers often point to large individual differences in the behavioral responses under caffeine to explain inconsistencies among experimental findings. These individual differences are often attributed to differences in subjects' regular caffeine use. It has been thought that consumers of high

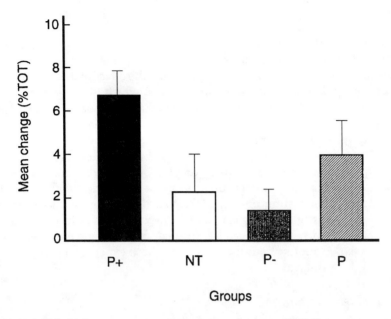

Figure 12.3   Mean change in percentage of time on target (%TOT) from pretreatment baseline for four groups (group n = 14). A positive change indicates improvement relative to pretreatment baseline %TOT. Vertical bars indicate standard error of the mean. (From Fillmore, Mark T., Investigating the behavioral effects of caffeine, *Pharmacopsychoecologia*, 1994. With permission.)

daily amounts of caffeine might have acquired tolerance to the behavioral effects of caffeine and are therefore less sensitive to its effect on behavior. However, little is known about the relationship between subjects' typical caffeine intake and their behavioral sensitivity to an acute dose of caffeine. There is some evidence that daily administration of 900 mg of caffeine can produce tolerance to its subjective effects in humans.[37] However, no evidence to date has demonstrated that daily consumption levels of caffeine relate to acute behavioral responses in humans. Fillmore and Vogel-Sprott[30] examined subjects' reports of typical daily consumption in their study of individual differences in response to caffeine. They concluded subjects' caffeine use could not account for individual differences, but subjects' expectancies about the effect of caffeine did predict individual differences in response to the drug. The relationship between expectancies and behavioral responses to caffeine is important because it suggests an explanation for the lack of consistency and null findings concerning caffeine effects on behavior. If the observed caffeine response consists of a pharmacological effect plus an expectancy effect, variability in expectancies could increase individual differences and mask any overall average response in a group. However, if subjects' expectations are measured, they could be isolated as a distinct source of response variability, and a more precise measure of the residual pharmacological effect could be obtained.

B. Placebo control groups

The evidence that expectancies affect responses to placebos has implications for the design of studies that assess the behavioral effect of caffeine or other psychoactive drugs. Most studies that examine the effect of a drug on behavior include a placebo group as a control against which to compare the effect of receiving caffeine. Given that subjects' expectations about the type of drug effect influence their behavior, two-group (caffeine vs placebo) comparisons provide only limited information because they lack the necessary control measure of performance when no drug effect is expected. A no-treatment control group that receives no drug or placebo is needed to provide this additional information. This group's behavior provides a control for task practice and other testing effects. Moreover, it is uncontaminated by any expectation of a drug or a drug effect. Placebo conditions cannot truly isolate testing and practice effects because subjects may enter experiments with their own expectations about the drug, and these expectancies could influence their performance.

C. Caffeine consumption

Much research attention has focused on the long-term effects and health risks associated with excessive caffeine intake.[38] Recently, there has been interest in whether caffeine should be considered to have abuse potential.[39,40] There are several reports that tobacco users consume higher levels of caffeine compared to the general population.[41,42] Also, psychiatric inpatients consume very high amounts of caffeine.[16,40]

However, despite concern over excessive caffeine consumption, surprisingly little is known about what factors might contribute to excessive use in the general population. No research has tested the degree to which expectancies about caffeine predict caffeine consumption. However, the research presented in this chapter shows that individuals do differ in their expectancies about caffeine effects on psychomotor performance. It is possible that individuals hold expectancies about caffeine effects on other types of behaviors as well (e.g., cognitive ability, alertness, affect, fatigue, endurance, etc.). Caffeine is well known as a stimulant, and·thus could evoke expectations of enhanced cognitive ability, increased work output, and general elevation in mood. These positive expectations or beliefs about caffeine effects could act to initiate and maintain heavy caffeine consumption. Some support for this notion comes from a study that surveyed reasons why people consume caffeine.[43] The study found that heavy users tended to report using caffeine to achieve particular behavioral effects or affective states (e.g., to pep up, to concentrate), and to alleviate adverse states (e.g., depression and fatigue).

The possible relationship between caffeine use and expectancies about the behavioral effect of caffeine would be consistent with research concerning alcohol use and expectancies.[27] That research shows that drinkers' expectancies about alcohol effects relate to their degree of alcohol use.

V. Conclusion

Research investigating expectancies about caffeine is a relatively new undertaking. This chapter presents caffeine research that draws on expectancy concepts and experimental techniques from alcohol research. The research presented shows that expectancies about caffeine effects can predict behavioral responses to the drug. These findings have important implications for understanding individual differences in the behavioral effect of caffeine. The findings also address the importance of using appropriate experimental designs and control groups for investigating the behavioral effect of caffeine and other psychoactive drugs.

Acknowledgment

This work was supported in part by a grant from the Alcoholic Beverage Medical Research Foundation and by fellowships from the Natural Science and Engineering Research Council of Canada.

References

1. Barone, J. J. and Roberts, H., Human consumption of caffeine, in *Caffeine: Perspectives from Recent Research*, Dews, P. B., Ed., Springer-Verlag, Berlin, 1984, 59.
2. Gilbert, R. M., Caffeine consumption, in *The Methylxanthine Beverages and Foods: Chemistry, Consumption, and Health Effects*, Spiller, G. A., Ed., Alan R. Liss Inc., New York, 1984, 185.
3. Gilbert, R. M., Caffeine as a drug of abuse, in *Research Advances in Alcohol and Drug Problems*, Vol. 3, Gibbons, R. J., Israel, Y., Kalant, H., Popham, R. E., Schmidt, W., and Smart, R. G., Eds., John Wiley & Sons, New York, 1976, 49.
4. Lieberman, H. R., Wurtman, R. J., Emde, G. G., and Coviella, I., The effects of caffeine and aspirin on mood and performance, *J. Clin. Psychopharmacol.*, 7, 315, 1987.
5. Smith, B. D., Rafferty, J., Lindgren, K., Smith, D. A., and Nespor, A., Effects of habitual caffeine use and acute ingestion: Testing a biobehavioral model, *Physiol. Behav.*, 51, 131, 1991.
6. Regina, E. G., Smith, G. M., Keiper, C. G., and McKelvey, R. K., Effects of caffeine on alertness in simulated automobile driving, *J. Appl. Psychol.*, 59, 483, 1974.
7. Franks, H. M., Hagedorn, H., Hensley, V. R., Hensley, W. J., and Starmer, G. A., The effect of caffeine on human performance, alone and in combination with ethanol, *Psychopharmacology*, 45, 177, 1975.
8. Kerr, J. S., Sherwood, N., and Hindmarch, I., Separate and combined effects of the social drugs on psychomotor performance, *Psychopharmacology*, 104, 113, 1991.
9. Flory, C. D. and Gilbert, J., The effects of benzedrine sulfate and caffeine citrate on the efficiency of college students, *J. Appl. Psychol.*, 27, 121, 1943.
10. Fagan, D., Swift, C., and Tiplady, B., Effects of caffeine on vigilance and other performance tests in normal subjects, *J. Psychopharmacol.*, 2, 19, 1988.

11. Swift, C. G. and Tiplady, B., The effects of age on the response to caffeine, *Psychopharmacology*, 94, 29, 1988.
12. Broverman, D. M. and Casagrande, E., Effect of caffeine on performances of a perceptual-restructuring task at different stages of practice, *Psychopharmacology*, 78, 252, 1982.
13. Zwyghuizen-Doorenbos, A., Roehrs, T. A., Lipschutz, L., Timms V., and Roth, T., Effects of caffeine on alertness, *Psychopharmacology*, 100, 36, 1990.
14. Weiss, B. and Luties, V. C., Enhancement of human performance by caffeine and the amphetamines, *Pharmacol. Rev.*, 14, 1, 1962.
15. Lader, M. H. and Bruce, M. S., The human pharmacology of the methylxanthines, in *Human Pharmacology: Measures and Methods*, Vol. 2, Hindmarch, I. and Stonier, P. D., Eds., John Wiley & Sons Ltd., Chichester, 1989, 179.
16. Nehlig, A., Daval, J., and Debry, G., Caffeine and the central nervous system: mechanisms of action, biochemical, metabolic and psychostimulant effects, *Brain Res. Rev.*, 17, 139, 1992.
17. Hindmarch, I., Kerr, J. S., and Sherwood, N., Psychopharmacological aspects of psychoactive substances, in *Addiction Controversies*, Warburton, D. M., Ed., Harwood Academic Publishers, Philadelphia, 1990, 36.
18. Billiland, K., Interactive effect of introversion-extroversion and alertness induced by caffeine on verbal performance, *J. Res. Person.*, 14, 482, 1980.
19. Anderson, K. J. and Revelle, W., The interactive effects of caffeine, impulsivity and task demands on a visual search task, *Person. Individ. Diff.*, 4, 127, 1983.
20. Bolles, R. C., Reinforcement, expectancy and learning, *Psychol. Rev.*, 79, 394, 1972.
21. Bolles, R. C., *Learning Theory*, 2nd ed., Holt, Rinehart, and Winston, New York, 1979.
22. Marlatt, G. A. and Rohsenow, D., Cognitive processes in alcohol use: Expectancy and the balanced placebo design, in *Advances in Substance Abuse*, Mello, N. K., Ed., J.A.I. Press Inc., Greenwich, Connecticut, 1980, 159.
23. Rohsenow, D. and Marlatt, G. A., The balanced placebo design: Methodological considerations, *Addict. Behav.*, 6, 107, 1981.
24. Vogel-Sprott, M., *Alcohol Tolerance and Social Drinking: Learning the Consequences*, Guilford Press, New York, 1992.
25. Vogel-Sprott, M. and Fillmore, M. T., Learning theory and research, in *Psychological Theories of Drinking and Alcoholism*, 2nd ed., Leonard, K. E. and Blane, H. T., Eds., Guilford Press, New York, in press.
26. Vogel-Sprott, M. and Fillmore, M. T., Expectancy and behavioral effects of socially-used drugs, in *Expectancy, Experience and Behavior*, Kirsch, I., Ed., American Psychological Association, Washington, D.C., in press.
27. Goldman, M., Brown, S., and Christiansen, B., Expectancy theory: thinking about drinking, in *Psychological Theories of Drinking and Alcoholism*, Blane H. T. and Leonard, K., Eds., Guilford Press, New York, 1987, 181.
28. Leigh, B. C., In search of the seven dwarves: Issues of measurement and meaning in alcohol expectancy research, *Psychol. Bull.*, 105, 361, 1989.
29. Fillmore, M. T. and Vogel-Sprott, M., Expectancies about alcohol-induced motor impairment predict individual differences in responses to alcohol and placebo, *J. Stud. Alcoh.*, 56, 90, 1995.
30. Fillmore, M. T. and Vogel-Sprott, M., Psychomotor performance under alcohol and under caffeine: Expectancy and pharmacological effects, *Exp. Clin. Psychopharmacol.*, 2, 319, 1994.

31. Fillmore, M. T., Carscadden, J., and Vogel-Sprott, M., Alcohol, cognitive impairment, and expectancies, *J. Stud. Alcoh.*, 59, 174, 1998.

32. Brown, S. A., Drug effect expectancies and addictive behavior change, *Exp. Clin. Psychopharmacol.*, 1, 55, 1993.

33. Fillmore, M. and Vogel-Sprott, M., Expected effect of caffeine on motor performance predicts the type of response to placebo, *Psychopharmacology*, 106, 209, 1992.

34. Fillmore, M. T., Mulvihill, L. E., and Vogel-Sprott, M., The expected drug and its expected effect interact to determine placebo responses to alcohol and caffeine, *Psychopharmacology*, 115, 383, 1994.

35. Hindmarch, I., Psychomotor function and psychoactive drugs, *Br. J. Clin. Pharmacol.*, 10, 189, 1980.

36. Goldstein, A., Kaizer, S., and Warren, R., Psychotropic effects of caffeine in man. II. Alertness, psychomotor coordination, and mood, *J. Pharmacol. Exp. Therap.*, 150, 146, 1965.

37. Evans, S. M. and Griffiths, R. R., Caffeine tolerance and choice in humans, *Psychopharmacology*, 108, 51, 1992.

38. Ernster, V. L., Epidemiologic studies of caffeine and human health, in *The Methylxanthine Beverages and Foods: Chemistry, Consumption, and Health Effects*, Spiller, G. A., Ed., Alan R. Liss, Inc., New York, 1984, 377.

39. Griffiths, R. R. and Woodson, P. P., Caffeine physical dependence: A review of human and laboratory studies, *Psychopharmacology*, 94, 437, 1988.

40. Hughes, J. R., Oliveto, A. H., Helzer, J. E., Higgens, S. T., and Bickel, W. K., Should caffeine abuse, dependence, or withdrawal be added to DSM-IV and ICD-10?, *Am. J. Psychiat.*, 149, 33, 1992.

41. Matarazzo, J. D. and Saslow, G., Psychological and related characteristics of smokers and nonsmokers, *Psychol. Bull.*, 57, 493, 1960.

42. Parsons, W. D. and Neims, A. H., Effect of smoking on caffeine clearance, *Clin. Pharmacol. Therap.*, 24, 40, 1978.

43. Graham, K., Reasons for consumption and heavy caffeine use: Generalization of a model based on alcohol research, *Addict. Behav.*, 13, 209, 1988.

chapter thirteen

Caffeine and cognitive performance: effects on mood or mental processing?

Jennifer Rusted

Contents

I. Introduction 221
II. The effects of caffeine ingestion 223
   A. Caffeine and mood 223
   B. Caffeine and cognitive performance 223
III. Differentiating mood and cognitive performance effects 224
   A. An experimental study 224
   B. Mood and cognitive performance revisited 226
IV. Concluding comments 227
References 227

I. Introduction

The effects of caffeine on human performance have been the focus of a great deal of research, but establishing the cognitive consequences of caffeine consumption has not been easy. A retrospective effort to extract some general pattern of effects from the literature is no easier, since it quickly becomes apparent that the methodological diversity across the studies defies even the most determined exploration of the data. These differences can basically be classified into three groups.

First, differences exist between studies on such basic design factors as dosage of caffeine employed, whether habitual patterns of intake among participants and baseline levels of caffeine are considered, and whether other conditions (such as diet) or habits (such as smoking) which affect metabolism of the compound are controlled in the pre-experimental period. Perhaps more seriously, most studies of the effects of caffeine used overnight deprivation as a tool to ensure that the effects measured in the laboratory were not influenced by differences in prior exposure to caffeine. James[1] was the first to point out that mood and performance changes may therefore reflect a reversal of the deleterious effects of deprivation in regular caffeine drinkers, rather than absolute benefits of caffeine usage.

Second, test selection varies considerably across studies. While the choice of vigilance or mood tasks is less critical to the demonstration of caffeine's effects, the choice of tests for monitoring effects on tasks of learning, memory, and other complex cognitive functions is both more varied and more problematic. Apart from the fact that many of the tests employed necessarily involve multifaceted cognitive processes, it is also true to say that the relatively small benefits which might be anticipated may be difficult to demonstrate unless rigorous measures are taken on all other potential confounding factors.

Finally, a number of specific factors which are known to influence caffeine's effects are variously controlled or ignored across the published studies. These factors include sex,[2,3] personality attributes such as impulsivity/extroversion,[4-7] time of day,[6,8,9] and fatigue state.[10-13]

Notwithstanding these differences, certain consistencies do emerge from the caffeine literature with regard to its effects on mood and cognitive performance. In this chapter, I want to examine the changes in mood and cognition associated with caffeine consumption and ask whether these effects can be attributed with any degree of confidence to specific effects on neurotransmitter systems. Caffeine's potential for promoting performance has been widely exploited in selling caffeinated products to promote vigilance, improve workplace performance (particularly for nightshift workers), sustain attention, reduce appetite, and stave off effects of fatigue or the common cold. Surprisingly few empirical studies have been completed to validate the workplace claims, although experimental evidence for effects on performance tasks requiring sustained attention is certainly robust. However, in common with other well-documented compounds that have demonstrable effects on cognitive function (benzodiazepines, scopolamine, nicotine, alcohol), caffeine also affects changes in mood state. These changes in mood can induce changes both in capacity for and quality of information processing.[14] Thus, any effect which these drugs have on cognitive function may arise through induced changes in mood state, or through a direct action on information processing systems, or a combination of both. The question of specificity of drug effects is a highly topical one.[15-17] While caffeine has not been a target of these discussions, it is a particularly interesting compound in this regard.

II. The effects of caffeine ingestion

A. Caffeine and mood

Caffeine's effects on mood have most commonly been examined with respect to measures of arousal. Numerous studies have reported increased alertness following caffeine consumption,[9,18-25] with alertness increasing in a linear fashion with dose. The extent to which these stimulant effects of caffeine are perceived positively may, however, relate both to the dose used and to the precaffeine arousal state of the individual. Greden,[26] Loke and Meliska,[27] Roach and Griffiths,[21] and Loke[28] have all reported subjective increases in anxiety levels in response to high (+400 mg) doses of caffeine, and work by Greden et al.[29] and by Gilliland and Adress[30] provided early evidence of the increased scores on the State–Trait anxiety inventory for nonclinical populations who habitually consumed large quantities of coffee over the course of the day. In clinical populations, caffeine has been found to increase the subjective experience of anxiety,[31] while decreased consumption of coffee reduces reported anxiety and irritability.[32] Consistent with these anxiogenic properties, caffeine has been shown to antagonize the effect of the anxiolytic benzodiazepines.[33-35]

Interestingly, studies which report a relationship between anxiety and coffee consumption have found that the same sample may score highly on measures of depression, too.[29,30] The co-occurrence of anxiety and depression is an issue that has received considerable attention in the clinical literature.[36] That caffeine should increase the subjective experience of both is consistent with nondifferentiation of the two mood disorders on clinical scales, and the pharmacological action of caffeine on serotonin and noradrenalin neurotransmitter systems.

B. Caffeine and cognitive performance

On measures of memory, caffeine effects are far from robust and are most frequently entangled in interactions with other variables. As stated earlier, test sensitivity and selection criteria make it difficult to determine whether this represents a genuine absence of effect on such cognitive processes or the insensitivity of tests in picking up small changes in multifaceted tasks. However, many studies have examined and failed to find effects of caffeine on acquisition measures of tests of spatial learning,[37] short-term verbal memory,[18,33] and immediate or delayed recall of word lists,[21,24,28,38] although an early study by Paroli, cited in Sawyer et al.,[39] reported increased speed of correct response in completing math problems, a task which loads heavily on working memory. When caffeine effects do emerge on memory tasks, they are associated with differential effects across caffeine usage[28,30] and time of day at which the testing takes place.[8] Performance of introverts is less affected by caffeine than that of extroverts,[40-42] and sex differences have also been reported (word list recall facilitated by caffeine for females but not males[2,3]).

In contrast, simple cognitive performance measures are consistently found to be enhanced by caffeine, over a wide range of doses. Thus, caffeine improves simple[8,43] and choice[21,23,44] reaction time measures, although this effect may again be dose-dependent.[20,43,45]

In addition, caffeine reliably enhances selective attention,[46] stimulus identification,[47] and vigilance.[6,9,13,18,20,23,48,49] While some of these studies have indicated that the effects increase over time,[20] Bättig and Buzzi[13] completed a detailed examination of the different indices of performance change on the Rapid Visual Information Processing task (RVIP: Wesnes and Warburton[50]) and found no evidence of an interaction with time on task for any of the indices used. No interaction with time was found in the selective attention task reported by Lorist and Snel.[46]

Swift and Tiplady[25] have reported differential effects in young and elderly adults, suggesting that only the young group showing enhanced alertness following caffeine, but Yu et al.[51] found positive effects of caffeine on both alertness and cognitive performance measures in the healthy elderly. More recently, neither Lorist et al.[47] nor Rogers and Dernoncourt[52] found substantial age-related differences in the effects of caffeine on attention and psychomotor performance measures.

III. Differentiating mood and cognitive performance effects

A. An experimental study

As indicated earlier, at a pharmacological level, there is evidence that caffeine modulates release of neurotransmitters which have been linked with mood and neurotransmitters linked with information processing. Since we are unable to determine either the absolute or the relative impact of a given dose of caffeine on these two systems directly, evidence for independent action on the two systems has to be inferred from either differential or dissociative effects on behavioral measures.

1. Design and methods

In an unpublished study, we have examined the temporal relationship between caffeine's effects on mood and vigilance. In a double-blind crossover design, 10 young adult volunteers (mean age 22 years) completed two experimental sessions, receiving either 0 or 3 mg/kg caffeine in respective sessions, with order of treatment counterbalanced across the group. Average weight of volunteers was 68 kg, making the average dose of caffeine 225 mg. Volunteers were selected to fall within the "average" range on a scale of heavy–light coffee drinkers, and all scored within the median range on measures on the introversion/extroversion scale of the Eysenck Personality Inventory. Each was given a series of three familiarization sessions on the

10-minute vigilance task (RVIP) to ensure that practice effects were minimized across the two subsequent experimental sessions. Visual analog mood scales (Bond and Lader[53]) were also completed in practice sessions.

The experimental sessions always took place in the morning, and both began at the same time for a given individual; the two sessions were separated by a week. Prior to dosing, all volunteers completed a baseline run on the two experimental parameters, namely mood scales and the RVIP. The same tasks were repeated at 30, 75, and 120 minutes after caffeine administration.

2. Results

On the RVIP task, volunteers performed better following caffeine at all time points. This was true for both the number of correctly identified targets and response time to those targets. These differences between caffeine and placebo only achieved statistical significance at 75 minutes after administration. Interestingly, the placebo condition resulted in performance levels below baseline for that day at all time points, while the caffeine condition improved performance above baseline measures at all points.

On the mood scales, caffeine did not affect subjective judgments on either the contentedness factor nor the sociability factor, but it did affect ratings on the alertness factor. Specifically, both caffeine and placebo treatment resulted in higher ratings of alertness by volunteers at 30 minutes after administration. At subsequent time points, the placebo condition produced decreasing alertness ratings, falling around baseline levels, while in the caffeine condition, enhanced alertness ratings were maintained across all time points.

In other words, the reported improvements in alertness at just 30 minutes post drug were in fact unrelated to actual consumption of caffeine and represent expectation effects. The real effects of caffeine, observed on alertness measures, emerged later and remained for longer than subjective expectancy effects. Questioned afterwards our volunteers all indicated that they anticipated fairly rapid onset of caffeine effects, and that they would not expect to still be feeling the benefits of caffeine two hours after drinking a coffee (the doses administered here fell within the range an individual might expect to receive from one mug of real coffee).

3. Summary

To summarize the results of this pilot study, the performance benefits of caffeine appeared within 30 minutes and increased up to 75 minutes post drug. In contrast, the short-term mood effects reported under caffeine were indistinguishable from those reported under placebo, with real differences in alertness emerging later and being maintained beyond the period of peak performance benefits.

So, the temporal relationship between cognitive change and mood change in this study is consistent with independence between the two measures; the enhancement of performance on the vigilance task cannot be

explained simply as a consequence of enhanced mood, but reflects independent effects of caffeine on cognition. Somewhat counterintuitively, the mood
effects emerge *after* the effects on the performance measure.

B. *Mood and cognitive performance revisited*

How does this result fit with the published literature? Surprisingly few
studies on the effects of caffeine have examined the temporal relationship
between cognitive and mood factors side by side. Many involve single measures on multiple tasks, rather than repeated measures on the same tasks.

Some measure of the independence of caffeine's effects on mood and
cognitive performance can be inferred from studies which have reported
changes in only one of the two parameters manipulated. For example, Loke
et al.[38] reported mood effects in the absence of changes on performance
parameters. Alternatively, changes in cognitive performance may be
observed in the absence of mood effects.[13,23] Bättig and Buzzi[13] measured
caffeine effects after a relatively short post-administration interval (30 minutes), while Lieberman et al.[23] were using relatively low doses of caffeine. In
both instances, the absence of mood effects suggests more robust and more
immediate performance effects of caffeine. Dissociative effects involving null
findings are problematic, however, because non-effects may indicate task
insensitivity.

Of the studies that have examined the time course of caffeine's effects
on both mood and cognitive performance together, there is some consensus
that, as in the study reported in this chapter, the "real" mood effects emerge
later in the time course of caffeine action, rather than earlier. For example,
Frewer and Lader[49] examined performance on two vigilance tasks and visual
analog mood scales at 45 and 165 minutes after administration of caffeine.
While analysis of performance on the vigilance tasks was complicated by
differences across blocks within each session, the pattern of change over
blocks indicated performance differences emerging at 45 minutes. Mood
changes were minimal across all sessions, emerging most strongly on the
calmness scale at 165 minutes post drug. Smith et al.[6] also reported performance effects on the RVIP task at 30 and 75 minutes post drug, while mood
effects (increased alertness ratings) between placebo and caffeine conditions
diverged only at the 75-minute test session. Fagan et al.[20] looked at the use
of extended vigilance tasks and reported changes in tapping performance
at 50 minutes post drug, while mood effects emerged 1 to 2 hours after
administration. In their extended auditory vigilance task, however, the time
course of caffeine's effects fell closer to the mood changes than to the tapping
changes.

Interestingly, studies concerned with the effects of caffeine deprivation
in regular users also argue for dissociative effects of caffeine on mood and
performance. In a review of the literature, Rogers et al.[54] concluded that while
there was substantial evidence for adverse mood changes following overnight deprivation from caffeine, impairments in psychomotor performance

were not often found after overnight deprivation. Lane[55] has reported a similar dissociation following short-term deprivation. Regular coffee drinkers who abstained for a minimum of 4 hours reported negative mood changes (decreased vigor, increased fatigue) but showed no deterioration on psychomotor measures.

IV. Concluding comments

Specificity of effects is an important issue in the human psychopharmacology of psychoactive compounds. While the source of any beneficial effects of caffeine may be of little importance to the healthy adult coffee drinker, it is critical to those interested in the potential of a given compound to improve cognitive function. The caffeine studies reviewed in this chapter provide evidence of cognitive-enhancing effects of this compound that can be observed independent of changes in mood. Counterintuitively, the mood effects seem to emerge later and less strongly than the effects on tasks of vigilance and attention. Future studies are needed to examine the boundaries and the temporal relationship of these independent effects more systematically. A greater understanding of caffeine's action on mood and memory, and on the underlying neurotransmitter systems, will have relevance not only for those interested in the cognitive-enhancing qualities of the compound, but also for the those concerned with the health consequences, both mental and physical,[56] of caffeine usage. The literature suggests that the negative consequences of caffeine usage, namely adverse effects of deprivation, may in fact be limited to caffeine-induced mood changes and not cognitive performance effects.

References

1. James, J. E., Does caffeine enhance or merely restore degraded psychomotor performance? *Neuropsychobiology*, 30, 124, 1994.
2. Erikson, G.C., Hager, L., Houseworth, C., Dungan, J., Petros, T., and Beckwith, B., The effects of caffeine on memory for word lists, *Physiol. Behav.*, 35, 47, 1985.
3. Arnold, M.E., Petros, T.V., Beckwith, B.E., and Coons, G., The effects of caffeine, impulsivity, and sex on memory for word lists, *Physiol. Behav.*, 41, 25, 1987.
4. Anderson, K.J. and Revelle, W., Interactive effects of caffeine, impulsivity, and task demands on a visual search task, *Person. Individ. Diff.*, 4, 127, 1983.
5. Revelle, W., Anderson, K.J., and Humphreys, M.S., Empirical tests and theoretical extensions of arousal-based theories of personality, in *Personality Dimensions and Arousal*, Strelau, J. and Eysenck, H.J. , Eds., Plenum Press, London, 1987, 17.
6. Smith, A.P., Rusted, J.M., Savory, M., Eaton-Williams, P., and Hall, S.R., The effects of caffeine, impulsivity and time of day on performance, mood and cardiovascular function, *J. Psychopharmacol.*, 5, 120, 1991.
7. Gupta, U., Effects of impulsivity and caffeine on human cognitive performance, *Pharmacopsychoecologia*, 1, 33, 1988.

8. Revelle, W., Humphreys, M.S., Simon, L., and Gilliland, K., The interactive effect of personality, time of day, and caffeine: A test of the arousal model, *J. Exp. Psychol.*, 109, 1, 1980.
9. Smith, A.P., Rusted, J.M., Eaton-Williams, P., Savory, M., and Leathwood, P., Effects of caffeine given before and after lunch on sustained attention, *Neuropsychobiology*, 23, 160, 1990.
10. Borland, R.G., Rogers, A.S., Nicholson, A.N., Pascoe, P.P., and Spencer, M.B., Performance overnight in shiftworkers operating a day-night schedule, *Aviat. Space Environ. Med.*, March, 241, 1986.
11. Walsh, J.K., Muehlbach, M.J., Humm, T.M., Dickins, Q.S., Sugerman, J.L., and Schweitzer, P.K., Effect of caffeine on physiological sleep tendency and ability to sustain wakefulness at night, *Psychopharmacology*, 101, 271, 1990.
12. Rosenthal, L., Roehrs, T., Zwyghuizen-Doorenbos, A., Plath, D., and Roth, T., Alerting effects of caffeine after normal and restricted sleep, *Neuropsychopharmacology*, 4, 103, 1991.
13. Bättig, K. and Buzzi, R., Effect of coffee on the speed of subject-paced information processing, *Neuropsychobiology*, 16, 126, 1986.
14. Blaney, P.H., Affect and mood: a review, *Psychological Bulletin*, 99, 229, 1986.
15. Sahakian, B.J., Cholinergic drugs and human cognitive performance, in *Handbook of Psychopharmacology*, Volume 20, Iversen, L.L., Iversen, S.D., and Snyder, S.H., Eds., Plenum Press, London, 1988, 393.
16. Curran, H.V., Benzodiazepines, memory and mood: a review, *Psychopharmacology*, 105, 1, 1991.
17. Rusted, J.M., Cholinergic blockade and human information processing: are we asking the right questions? *J. Psychopharmacol.*, 8, 54, 1994.
18. Clubley, M., Bye, C.E., Henson, T.A., Peck, A.W., and Riddington, C.J., Effects of caffeine and cyclizine alone and in combination on human performance, subjective effects and EEG activity, *Br. J. Clin. Pharmacol.*, 7, 157, 1979.
19. Leathwood, P. and Pollet, P., Diet-induced mood changes in normal populations, *J. Psychiat. Res.*, 17, 147, 1982.
20. Fagan, D., Swift, C.G., and Tiplady, B., Effects of caffeine on vigilance and other performance tests in normal subjects, *J. Psychopharmacol.*, 2, 19, 1988.
21. Roach, J.D. and Griffiths, R.R., Interactions of diazepam and caffeine: behavioural and subjective dose effects in humans, *Pharmacol. Biochem. Behav.*, 26, 801, 1987.
22. Zahn, T.P. and Rapoport, J.L., Acute autonomic nervous system effects of caffeine on prepubertal boys, *Psychopharmacology*, 91, 40, 1987.
23. Lieberman, H.R., Wurtman, R.J., Emde, G.G., and Coviella, I.L.G., The effects of caffeine and aspirin on mood and performance, *J. Clin. Psychopharmacol.*, 7, 315, 1987.
24. Leiberman, H.R., Beneficial effects of caffeine, in *Twelfth International Scientific Colloquium on Coffee*, ASIC, Paris, 1988.
25. Swift, C.G. and Tiplady, B., The effect of age on the response to caffeine, *Psychopharmacology*, 94, 29, 1988.
26. Greden, J.F., Anxiety and caffeinism: a diagnostic dilemma, *Am. J. Psychiat.*, 131, 1089, 1974.
27. Loke, W.H. and Meliska, C.J., Effects of caffeine use and ingestion on a protracted visual vigilance task, *Psychopharmacology*, 84, 54, 1984.
28. Loke, W.H., Effects of caffeine on mood and memory, *Physiol. Behav.*, 44, 367, 1988.

29. Greden, J.F., Fontaine, P., Lubetsky, M., and Chamberlin, K., Anxiety and depression associated with caffeinism among psychiatric inpatients, *Am. J. Psychiat.*, 135, 963, 1978.
30. Gilliland, K. and Andress, D., Ad lib caffeine consumption, symptoms of caffeinism, and academic performance, *Am. J. Psychiat.*, 138, 512, 1981.
31. Charney, D.S., Heninger, G.R., and Jatlow, P.I., Increased anxiogenic effects of caffeine in panic disorders, *Arch. Gen. Psychiat.* , 42, 233, 1985.
32. Shisslak, C.M., Beutler, L.E., Scheiber, S., Gaines, J.A., La Wall, J., and Crago, M., Patterns of caffeine use and prescribed medications in psychiatric inpatients, *Psychol. Rep.*, 57, 39, 1985.
33. File, S.A., Bond, A.J., and Lister, R.G., Interaction between effects of caffeine and lorazepam on performance tests and self-ratings, *J. Clin. Psychopharmacol.*, 2, 102, 1982.
34. Lee, M.A., Cameron, O.G., and Greden, J.F., Anxiety and caffeine consumption in people with anxiety disorders, *Psychiat. Res.*, 15, 211, 1985.
35. Johnson, L.C., Spinweber, C.L., and Gomez, S.A., Benzodiazepines and caffeine: effect on daytime sleepiness, performance, and mood, *U.S. Naval Health Research Center Report*, Rep. No. 88-51, 1988, 1.
36. Stavrakaki, C. and Vargo, B., The relationship of anxiety and depression: A review of the literature, *Br. J. Psychiat.*, 149, 7, 1986.
37. Bättig, K., Buzzi, R., Martin, J.R., and Feierabend, J.M., The effects of caffeine on physiological functions and mental performance, *Experientia*, 40, 1218, 1984.
38. Loke, W.H., Hinrichs, J.V., and Ghonheim, M.M., Caffeine and diazepam: separate and combined effects on mood, memory and psychomotor performance, *Psychopharmacology*, 87, 344, 1985.
39. Sawyer, D.A., Julia, H.L., and Turin, A.C., Caffeine and human behaviour: arousal, anxiety, and performance effects, *J. Behav. Med.*, 5, 415, 1982.
40. Keister, M. and McLaughlin, R., Vigilance performance related to extroversion-introversion and caffeine, *J. Exp. Res. Person.*, 6, 5, 1972.
41. Gilliland, K., The interactive effects of introversion extroversion with caffeine induced arousal on verbal performance, *J. Exp. Res. Person.*, 14, 482, 1980.
42. Terry, W.S. and Phifer, B., Caffeine and memory performance on the AVLT, *J. Clin. Psychol.*, 42, 860, 1986.
43. Lieberman, H.R., Wurtman, R.J., Emde, G.G., Roberts, C., and Coviella, I.L.G., The effects of low doses of caffeine on human performance and mood, *Psychopharmacology*, 92, 308, 1987.
44. Smith, D.L., Tong, J.E., and Leigh, G., Combined effects of tobacco and caffeine on the components of choice reaction time, heart rate, and hand steadiness. *Percept. Mot. Skills*, 45, 635, 1977.
45. Kuznicki, J.T. and Turner, L.S., The effects of caffeine on caffeine users and nonusers, *Physiol. Behav.*, 37, 397, 1986.
46. Lorist, M. M. and Snel, J., Caffeine effects on perceptual and motor processes, *Electroencephalogr. Clin. Neurophysiol.*, 102, 401, 1997.
47. Lorist, M.M., Snel, J., Mulder, G., and Kok, A., Aging, caffeine and information processing: An event-related potential analysis: Evoked potentials, *Electroencephalogr. Clin. Neurophysiol.*, 96, 453, 1995.
48. Baker, W.T. and Theologus, G.C., Effects of caffeine on visual monitoring, *J. Appl. Psychol.*, 56, 422, 1972.
49. Frewer, L.J. and Lader, M.H., The effects of caffeine on two computerized tests of attention and vigilance, *Hum. Psychopharmacol.*, 6, 119, 1991.

50. Wesnes, K. and Warburton, D.M., The effects of cigarette smoking and nicotine tablets upon human attention, in *Smoking Behaviour: Physiological and Psychological Influences*, Thornton, R.E., Ed., Churchill-Livingston, London, 1978, 131.
51. Yu, G., Maskray, V., Jackson, S.H.D., Swift, C.G., and Tiplady, B., A comparison of the central nervous system effects of caffeine and theophylline in elderly subjects, *Br. J. Clin. Pharmacol.*, 32, 341, 1991.
52. Rogers, P. J. and Dernoncourt, C., Regular caffeine consumption: a balance of adverse and beneficial effects for mood and psychomotor performance, *Pharmacol. Biochem. Behav.*, 59, 1039, 1998.
53. Bond, A.J. and Lader, M.H., The use of analogue scales in rating subjective feelings, *Br. J. Med. Psychol.*, 47, 211, 1974.
54. Rogers, P. J., Richardson, N.J., and Dernoncourt, C., Caffeine use: is there a net benefit for mood and psychomotor performance, *Neuropsychobiology*, 31, 195, 1995.
55. Lane, J.D., Effects of brief caffeinated beverage deprivation on mood, symptoms, and psychomotor performance, *Pharmacol. Biochem. Behav.*, 58, 203, 1997.
56. Benowitz, N.L., Clinical pharmacology of caffeine, *Annu. Rev. Med.*, 41, 277, 1990.

chapter fourteen

Caffeine and attention

Odin van der Stelt

Contents

I. Introduction 231
II. Caffeine and attentional performance...231
   A. Selective attention..231
   B. Sustained attention 235
III. Conclusion...237
References...238

I. Introduction

For centuries, caffeine-containing substances have been believed to enhance human alertness and mental efficiency.[1] Indeed, many laboratory studies conducted over the past 80 years provide experimental evidence confirming this belief, although the exact nature of the effects of caffeine on performance remains to be specified.[2-5] The present paper evaluates some recent studies concerned with the effects of caffeine on the performance of selective-attention and sustained-attention tasks. The aim is to provide a brief account of our current understanding of the effects of caffeine on human attention and information processing.

II. Caffeine and attentional performance

A. Selective attention

The study of selective attention involves either focused attention or divided attention.[6,7] In focused-attention tasks, there are usually two concurrent sources of information, and the subject is asked to direct attention to one source of information while ignoring and/or excluding the other source.

0-8493-1166-7/99/$0.00+$.50

This task paradigm is used to evaluate the subject's susceptibility to distraction and to determine the levels of processing at which different kinds of information are selected for further analysis.[8] In contrast, divided-attention tasks require the subject to divide attention among multiple concurrent sources of information. This paradigm is used to assess limitations in information processing and to study the efficiency with which two tasks can be performed simultaneously.[9] Thus, whereas focused-attention tasks are used to assess how effectively people select some information in favor of other, divided-attention tasks are used to establish limits on the amount of information people can select. Succeeding sections consider the behavioral effects of caffeine in both task situations.

1. Focused attention

A deficit in focused attention is thought to occur when one attempts to attend to just one stimulus attribute, stimulus, or type of stimuli, but finds performance impaired because of distraction from other inputs. The classical experimental example of the inability of subjects to ignore irrelevant information is the Stroop color–word test.[10] The stimulus materials of this test consist of three types of cards. On the first card, the subject is required to read a list of color words that are printed in black. On the second card, the subject must name the color of a set of color patches. On the third card, the subject is asked to name the color of a list of color words while the words are printed in incongruent colors (e.g., to say "green" when presented with the word RED printed in green). The basic findings are that word reading is faster than color naming, and that naming the incongruently printed color words takes much more time than naming the color patches. The time difference between naming the colors in the nonconflicting and conflicting condition defines the Stroop interference effect, which signifies the subject's ability to focus attention on a relevant stimulus attribute of color while ignoring an irrelevant, semantic one.

Several studies have examined the effects of caffeine on Stroop performance. The results, however, are not consistent. Nash[11] reported that a dose of 100 mg of caffeine tended to speed both word reading, color naming, and Stroop reading, as well as tending to reduce the Stroop interference effect. On the other hand, Borland et al.,[12] utilizing shift workers operating a day–night schedule, evaluated the effects of 300 mg caffeine on a computerized version of the Stroop color–word test at night. Although caffeine was found to improve overnight performance of a number of other tasks as compared to a placebo, performance of the Stroop test was not affected. Foreman et al.[13] investigated the effects of two doses of caffeine (125 and 250 mg) using a computer-controlled numerical version of the Stroop test. The results showed that when subjects were administered the high dose of caffeine, the Stroop interference effect was significantly larger than when subjects ingested a placebo. The interference effect when subjects ingested the low dose of caffeine failed to differ significantly from either of the two other treatments. Subsequently, Hasenfratz and Bättig[14] used two versions of the numerical Stroop test. In the relatively easy version of the task, when the stimuli were presented with a post-response delay of 1 sec, a dose of 250 mg caffeine was found to reduce the reaction times to both nonconflicting and conflicting information, but the Stroop interference effect was not affected by caffeine. However, in the difficult version of the task, when the stimuli were presented without a delay, caffeine failed to affect the speed of reactions to both types of information, but did reduce the Stroop interference effect. These results indicate that, in addition to the dose of caffeine, task parameters need careful consideration in the evaluation of caffeine on Stroop performance.

Broverman and Casagrande[15] investigated the effects of caffeine on the Embedded Figures Task (EFT) at different stages of practice. This task required subjects to search simple geometric figures embedded within complex, distracting patterns of lines and colors. In order to assess the effects of practice, two different but equivalent sets of five figures were used. The total time needed to detect each of the target figures in a given set served as the dependent variable. Three groups of subjects were tested on the EFT twice, in one of the following three treatment sequences: caffeine–placebo, placebo–caffeine, and placebo–placebo. Caffeine was administered in the form of two cups of caffeinated instant coffee, containing approximately 113 mg caffeine. Two cups of decaffeinated coffee, containing approximately 3.6 mg caffeine, was used as placebo. Caffeine was hypothesized to impair performance at the first test session when the task was novel, but to facilitate practice-induced gains in performance from the first to the second session as the task became less novel and automatized. These predictions were confirmed. First, the subjects who received caffeine at the first test session performed more slowly than those who received placebo (means of 242.1 sec and 193.6 sec, respectively). Further, the group of subjects under the placebo–placebo treatment condition performed better at the second test session than at the first session; that is, there was a practice effect. Finally, the subjects receiving caffeine at the second test session performed better and had about a 2.5 times larger practice effect than those who did not receive caffeine. In terms of automatic-control processing theory[7,16,17] these results suggest that caffeine enhanced the automatic information processing, which may have overruled the subject-controlled processing mode required to deal with novel information. The automatic-controlled processing distinction deserves more consideration in the study on the performance effects of caffeine, since it provides a useful framework for analysis and interpretation.

These findings indicate that the ingestion of caffeine is sometimes associated with detrimental effects on performance, presumably due to distraction. Smith et al.,[18] however, did not find an effect of 4 mg/kg caffeine on the performance of a visual focused-attention task in which subjects had to press a button in response to a centrally presented single letter that was occasionally accompanied by other, distracting letters. It should be noted, as the authors acknowledged, that only the main effect of caffeine on performance was presented, whereas the possible interaction with task variables (e.g., the separation between targets and distracters) would be the most informative. Yet, in a recent event-related brain potential (ERP) study, using a combined visual focused-attention and memory-search task paradigm, a dose of 250 mg caffeine was found to enlarge the P3 ERP component in response to task-irrelevant stimuli,[19] suggesting that caffeine enhanced the involuntary, reflexive aspects of attention[20] and facilitated the processing of the irrelevant information.[21]

In conclusion, although the existing database is extremely small, the overall result suggests that caffeine does not affect consistently nor specifically the subject's ability to differentiate relevant from irrelevant information. Rather, a more general facilatory effect on the processing of both relevant and irrelevant stimulus information may occasionally be observed with caffeine.

2. Divided attention

Investigators interested in the effects of caffeine on divided attention have generally employed the dual task paradigm. Putz-Anderson et al.[22] examined the effects of caffeine, alcohol, and methylchloride on the performance of a 25-min. dual task involving compensatory tracking and auditory vigilance. The subjects were instructed to use a joystick to align a cursor with a moving target while simultaneously detecting deviant (slightly higher) tones occurring infrequently and randomly within a series of standard tones. As compared with pretreatment scores, the subjects, after ingestion of 3 mg/kg caffeine, showed a significant improvement (approximately 10% fewer errors) in tracking performance, with no significant effect on the speed of detections in the vigilance task. Alcohol was found to impair both aspects of dual task performance; methylchloride had no influence.

Borland et al.[12] and Rogers et al.[23] aimed to assess the usefulness of caffeine in sustaining mental performance overnight. Several tasks were used, including a 20-min dual task that combined compensatory tracking with auditory vigilance. The dual task required subjects to use a joystick to maintain the cross pointers of an instrument landing system meter at the zero position. At the same time, they had to detect a pause of 1.5 sec duration within a series of tone–pause–tone sequences presented binaurally through headphones. In both studies, a dose of 300 mg caffeine administered shortly before midnight was found to attenuate significantly the decrement in overnight performance of most of the tasks, including the dual task, that occurred with placebo. The enhancing effect of caffeine on dual task performance took the form of an improvement of tracking accuracy and/or a reduction of the number of misses in the auditory vigilance task. Its efficacy was already evident around midnight and persisted throughout the night until the next morning.

Zwyghuizen-Doorenbos et al.[24] used a 15-min dual task that involved tracking a moving target across a video screen using a joystick, and detecting the occurrence of a white circle on the periphery of the screen or at the center of the moving target. The investigators also used an auditory vigilance task and the Multiple Sleep Latency Test (MSLT) — a physiological measure of daytime sleepiness/alertness. The results showed that while 250 mg of caffeine increased daytime alertness and vigilance performance relative to a placebo, dual task performance was not influenced by caffeine.

Kerr et al.[25] examined the separate and combined effects of nicotine, alcohol, and caffeine on performance. One of the tasks was a dual task involving compensatory tracking and detection of peripheral visual stimuli. In addition, the subject's critical flicker fusion threshold (CFF) was measured and used as an index of the state of CNS arousal. Tracking accuracy over four trials of 60 sec and the reaction time to 40 lights (10 per trial) were assessed prior to treatment (baseline test) and at 1, 2, 3, and 4 hours after administration of treatment. Alcohol alone was found to impair significantly tracking accuracy, whereas all drug combinations which included nicotine, or 250 mg of caffeine alone, produced an improvement of tracking performance. There were no drug effects on the reaction time to the simultaneously presented visual stimuli. The effect of caffeine on tracking performance did not vary as a function of the time of administration. Because caffeine did not affect the CFF, the authors suggested that the enhancing effects of caffeine on tracking performance were not related to an increase in the general level of arousal. Instead, a more specific action of caffeine on information processing might be involved.

Indeed, these last results are in accordance with most prior research[2] and point to the conclusion that caffeine has properties for the facilitation of dual task performance. Although the available evidence is limited, the results may indicate that caffeine can enhance the ability to divide attention, although this may be gained at the expense of selection efficacy,[26] as implicated in the previous section.

B. Sustained attention

In contrast to the study of selective attention, another area of attention research involves the study of long-term attentive behavior or vigilance, that is, of sustained attention. Sustained attention research is concerned with the ability to maintain attention and performance over prolonged, unbroken periods of time. Although a wide variety of tasks, such as tracking tasks, reaction-time tasks, and problem-solving tasks, have been used in laboratory studies on sustained attention, the vigilance task has been more widely used than any other. According to Jerison,[27] the vigilance task can be seen "to provide a fundamental paradigm for defining sustained attention as a behavioral category." In a vigilance task, the subject usually has to maintain attention on one or more sources of information in order to detect the occurrence of infrequent but critical signals. Two principal aspects of vigilance performance need be distinguished.[28,29] One aspect is the overall *level* of vigilance performance, which is closely related to the general level of arousal. The second aspect is the *decrement* in vigilance performance over time, which may be seen as the hallmark of attention failure in these tasks. Much research effort has been devoted to the environmental and task factors that determine vigilance performance, and in particular to the factors that are responsible for its breakdown.[28-30]

Several vigilance tasks have been used in the study on the effects of caffeine on sustained attention, but two types of vigilance tasks have probably been most often used. These are the auditory vigilance task (AVT) adapted from Wilkinson,[31] and auditory and visual variants of the vigilance task originally developed by Bakan.[32] Fagan et al.[33] examined the effects of caffeine on the performance of the AVT. In this vigilance task, lasting 1 hour, the subjects listened through headphones to a series of tones that were presented against a background of white noise. The tones were presented every 2 s, and had a duration of 500 ms, except that occasionally a tone occurred with a duration of 400 ms. The subject's task was to detect the short tones by pressing a button. In their first experiment, a dose of 200 mg caffeine was found to increase significantly the overall number of correct detections as compared to a placebo, confirming the results of an earlier study.[34] In their second experiment the dose of caffeine did not affect the overall number of hits. However, whereas the placebo group showed a decline in the number of detections from the first half of the task to the second half, vigilance performance of the caffeine group remained stable. In both experiments, there was a slight, but nonsignificant increase observed in the number of incorrect detections with caffeine. Lieberman et al.[35] aimed to assess the behavioral effects of different doses of caffeine. In this study, caffeine in doses of 256, 128, 64, and even 32 mg (less than that found in a single cup of coffee) as compared with a placebo, significantly increased the overall number of correct detections in a modified version of the AVT, while the false-alarm rate was not affected by caffeine. In a subsequent study,[36] a dose of 64 mg caffeine when given alone as well as when combined with a dose of 800 mg aspirin also significantly improved the level of vigilance in the AVT. Likewise, Zwyghuizen-Doorenbos et al.[24] observed that 250 mg caffeine, relative to a placebo, improved the reaction times to the signals as well as detection accuracy in a 40-min version of the AVT. Unfortunately, analyses of AVT performance as a function of time of task were not reported in this study nor in the two last studies.

Keister and McLaughlin[37] investigated the effects of caffeine on the performance of an auditory version of the Bakan task. In this task, the subjects wore headphones and listened to a 48-min tape recording of digits, which were spoken at a rate of one per second. The subjects were instructed to write down on a sheet of paper the occurrence of odd–even–odd sequences of digits. Two groups of subjects were formed on the basis of scores on the extremes of the introversion–extraversion dimension of the Eysenck Personality Inventory,[38] because caffeine was assumed to produce differential effects on the performance of the two personality groups. Indeed, the results showed that four 50 mg capsules of caffeine abolished the decrement in the number of correct detections seen in extroverts under a nondrug condition.

Introverts exhibited no vigilance decrement, and caffeine exerted no effect on their performance.

Frewer and Lader[39] used a visual variant of the Bakan task. In this vigilance task, lasting 15 minutes, single digits were presented at a rate of 100 digits per minute. The subject's task was to detect, by pressing a button, the occurrence of three successive odd or even digits. The main finding was that caffeine in doses of 250 mg and 500 mg significantly increased the overall detection rate, without affecting the false-alarm rate. Smith et al.[40] examined the influence of caffeine before and after lunch on the performance of a 10-min visual version of the Bakan task. A dose of 3 mg/kg caffeine was found to improve significantly the overall detection rate and reaction times as compared with a placebo, while the false-alarm rate was not affected. The improvement of performance with caffeine was observed before lunch but was significantly more pronounced after lunch, totally removing the post-lunch dip observed in the placebo condition. Caffeine, however, failed to affect the vigilance decrement. Bättig and Buzzi[41] employed a subject-paced and monetarily reinforced 30-min visual version of the Bakan task. The stimulus presentation rate increased after each correct response and decreased after each error and was continuously recorded as the index of performance. The main result was that caffeine in doses of 150 mg and 450 mg significantly improved overall performance but without affecting the decline in processing rate that occurred as a function of time on task. This result was confirmed by Hasenfratz et al.[42] in two other experiments using 250 mg caffeine.

These findings show that caffeine consistently improves the level of vigilance in AVT and Bakan types of vigilance tasks. These benefits can be observed in both the detection rate and the speed of detections, while the false-alarm rate is usually not affected by caffeine. This observation suggests a real effect on detection efficiency with caffeine, rather than a change in task strategy. In addition, caffeine seems to have the capacity to reduce vigilance decrements, although this caffeine effect appears to be less robust and may be more task specific. Because investigators using the AVT have focused more on the level of vigilance than on the vigilance decrement, this issue needs further research.

III. Conclusion

This chapter considered the effects of caffeine on human attention and performance. Much more research is needed, however, before any firm conclusions can be drawn. For instance, studies on caffeine and dual task performance are needed that manipulate the structure of the tasks and the allocation of the subject's attention.[9] In addition, studies on the effects of caffeine on sustained attention may profit from manipulation of task variables and also, when carefully applied, from signal detection methods of analysis, which can provide a comprehensive framework within which several vigilance performance measures can be interpreted.[28]

Because the available evidence is limited, the present conclusion is necessarily tentative and rather general. Attention is not a unitary function of the brain but can be divided into a *selective* (or channel) function and an *intensive* (or state) function.[28,43,44] The former function is concerned with the selectivity of information processing, whereas the latter function refers to the amount of available processing capacity, resources, or effort, and is closely related to general arousal. It may be concluded that the intensive aspects of attention and information processing may be more sensitive to the effects of caffeine than the selective aspects. That is, the impact of caffeine on human performance may be conceptualized in terms of an increase in processing resources with caffeine, which may not directly relate to its action on the general level of arousal. The precise mechanisms by which this hypothesized caffeine-induced increase in processing resources results in improved performance remain to be identified.

References

1. Rall, T. W., Drugs used in the treatment of asthma, in *The Pharmacological Basis of Therapeutics*, Goodman, L. S., Gilman, A., Rall, T. W., Nies, A. S., and Taylor, P., Eds., Pergamon Press, New York, 1990, 618.
2. Weiss, B. and Laties, V. G., Enhancement of human performance by caffeine and the amphetamines, *Pharmacol. Rev.*, 14, 1, 1962.
3. Snel, J., Coffee and caffeine: Sleep and wakefulness, in *Caffeine, Coffee and Health*, Garattini, S., Ed., Raven Press, New York, 1993, 255.
4. Van der Stelt, O. and Snel, J., Effects of caffeine on human information processing: A cognitive-energetic approach, in *Caffeine, Coffee and Health*, Garattini, S., Ed., Raven Press, New York, 1993, 291.
5. Van der Stelt, O. and Snel, J., Caffeine and human performance, in *Nicotine, Caffeine, and Social Drinking. Behaviour and Brain Function*, Snel, J. and Lorist, M. M., Eds., Harwood Academic Publishers, Reading, 1998, 167.
6. Treisman, A. M., Strategies and models of selective attention, *Psychol. Rev.*, 76, 282, 1969.
7. Schneider, W., Dumais, S. T., and Shiffrin, R. M., Automatic and control processing and attention, in *Varieties of Attention*, Parasuraman, R. and Davies, D. R., Eds., Academic Press, Orlando, 1984, 1.
8. Johnston, W. A. and Dark, V. J., Selective attention, *Annu. Rev. Psychol.*, 37, 43, 1986.
9. Wickens, C. D., Processing resources in attention, in *Varieties of Attention*, Parasuraman, R. and Davies, D. R., Eds., Academic Press, Orlando, 1984, 63.
10. Stroop, J. R., Studies of interference in serial verbal reactions, *J. Exp. Psychol.*, 18, 643, 1935.
11. Nash, H., *Alcohol and Caffeine*, Thomas, Springfield, 1962.
12. Borland, R. G., Rogers, A. S., Nicholson, A. N., Pascoe, P. A., and Spencer, M. B., Performance overnight in shiftworkers operating a day-night schedule, *Aviat. Space Environ. Med.*, 57, 241, 1986.
13. Foreman, N., Barraclough, S., Moore, C., Mehta, A., and Madon, M., High doses of caffeine impair performance of a numerical version of the Stroop task in men, *Pharmacol. Biochem. Behav.*, 32, 399, 1989.

14. Hasenfratz, M. and Bättig, K., Action profiles of smoking and caffeine: Stroop effect, EEG, and peripheral physiology, *Pharmacol. Biochem. Behav.*, 42, 155, 1992.
15. Broverman, D. M. and Casagrande, E., Effect of caffeine on performances of a perceptual-restructuring task at different stages of practice, *Psychopharmacology*, 78, 252, 1982.
16. Schneider, W. and Shiffrin, R. M., Controlled and automatic human information processing: I. Detection, search, and attention, *Psychol. Rev.*, 84, 1, 1977.
17. Shiffrin, R. M. and Schneider, W., Controlled and automatic human information processing: II. Perceptual learning, automatic attending, and a general theory, *Psychol. Rev.*, 84, 127, 1977.
18. Smith, A. P., Kendrick, A., and Maben, A., Effects of caffeine, lunch and alcohol on human performance, mood and cardiovascular function, *Proc. Nutr. Soc.*, 51, 325, 1992.
19. Lorist, M. M., Snel, J., Kok, A., and Mulder, G., The influence of caffeine on selective attention in well rested and fatigued subjects, *Psychophysiology*, 31, 525, 1994.
20. Kok, A., Internal and external control: A two-factor model of amplitude change of event-related potentials, *Acta Psychologica*, 74, 203, 1990.
21. Johnson, R., Jr., A triarchic model of P300 amplitude, *Psychophysiology*, 23, 367, 1986.
22. Putz-Anderson, V., Setzer, J. V., and Croxton, J. S., Effects of alcohol, caffeine and methylchloride on man, *Psychol. Rep.*, 48, 715, 1981.
23. Rogers, A. S., Spencer, M. B., Stone, B. M., and Nicholson, A. N., The influence of 1 h nap on performance overnight, *Ergonomics*, 32, 1193, 1989.
24. Zwyghuizen-Doorenbos, A., Roehrs, T. A., Lipschutz, L., Timms, V., and Roth, T., Effects of caffeine on alertness, *Psychopharmacology*, 100, 36, 1990.
25. Kerr, J. S., Sherwood, N., and Hindmarch, I., Separate and combined effects of the social drugs on psychomotor performance, *Psychopharmacology*, 104, 113, 1991.
26. Johnston, W. A. and Heinz, S. P., Flexibility and capacity demands of attention, *J. Exp. Psychol. Gen.*, 107, 420, 1978.
27. Jerison, H. J., Vigilance: Biology, psychology, theory and practice, in *Vigilance: Theory, Operational Performance and Physiological Correlates*, Mackie, R. R., Ed., Plenum, New York, 1977, 27.
28. Davies, D. R. and Parasuraman, R., *The Psychology of Vigilance*, Academic Press, London, 1982.
29. Parasuraman, R., Sustained attention in detection and discrimination, in *Varieties of Attention*, Parasuraman, R. and Davies, D. R., Eds., Academic Press, Orlando, 1984, 243.
30. Broadbent, D. E., *Decision and Stress*, Academic Press, London, 1971.
31. Wilkinson, R. T., Sleep deprivation: Performance tests for partial and selective sleep deprivation, *Prog. Clin. Psychol.*, 8, 28, 1968.
32. Bakan, P., Extraversion-introversion and improvement in an auditory vigilance task, *Br. J. Psychol.*, 50, 325, 1959.
33. Fagan, D., Swift, C. G., and Tiplady, B., Effects of caffeine on vigilance and other performance tests in normal subjects, *J. Clin. Psychopharmacol.*, 2, 19, 1988.
34. Clubley, M., Bye, C. E., Henson, T. A., Peck, A. W., and Riddington, C. J., Effect of caffeine and cyclizine alone and in combination on human performance, subjective effects and EEG activity, *Br. J. Clin. Pharmacol.*, 7, 157, 1979.

35. Lieberman, H. R., Wurtman, R. J., Emde, G. G., Roberts, C., and Coviella, I. L. G., The effects of low doses of caffeine on human performance and mood, *Psychopharmacology,* 92, 308, 1987.
36. Lieberman, H. R., Wurtman, R. J., Emde, G. G., and Coviella, I. L. G., The effects of caffeine and aspirin on mood and performance, *J. Clin. Psychopharmacol.,* 7, 315, 1987.
37. Keister, M. E. and McLaughlin, R. J., Vigilance performance related to extraversion–introversion and caffeine, *J. Exp. Res. Person.,* 6, 5, 1972.
38. Eysenck, H. J. and Eysenck, S. B. G., *Eysenck Personality Inventory,* Educational and Industrial Testing Services, San Diego, 1964.
39. Frewer, L. J. and Lader, M., The effects of caffeine on two computerized tests of attention and vigilance, *Hum. Psychopharmacol.,* 6, 119, 1991.
40. Smith, A. P., Rusted, J. M., Eaton-Williams, P., Savory, M., and Leathwood, P., Effects of caffeine given before and after lunch on sustained attention, *Neuropsychobiology,* 23, 160, 1990.
41. Bättig, K. and Buzzi, R., Effect of coffee on the speed of subject-paced information processing, *Neuropsychobiology,* 16, 126, 1986.
42. Hasenfratz, M., Jaquet, F., Aeschbach, D., and Bättig, K., Interactions of smoking and lunch with the effects of caffeine on cardiovascular functions and information processing, *Hum. Psychopharmacol.,* 6, 277, 1991.
43. Kahneman, D., *Attention and Effort,* Prentice-Hall, Englewood Cliffs, NJ, 1973.
44. Mesulam, M.-M., Attention, confusional states, and neglect, in *Principles of Behavioral Neurology,* Mesulam, M.-M., Ed., F. A. Davies Company, Philadelphia, 1985, 125.

chapter fifteen

Caffeine and fatigue

Jan Snel and Monicque M. Lorist

Contents

I. Introduction 241
II. Types of fatigue...242
   A. Natural fatigue or sleepiness...242
   B. Experimentally induced fatigue...244
III. Discussion 251
IV. Summary 252
References..252

I. Introduction

Experimental evidence suggests that effects of caffeine on task performance, subjective reactions, and psychophysiological measures are state dependent. Caffeine is said to have, nonspecific arousal-enhancing effects and specific cognitive effects.[1,2] If true, especially in states of fatigue, energizing effects of caffeine should be found more prominently. Some studies could or do show that caffeine may improve performance beyond optimum levels.[3-6] However, in states of overarousal, such as agitation, excitement, or strenuous exercise, caffeine should deteriorate performance. In this chapter, the relationship of caffeine and fatigue will be discussed.

When a certain level of extrinsic and/or intrinsic stimulation is present, the subject will usually stay awake and be conscious of his environment. In a state of wakefulness, when sufficient stimulation is offered, the subject is said to be aroused or activated. In the case of an absence of stimulation, the subject will probably fall asleep. In this sense, one might say that arousal is the inverse probability of falling asleep.

The antihypnoid or fatigue-compensating effects of moderate amounts of caffeine, as habitually used, appear to be caused predominantly by the blocking of inhibitory adenosine receptors in the medial reticular formation (mRF). This activity of the mRF in the brain stem is gated, presumably[7,8] through the reticular nucleus of the thalamus to cortical areas. It implies that effects of caffeine depend on the level of cortical arousal, hence the term "state-dependent." Indeed, caffeine induced increased cortical arousal is manifested as a speeding up of frequency and amplitude of the EEG.[9-11] However, since these effects are nonspecific and depend on the effects of many concurrently running facilitory and inhibitory systems of the nervous system a decrease of arousal may also be found.[12] As a rule however, after ingesting doses to 250 mg caffeine, subjects report a decrease of fatigue, sleepiness, or drowsiness, and increased alertness or more vigor.[3,13-19]

II. Types of fatigue

Since effects of caffeine on fatigue might diverge, depending on the way fatigue has been defined, a first step is to distinguish types of fatigue. Two broad classes of fatigue will be discerned: "natural" fatigue or sleepiness and fatigue induced by experimental manipulation. The latter category is subdivided into three subtypes: fatigue induced by sleep deprivation; cognitive fatigue induced by performing mentally demanding tasks; physical fatigue induced by physical activity.

A. Natural fatigue or sleepiness

The occurrence of fatigue without any experimental manipulation may be said to represent natural wakefulness or the natural tendency to fall asleep. A well-known example in this category is the so-called after-lunch dip in alertness[20] or postprandial sleepiness.[21] If it is true that caffeine enhances physiological activation, caffeine should have positive effects on alertness at this time of the day. Indeed Smith and colleagues[22,23] could show that after administering 3 mg/kg of caffeine to 32 university students after abstaining from caffeine for 3 hours, caffeine improved performance on a vigilance task and removed the post-lunch dip in alertness. In order to study the nature of this post-lunch dip more thoroughly, Mavlee[20] combined conditions of high (29°C) and low temperature (18°C) and a boring vs. highly interesting situation. He found that reaction time and fatigue peaked at around 15 00 h and showed a linear trend in the evening. Taking into account the role of food made the author concluded that the curvilinear relation of reaction time and fatigue could be better referred to as the Afternoon Pressure for Sleep rather than the post-lunch dip in alertness.

Effects of caffeine on sedation–stimulation were studied by Leathwood and Pollet[24] in a setting where subjects were sitting quietly. Sixty subjects ranked, double-blind, the mood effects of 100 mg caffeine, 500 mg tryptophan, 500 mg tyrosine, or placebo. Caffeine increased wakefulness, clarity of mind, energy, vigor, and efficiency and was found to be the most stimulating compound compared to the other substances.

To evaluate the effects of caffeine on wakefulness and alertness more precisely, the circadian fluctuation in arousal and the habitual use of caffeine should be taken into account. The relevance of this is nicely illustrated by Kole et al.,[25] who examined the effects of 250 mg caffeine on event-related potentials (ERPs) in selective attention in extreme morning and evening types. In the early evening at 18 30 h in the extreme morning types, caffeine caused larger changes in the amplitude of the P3 compared to placebo than in evening types, indicating a higher cortical arousal level. Unfortunately, no attempt was made to determine the level of arousal of each subject at the time of measurement to determine the role of state-dependency in caffeine effects even further.

The important role of the circadian activity rhythm in the effects of caffeine on daytime sleepiness was further demonstrated by Johnson et al.[26] in 20.3 ± 2.7-yr-old marine personnel. Eighty nonsmokers with a daily caffeine intake of ≤3 cups received a capsule, always filled with placebo at 21 45 h, and either filled with 250 mg caffeine or placebo at 05 15 h. EEG recordings as part of the Multiple Sleep Latency Test (MSLT) were made every 2 h between 07 00 and 17 00 h. Subjects who got caffeine had longer falling-asleep latencies and a decreased subjective sleepiness throughout the day, especially in the first part of the day. Apparently, the relationship between these two measures depends on the arousal level at a specific point of the day. Diurnal variation of caffeine effects on alertness and cognitive function was assessed in 10 healthy regular users (age 19 to 28) by Lipschutz et al.[27] Seven days prior to the study they were asked to maintain a regular sleep schedule and to abstain from caffeine. From day 1 to 13 they got a 250 mg caffeinated beverage or a placebo at 08 30 h. Auditory evoked potentials (AEP) were recorded at 08 00, 14 30, and 16 30 h; and the MSLT was filled every second hour between 10 00 and 18 00 h. The subjects were less sleepy from day 1 on. This activation was also evident after repeated exposure on day 13 at a reduced, but still significantly higher level than the baseline. The lower alertness at mid afternoon, compared to other times on day 13, suggested the development of tolerance at a susceptible time of day. Zwyghuizen-Doorenbos et al.[15] used 24 male normal sleepers, nonsmokers, 21 to 36 years old, and with a daily caffeine intake of less than 250 mg to assess the alerting effect of caffeine at different times of the day. Caffeine (250 mg) was given at 09 00 and 13 00 h. The MSLT was given at 10 00, 12 00, 14 00 and 16 00 h. Caffeine, compared to placebo, improved alertness and auditory vigilance. Although the sleep-latency in the caffeine group diminished 3 h post-treatment, the delay in falling asleep remained above that of the placebo treatment. It indicates that caffeine increases wakefulness, although this effect diminishes over time.

The role of habitual use of caffeine is illustrated by Regestein.[28] He described 6 heavy coffee drinkers (5 to 12 cups daily) who felt very drowsy and confused in the morning and complained of excessive sleepiness during daytime. After stopping their caffeine consumption, sleepiness decreased or remitted. The explanations for the pathological sleepiness were the nocturnal excitation by caffeine with the subsequent daily sleepiness, a direct sedative action of caffeine, or idiosyncrasies of the subjects. Bonnet and Arand[29] used caffeine to develop a physiological arousal model of chronic insomnia in a group of healthy adults. The subjects received 400 mg caffeine three times daily for 7 days and nights. In addition to a reduced sleep efficiency, sleepiness during daytime measured with the MSLT was significantly increased.[14] Obviously, physiological arousing effects of caffeine at night may cause the increased sleepiness during daytime. Only a few studies are known in which subjects were assessed during their sleep for the role of caffeine in wakefulness or alertness. The effect of 4 mg caffeine/kg administered at 23 20 to 10 normal sleepers with a mean age of 30.3 yrs was assessed by Walsh et al.[30] Caffeine reduced sleepiness, as measured with the MSLT and promoted alertness. Bonnet and co-workers[31,32] used auditory stimuli of 1,000 Hz stimuli 5 to 8 times during sleep to measure awakening thresholds and sleep latencies of 6 normal sleepers (aged 21 to 23 yrs). The 400 mg caffeine or placebo was given 15 min before going to sleep at 23 30 h. The sleep latencies increased throughout the night, although the efficacy of caffeine became less with time. Caffeine taken during awakenings from sleep at night is associated with increased wakefulness and alertness. Indeed, caffeine in moderate doses might be a useful antidote to fatigue. However, the evidence so far does not indicate whether caffeine increases alertness directly or does so indirectly through a compensation of fatigue. Also, the sometimes negative repercussion effects of caffeine on sleep, its subsequent daytime fatigue and decreased vigor, ask for a more thorough study of the efficacy of caffeine on fatigue.

B. Experimentally induced fatigue

1. Sleep deprivation (SD)

One obvious way to induce fatigue is to disrupt the sleep–wake cycle by keeping the subject awake for some time. The assumption is, the more hours of SD, the more fatigue, and vice versa. However, self-reporting on fatigue may be biased by expectancy. For example, when keeping subjects awake until the early morning or after a hard day's work, they expect to be tired, and after a long rest one would expect to be well-rested. In particular, the latter assumption should be treated with caution, since sleep restriction up to a certain limit may have arousing effects.[33] This also appears from studies using SD as a therapeutic treatment in depression.[34] Extending sleep to more than 10 hours rather than the usual 7.45 h of sleep may result in changes indicating increased arousal, such as jitteriness, agitation, headache, and so on, sometimes called postdormital confusion. Put in more general terms, a regular, habitual sleep length, considered by the subject as normal, pleasant, and sufficiently efficient to recover, might be indicative during daytime of a more or less optimum level of wakefulness and functioning, while a reduction of sleep may effectuate a higher than usual level of arousal, and an extension of sleep beyond its normal length may result in lower than habitual arousal levels.

So positive effects of caffeine found on mood, task performance, and EEG-activity in fatigued compared to well-rested people do not necessarily mean that these effects are caused exclusively by caffeine. They are, however, at least the outcome of the interaction of the effect of caffeine, the effort to mobilize energetical resources, and possibly relief of caffeine withdrawal effects.[35] This confounding of caffeine effects with withdrawal effects is probably not the case in the study by Penetar et al.[16] in 15 healthy, nonsmoking 18 to 32-yr-old men. A caffeine and alcohol abstinence period of 72 hours was used. The aim of the study was to determine the ability of caffeine (150, 300, or 600 mg/70 kg) to restore changes in mood and alertness after 49 hours of sleep deprivation. Caffeine restored the increase of subjective sleepiness to near rest values, while the changes of vigor, fatigue, and confusion were reversed. A similar study was done by Wright et al.[36] who studied 46, 18, to 28-year old, healthy female college students who were sleep deprived for 48 hours, using a constant routine procedure. Caffeine alone (a dose of 100 mg given 4 times at 20 00, 23 00, 02 00, and 05 00 h) improved performance and alertness. These effects were greatest in the morning hours between 02 00 and 08 00 h. Apparently, caffeine does produce alerting and long-lasting beneficial mood effects in subjects deprived of sleep for long periods. Since this period is long enough not to be confounded with caffeine withdrawal effects, the improvement of mood, performance and alertness might be described to caffeine only.

Whether abstinence from caffeine was applied in the early Muehlbach and Walsh' studies (1991, 1993)[37,38] was not mentioned, but in their 1995 study abstinence began 7 days prior to the study.[39] Performances in a simulated assembly line task were improved after 2 mg/kg caffeine at 22 20 and 01 20 for 5 consecutive nights in 30 young adults (age m = 24.7 yrs). Physiological sleepiness (MSLT) was measured every 2 hours from 23 00 to 07 00 h. The data on sleepiness showed that the caffeine group was more alert on nights 1 and 3 compared to the controls. Caffeine clearly decreased physiological sleepiness compared to placebo, but failed to completely overcome the dip in the sleepiness/alertness circadian rhythm. Also, there were no differences between the groups on the POMS-mood scales or on any objective measure of daytime sleepiness. The conclusion was that during night shifts caffeine appears to be beneficial in improving alertness, without detrimental effects on mood or subsequent daytime sleep. However, night time performance was not significantly improved, and sleepiness at the circadian trough (05 00 to 07 00 h) could not be compensated fully by the caffeine dose of 2 mg/kg. Similar positive effects of caffeine on task performance and fatigue during overnight working were found by other investigators.[14,30,40-43]

The summarizing conclusion is that caffeine in doses up to 300 mg in overnight work apparently has dose-related beneficial effects on task performance and fatigue. Deviating findings are reported by Plath et al.[44] Sleep restriction until 02 30 h was introduced by Plath et al.[44] to 27 19 to 35-yr-old men. Caffeine up to 300 mg produced the same amount of increased wakefulness when the basal sleepiness level is normal as when it is increased.[4,45] As argued before, it suggests that the level of sleepiness does not always interact with the alerting effects of caffeine.

After a night of sleep deprivation with a maximum of 3 hours of sleep, Engleman and co-workers[46] gave 11 medical students a dose of 200 mg every 2 hours between 07 00 h to 17 00 h. In the late afternoon, caffeine compared to placebo improved cognitive performance for all tests. After a recovery night the task performance scores of the next morning were still better in the caffeine group. The latter suggests that caffeine intake during a period of sleep deprivation need not go at the cost of a good night's sleep, but may increase the efficiency of the arousal system, such that spared capacity of the system can be applied to subsequent tasks. Arousal, a concept ranging from extreme fatigue at one end to extreme excitement at the other, influences the stimulus evaluation system according to Humphreys and Revelle.[47] The stimulus evaluation system and the response activating system represent two subsystems in the human information processing system.[48] Since arousal helps rapid information processing, sleep deprivation or fatigue will affect the encoding and response activation negatively.[49] Consequently, caffeine as a hypothesized antidote for fatigue must have beneficial effects on these stages in particular. This hypothesis was tested extensively by Lorist and co-workers.[3] To ascertain whether caffeine would affect the specific stages of information processing, special attention was paid to manipulation of task variables. Task variables were stimulus-degradation, stimulus-response compatibility, and time-uncertainty, representing the encoding, response selection, and motor activation stages, respectively. After 12 hours of caffeine abstinence, the 30 regular coffee users were kept awake until 04 00 h. After 250 mg of caffeine, mood and task performance were assessed and event related potentials (ERPs) were recorded. The ERP data showed that in 24 of the 30 subjects, caffeine improved mood and task performance and had its effect predominantly on stimulus evaluation processes and response activation. This effect of caffeine was more pronounced after sleep deprivation. The results supported the hypothesis that sleep deprivation affects these stages of information processing especially.

At first glance, SD is an attractive and easy method to induce fatigue. Inspecting SD more carefully, it has serious methodological shortcomings. First, SD knows no placebo treatment and cannot be applied unnoticed to the subject. Second, keeping subjects awake disturbs the activity–rest cycle relative to other rhythms. Since desynchronization of endogenous rhythms as evidenced from night and shift work is an intensive stressor, the effects of caffeine in sleep deprivation studies must be the result of the combined effects of fatigue, changes in arousal, and desynchronization of circadian rhythms; the latter is sometimes used as a definition of stress. Another problem inherent to SD is how to keep subjects awake. The usual approach is to keep subjects physically and mentally inactive. Only low-intensity activities are permitted, such as reading, playing cards, small talk, etc. Small moments of sleep, so-called *micro sleeps* may intrude upon the waking state and remain unnoticed, and bouts of extreme sleepiness may occur. To prevent subjects from falling asleep, manipulations have to be made, self-evidently of an arousing nature, such as physical activity, walking, ball games, or pedaling on an ergometer, which may unintentionally induce fatigue.

Keeping these comments in mind, the cautious conclusion is that caffeine may compensate SD-induced fatigue, which is probably achieved by an interaction of factors not controlled for. Hence, in order to assess the exact contribution of caffeine to the alleviation of sleep loss is a task still to be done.

2. Mental fatigue

An important factor in the fatiguing potential of cognitive tasks is task complexity. In general, tasks which show straightforward deterioration over time are mostly simple, monotonous tasks, such as Wilkinson vigilance-like tasks or simple reaction time tasks. In more complex and perhaps arousing, intrinsically interesting tasks, fatiguing effects are often less direct. Application of the Easterbrook-hypothesis[50] on caffeine implies that in a state of heightened arousal, performance of simple tasks will benefit more from caffeine than very complex tasks, since cue utilization will be hindered by increased arousal. As for the capacity to attract interest, more complex tasks may be more interesting than monotonous, simple tasks such as vigilance tasks, simple reaction time tasks, etc. Roughly, two types of mental tasks can be discerned: simple tasks provoking boredom or understimulaton and complex tasks provoking interest or overstimulation.

The role of caffeine in performing cognitive tasks is often interpreted as changes in subjective energy. Indeed Bruce et al.[10] and Zwyghuizen-Doorenbos et al.[15] did show that caffeine diminished tiredness and increased alertness, which could be interpreted as a caffeine-induced increased level of arousal. Another line of reasoning comes from theories stressing that the energetical states of the subject play a controlling role in information processing.[13,47,49] Support for this view comes from studies showing that caffeine increases EEG power.[3,10,17,25,51,52] According to the Humphreys–Revelle model, the information transfer component of cognitive tasks should be facilitated by an increase of arousal, and since caffeine is supposed to do so, numerous studies have been performed to verify this assumption. Although there are inconsistent findings, the position still holds that the effects of caffeine on cognitive function are mediated predominantly by arousal or the energetical state of the subject. It is obvious that many factors play a role in this state, such as time of day, personality, age, etc. It is beyond the scope of this discussion to ascertain the role of each factor. In an attempt to interpret information processing more satisfactorily and in more detail, multiple energy resources have been assumed, which supposedly perform a modulating role in cognitive operations.[49,53,54] Studies have been performed recently to determine the role of caffeine in these energy resources.[3,17,55,56]

These show that caffeine has specific energetic effects on the input and output stages of information processing, but not on the central processing stage.

Effects of caffeine on vigilance, representing the simple cognitive task type have been reviewed by Koelega.[57] He left out those studies that for several reasons precluded comparison with the usual simple vigilance task. After caffeine, 6 studies showed no improvement, 3 showed improvement under special conditions (only in the first experiment, at the end of a 60 min task; in the elderly 3 h after treatment); 14 studies reported an improvement in hits. One of Koelega's conclusions was that caffeine improves the absolute level of stimulus detection. Since detection of stimuli represents a stimulus evaluation activity, it supports Lorist's conclusion[3] that caffeine produces effects on the input stage of the information processing system.

A more complex vigilance task has been used by Bättig, in which presenting digits occurred at a subject-paced rate rather than at a fixed rate. Twenty female subjects, aged ± 42.2 years were given a caffeinated beverage containing 150 or 450 mg caffeine or decaffeinated coffee.[4] Fatigue-induced performance decrements were significant across sessions. Interestingly, no interaction was found between arousal state and caffeine. Corresponding results were reported by Regina et al.,[58] Plath et al.,[44] and as referred to before by Lorist et al.[3,17] in well-rested and fatigued subjects, implying that caffeine may improve performance independent of state. It suggests that caffeine might be able to let subjects invest effort in a well-rested condition, when fatigued and consequently when physically exhausted (see II.B.3: Physical Fatigue). Beneficial effects of decaffeinated and caffeinated (300 mg) coffee on a 21-item free word association test were found by Pons et al.[59] in 19 normal subjects and two groups of 40 controls (age ± 21 yrs). Caffeine was found to increase attention and vigilance more than the placebo as shown from the more numerous response repetitions. Bruce and colleagues[10] assessed the effects of 250 and 500 mg caffeine in 9 healthy moderate coffee-drinkers (18 to 40 yrs) who abstained from caffeine for 24 hours. No effect of caffeine was found on tapping rate, simple reaction time, digit symbol substitution, and symbol copying. The objective and subjective tiredness and alertness all pointed to an increased level of arousal.

Roache and Griffiths[60] assessed the interactive effects of diazepam and caffeine (0, 200, 400, 600 mg) alone and in combination in 9 healthy males. Caffeine improved performance in the rapid reaction speed tests and apparently decreased fatigue, as could be inferred from the increased restlessness, subjective tension, and alertness. The effects of repeated caffeine administration on cognitive and mood tests in 32 students 19 yrs ± 1.07 yrs old were investigated by Loke.[61] In general, caffeine produced no significant effects on addition, card rotation, sequence completion, and symbol cancellation; the 600 mg dose slowed the completion of additions relative to the other doses and also impaired the performance of the low caffeine users more than that of the heavy users. In conclusion in relatively complex tasks caffeine impairs performance due to an inability to select the relevant cues.

Although mostly beneficial effects of caffeine have been reported, detrimental effects of caffeine have been demonstrated on short-term memory.[60,62] Although fatigue is not a subject of the major part of studies on caffeine, a reasonable assumption is that the more complex a cognitive task is, the earlier it will fatigue the subject. This assumption may seem a simple way of reasoning, but it can be supported from the following argument based on our experimental findings. Energetical aspects of information processing can be described in terms of the Yerkes–Dodson law, which says that negative effects of caffeine occur in tasks that are rather complex or occur in combination with arousal-related factors such as dose, personality, circadian arousal rhythm, etc. What may occur under the influence of caffeine-induced changes in arousal are changes in attention. We could illustrate in our ERP-data[17] that caffeine enhanced the processing of irrelevant input that had to be ignored, but did not, as might be expected, adversely affect the incidence of hits or errors. In an attempt to explain this, the presumption was made that the caffeine effects on the ERPs to relevant stimuli, which had to be attended, were largely masked by increased processing requirements associated with the "attend" condition. This presumption was supported by the P3 amplitudes, an index of the energetic level of information processing. These were larger for the relevant than for the irrelevant stimuli. This was interpreted as a larger energy investment in attention to relevant information. In contrast to its larger effect on the ERP-amplitudes of the irrelevant stimuli, caffeine had only minor effects on the ERPs of relevant stimuli. Apparently, caffeine improves the signal–noise ratio by a more intense processing of irrelevant information, that is by a better ignorance, but contributes only marginally to the processing of relevant information.[1,2,17] Consequently, in a fatigued state, processing irrelevant information is less hampered than processing relevant information, and caffeine in such a state will exert its effect predominantly on the processing of relevant information. This speculation needs additional experimental verification.

3. Physical fatigue

Most literature on the role of caffeine in physical performance supports the fatigue hypothesis. This hypothesis says that caffeine as an ergogenic aid, delays fatigue and exhaustion by sparing the muscle glycogen depots or by enhancing free fatty acid utilization,[63,64] which will result in an enhanced metabolic efficiency.[65] However, the literature is not consistent. Although caffeine may have ergogenic effects for some activities, such as mid-term and long-term performance[66,67] at least at submaximal intensity,[68] it does not appear to improve short term performance convincingly and consistently.[68,69] Physical exercise in untrained sportsmen, not accustomed to exercising, results in a relatively lasting increase of physiological activation, which may be manifested as sleep disturbance, increased heart rate, and build up of lactic acid. In trained, fit subjects, habituated to exercise, a return to basic levels of arousal goes relatively fast due to their greater physiological reactivity, and hence effects of similar amounts of exercise may go unnoticed and will not result in physical fatigue. Hill and Smith[70] studied 6 healthy men age 22 ± 3 yr at 03 00, 09 00, 15 00, and 21 00 h with the POMS and a physical exertion test. The fatigue score on the POMS correlated nonsignificantly (−.28) with the anaerobic peak power. Time of day and fatigue together explained 51% of the variance in peak power. Their conclusion was that feelings of fatigue and subsequent performance are dependent on physical fitness. It suggests also that variation in physical performance, caffeine treatment effects, and mood are a function of fitness level and time of day. So the fitness level is an important determinant of the way in which fatigue induced by physical activity is expressed in physiological parameters.

There are only a few studies which show positive effects of caffeine on the early stages of physical exercise[58,71,72] or on acute performance, such as reaction time and muscle strength. Bättig and Buzzi[4] refer to Eichler, who found an improvement on the early stages of different types of physical performance after 250 mg of caffeine. The general picture is that studies regarding any ergogenic effect of caffeine in endurance, muscle strength, movement time, coordination, etc., have found confusing and confounding results.[73] If caffeine could postpone fatigue by mobilizing spare capacity, it would be interesting to assess the effects of caffeine on performance *after* straining physical performance. Unfortunately, there is hardly any evidence on caffeine effects in such a situation, although the study done by Falk et al.[74] might be considered. The effects of a 5 mg/kg dose of caffeine were determined on a 40 km march in two groups matched in aerobic capacity. After the march, a cycle ergometer test until exhaustion was performed at 90% of maximal oxygen consumption. It appeared that caffeine could not postpone exhaustion following prolonged exercise. The authors conclude that caffeine could not improve the ability to sustain intensive exercise. This conclusion can be reworded, saying that caffeine apparently is not able to compensate for physical fatigue for long periods.

An aspect of fatigue induced by exercise is that it may trigger changes similar to stress, such as increased blood pressure, heart, and respiratory rate, and sweating, and at the behavioral level, more agitation, tension, rivalry, or even aggression. The crucial factor in these changes seems to be the level of fitness. Highly trained, aerobically fit subjects are characterized in rest by a low cardiorespiratory activity, lowered muscle tonus, and a high cardiovascular reactivity, while unfit subjects show mostly the opposite. Due to these differences unfit people react to intensive or enduring physical exercise with long-lasting changes in the before-mentioned physiological parameters. Such changes are indicative of increased physiological arousal. So rather than lowered levels of those physiological parameters, unfit subjects during and after exercise can be described physiologically as being highly activated, restless, and fully awake. Hence, when performing on identical physical performance task, untrained, unfit subjects, due to the evoked higher arousal level, will profit less from caffeine than well-trained subjects.[68,75,76] Caffeine will help trained people more than untrained people to sustain performance by its greater capacity to compensate low arousal.

Apparently, caffeine does not act directly on sustained physical performance by increasing physiological arousal, but indirectly by its power to make the subject invest more effort in the task by allocating spare capacity. The current state of our poor knowledge of the role of caffeine on physical performance, while taking into account fitness level, does not permit firmer conclusions. One reason is that to our knowledge no studies have been designed with the explicit aim of assessing the effects of caffeine on physical fatigue. Other reasons are that most physical performance studies suffer from methodological flaws, such as poor standardization of protocol, small numbers of subjects, the use of either highly trained or untrained subjects[75] and different types of exercise of varying intensities and varying periods of time.[68,73] Since there is no a good understanding of the energy mechanisms (physical and/or physiological) involved in performing physical tasks in comparison with cognitive tasks, it is obvious that much work still has to be done before ascertaining the role of caffeine in physical performance studies.

III. Discussion

A serious problem (see also Reference 77) that arose in the present contribution was the absence of a reliable and valid definition of the concept of fatigue. Is fatigue the lessened capacity to perform? Is it a suboptimum physiological or cortical arousal, the reluctance to invest effort, or is it simply "hours on duty," hours of sleep loss, etc? To circumvent this problem, an attempt was made to use the concept of arousal as the ubiquitous factor which is related to fatigue.

Effects of caffeine seem to obey the law of initial values of arousal saying that caffeine effects are determined by the subject's arousal level. In states of high arousal caffeine might lead to overarousal and impair performance, while in states of underarousal caffeine may increase arousal to more optimum levels. In states of high arousal, caffeine is less able to mobilize additional arousal or energy than in states of low arousal. Since the efficacy of moderate amounts of caffeine depends predominantly on inhibition of adenosine receptors and since these receptors are widespread in the CNS, caffeine not only may have widespread effects, it may also affect the balance of inhibitory and facilitory activity. This notion may help explain sometimes the puzzling and unexpected findings of caffeine studies. For example, Klein and Saltzman[78] found that 2 h after taking 300 mg of caffeine the amplitude of auditory evoked potentials produced by sounds of 1000 Hz and 60 dB were smaller than under placebo conditions. Pollock et al.[79] showed that after caffeine intake the EEG activity was reduced during responding to stimulating modulated light. Since these results indicate a caffeine induced decrease of energy levels, instead of increases, the interpretation is that caffeine may have biphasic effects under certain circumstances.

Since fatigue as a construct is still hardly measured, the effects of caffeine on it were inferred from changes in physiological measures, performance, and mood. In general, the available evidence points out that caffeine may indeed decrease fatigue, although there are exceptions. It also pointed to the plausible possibility that the underlying arousal system for cognitive and physical fatigue is different, although the common factor in all types of fatigue might be the capacity of caffeine to increase effort, that is to mobilize spare energy capacity. Using this conception of fatigue implies an important role for motivation.

Considering these reflections together, the summarizing conclusion can only be preliminary. Caffeine in moderate doses has energizing effects and is capable of compensating fatigue, in spite of the absence of a definition of fatigue, the lack of standardization of protocols, and confounding factors such as withdrawal effects, personality, time of day, physical fitness, etc. In order to draw firmer conclusions on the effects of caffeine on fatigue, attention should first be paid to a valid and reliable measurement of fatigue.

IV. Summary

The role of caffeine in fatigue was evaluated by reviewing studies on cognitive and physical performance. Fatigue was divided into two categories: natural fatigue and experimentally induced fatigue.

The concept of arousal was used as the central concept in an attempt to interpret the efficacy of caffeine on fatigue. Since fatigue is not the main subject in the studies referred to, it is rather hard to assess its role in these studies, not to mention the effect of caffeine on fatigue.

With this point in mind, the cautious conclusion is that caffeine has the potential to alleviate fatigue, although its precise role needs clarification in research aimed exclusively at the relationship of caffeine and fatigue.

References

1. Van der Stelt, O. and Snel, J., Effects of caffeine on human information processing — A cognitive-energetic approach, in *Caffeine, Coffee and Health*, Garattini, S., Ed., Raven Press, New York, 1993, 291.
2. Van der Stelt, O. and Snel, J., Caffeine and human performance, in *Nicotine, Caffeine and Social Drinking — Behaviour and Brain Function*, Snel, J. and Lorist, M.M., Eds., Harwood Academic Publishers, Reading, U.K., 1998, 167.
3. Lorist, M. M., Snel, J., and Kok, A., Influence of caffeine on information processing stages in well rested and fatigued subjects, *Psychopharmacology,* 113, 411, 1994.
4. Bättig, K. and Buzzi, R., Effect of coffee on the speed of subject-paced information processing, *Neuropsychobiology,* 16, 126, 1986.
5. Weiss, B. and Laties, V.G., Enhancement of human performance by caffeine and the amphetamines, *Pharmacol. Rev.,* 14, 1, 1962.

6. Rogers, P. J., Richardson, N. J., and Dernoncourt, C., Caffeine use: Is there a net benefit for mood and psychomotor performance? *Neuropsychobiology,* 31, 195, 1995.
7. Skinner, J. E. and Yingling, C. D., Central gating mechanisms that regulate event-related potentials and behavior, in *Attention, Voluntary Contraction and Event-Related Cerebral Potentials,* Vol. 1., Desmedt, J. E., Ed., Karger, Basel, 1977, 30.
8. Yingling, C. D. and Skinner, J. E., Gating of thalamic input to cerebral cortex by nucleus reticularis thalami, in, *Attention, Voluntary Contraction and Event-Related Cerebral Potentials,* Vol. 1., Desmedt, J. E., Ed., Karger, Basel, 1977, 70.
9. Künkel, H., Vielkanal EEG- Spektralanalyse der Coffein-Wirkung, *Z. Ernährungswiss.,* 15, 71, 1976.
10. Bruce, M., Scott, N., Lader. M., and Marks. V., The psychopharmacological and electrophysiological effects of single doses of caffeine in healthy human subjects, *Br. J. Clin. Pharmacol.,* 22, 81, 1986.
11. Hasenfratz, M. and Bättig, K., Acute dose-effect relationships of caffeine and mental performance, EEG, cardiovascular and subjective parameters, *Psychopharmacology,* 114, 281, 1994.
12. Lorist, M. M., *Caffeine and Human Information Processing,* Ph.D. Thesis, University of Amsterdam, Faculty of Psychology, Dept. of Psychonomics, Amsterdam, 1995.
13. Clubley, M., Bye, C. E., Henson, T. A., Peck, A. W., and Riddington, C. J., Effects of caffeine and cyclizine alone and in combination on human performance, subjective effects and EEG activity, *Br. J. Clin. Pharmacol.,* 7, 157, 1979.
14. Koopmans, R. and Van Boxtel, C. J., The influence of caffeine on the adjustment to night shift, in *Chronopharmacology and Shift Work: Studies with Oxyprenolol, Midazolam, Terbutaline, Nityroglycerin, Prednisolone, Dexamethasone and Caffeine,* Koopmans, R., Ed., Unpublished doctoral dissertation, University of Amsterdam, Amsterdam, 1990, 171.
15. Zwyghuizen-Doorenbos, A., Roehrs, T. A., Lipschutz, L., and Timms, V., Effects of caffeine on alertness, *Psychopharmacology,* 100, 36, 1990.
16. Penetar, D., McCann, U., Thorne, D., Kamimori, G., Galinski, C., Sing, H., Thomans, M., and Belenky, G., Caffeine reversal of sleep deprivation effects on alertness and mood, *Psychopharmacology,* 112, 359, 1993.
17. Lorist, M. M., Snel, J., Kok, A., and Mulder, G., The influence of caffeine on selective attention in well rested and fatigued subjects, *Psychophysiology,* 31, 525, 1994.
18. Horne, J. A. and Reyner, L. A., Counteracting driver sleepiness: effects of napping, caffeine, and placebo, *Psychophysiology,* 33, 306, 1996.
19. Reyner, L. and Horne, J., Caffeine combined with a short nap effectively counteracts driver sleepiness, *Sleep Res.,* 26, 625, 1997.
20. Mavlee, V., Sleepiness, time of day, boredom and room temperature, in *Sleep '90,* Horne, J., Ed., Pontenagel Press, Bochum, 1990, 55.
21. Harnish, M. J., Orr, W. C., and Chard, S. R., *Postprandial Sleepiness: Is It All in the Head?* 10th Annual Meeting, May 28-June 2, Washington, D. C., Sleep Res. Society and the American Sleep Disorders Association, Abstractbook, 1996, 160.
22. Smith, A. P., Rusted, J. M., Eaton-Williams, P., Savory, M., and Leathwood, P., Effects of caffeine given before and after lunch on sustained attention, *Neuropsychobiology,* 23, 160, 1990.

23. Smith, A., Effects of caffeine on attention: Low levels of arousal, in *Nicotine, Caffeine and Social Drinking* — *Behaviour and Brain Function*, Snel, J. and Lorist, M.M., Eds., Harwood Academic Publishers, Reading, U.K., 1998, 215.

24. Leathwood, P. D. and Pollet, P., Diet-induced mood changes in normal populations, *J. Psychiat. Res.*, 17, 147, 1982-83.

25. Kole, A., Snel, J., and Lorist, M. M., Caffeine, morning-evening type and coffee odour, attention, memory search and visual event related potentials, in *Nicotine, Caffeine and Social Drinking* — *Behaviour and Brain Function*, Snel, J. and Lorist, M.M., Eds., Harwood Academic Publishers, Reading, U.K., 1998, 201.

26. Johnson, L.C., Freeman, C. R., Spinweber, C. L., and Gomez, S. A., Subjective and objective measures of sleepiness: effect of benzodiazepine and caffeine on their relationship, *Psychophysiology*, 28, 65, 1991.

27. Lipschutz, L., Berman, S., and Spielman, A. J., Acute and chronic caffeine administration: alertness, cognition, and diurnal variation, *Sleep Res.*, 19, 74, 1990.

28. Regestein, Q. R., Pathological sleepiness induced by caffeine, *Am. J. Med.*, 87, 586, 1989.

29. Bonnet, M. H. and Arand, D. L., Caffeine use as a model of acute and chronic insomnia, *Sleep*, 15, 526, 1992.

30. Walsh, J. K., Muehlbach, M. J., Humm, T. M., Dickins, Q. Stokes, R., Sugerman, J. L., and Schweitzer, P. K., Effect of caffeine on physiological sleep tendency and ability to sustain wakefulness at night, *Psychopharmacology*, 101, 271, 1990.

31. Bonnet, M. H., Webb, W.B., and Barnard, G., Effects of flurazepam, pentobarbital and caffeine on arousal threshold, *Sleep*, 1, 271, 1979.

32. Bonnet, M. H. and Webb, W. B., The return to sleep, *Biol. Psychol.*, 8, 225, 1979.

33. Hicks, R.A. and Guista, M., The energy level of habitual long and short sleepers, *Bull. Psychon. Soc.*, 19, 131, 1982.

34. King, D., Sleep deprivation therapy in depression syndromes, *Psychosomatics*, 21, 404, 1980.

35. Linde, L., Mental effects of caffeine in fatigued and non fatigued female and male subjects, *Ergonomics*, 38, 864, 1995.

36. Wright, K.P., Badia, P., Plenzler, S. C., Myers, B. L., and Drake, C. L., Enhancement of nighttime alertness and performance in women with bright light and caffeine during 48 h of sleep deprivation, *Sleep Res.*, 26, 635, 1997.

37. Muehlbach, M. J., Schweitzer, P. K., Stuckey, M. L., and Walsh, J. K., The effect of caffeine on continuous performance at night, *Sleep Res.*, 20, 464, 1991.

38. Muehlbach, M. J. and Walsh, J.K., The effects of caffeine on simulated night shift work, *Sleep Res.*, 22, 412, 1993.

39. Muehlbach, M.J. and Walsh J.K., The effects of caffeine on simulated night shift work and subsequent daytime sleep, *Sleep*, 18, 22, 1995.

40. Borland, R. G., Rogers, A. S., Nicholson, A. N., Pascoe, P. A., and Spencer, M.B., Performance overnight in shiftworkers operating a day-night schedule, *Aviat. Space Environ. Med.*, 57, 241, 1986.

41. Rogers, A. S., Spencer, M. B., Stone, B. M., and Nicholson, A. N., The influence of a 1 h nap on performance overnight, *Ergonomics*, 32, 1193, 1989.

42. Walsh, J. K., Muehlbach, M. J., and Schweitzer, P. K., Hypnotics and caffeine as countermeasures for shiftwork related sleepiness and sleep disturbance, *J. Sleep Res.*, 4(suppl. 2), 80, 1995.

43. Snel, J., Coffee and caffeine — sleep and wakefulness, in *Caffeine, Coffee and Health*, Garattini, S., Ed., Raven Press, New York, 1993, 255.

44. Plath, D., Roehrs, T. A., Zwyghuizen-Doorenbos, A., Sicklesteel, J., Wittig, R. M., and Roth, T., The alerting effects of caffeine after sleep restriction, *Sleep Res.*, 20, 124, 1989.

45. Lumley, M., Roehrs, T., Asker, D., Zorick, F., and Roth, T., Ethanol and caffeine effects on daytime sleepiness/alertness, *Sleep*, 10, 306, 1987.

46. Engleman, H., Ronald, P., and Shapiro, C. M., The effect of caffeine and sleep deprivation on daytime performance, Paper presented at the VIIIth Eur. Sleep Congr. Strasbourg, France, 1990.

47. Humphreys, M.S. and Revelle, W., Personality, motivation, and performance: a theory of the relationship between individual differences and information processing, *Psychol. Rev.*, 91, 153, 1984.

48. Gunter, T. C., Van der Zande, R.D., Wiethoff, M., Mulder, G., and Mulder, L. J. M., Visual selective attention during meaningful noise and after sleep deprivation, in *Current Trends in Event-Related Potential Research, (EEG Suppl. 40)*, Johnson, R., Jr., Rohrbaugh, J. W., and Parasuraman, R., Eds., Elsevier Science Publishers B. V., Amsterdam, 1987.

49. Sanders, A. F., Toward a model of stress and human performance, *Acta Psychologica*, 53, 61, 1983.

50. Easterbrook, J. A., The effect of emotion on cue utilization and the organization of behavior. *Psychol. Rev.*, 66, 183, 1959.

51. Münte, T.-F., Heinze, H.-J., Künkel, H., and Scholz, M., Personality traits influence the effects of diazepam and caffeine on CNV magnitude, *Neuropsychobiology*, 12, 60, 1984.

52. Goldstein, L., Muphee, H. B., and Pfeiffer, C. C., Quantitative electroencephalography in man as a measure of CNS stimulation, *Ann. N. Y. Acad. Sci.*, 107, 1045, 1963.

53. Hockey, G. R. J., A state control theory of adaptation to stress and individual differences in stress management, in *Energetics and Human Information Processing*, Hockey, G. R. J., Gaillard, A. W. K., and Coles, M. G. H., Eds., Nijhoff, Dordrecht, the Netherlands 1986, 285.

54. Pribram K. H. and McGuiness, D., Arousal, activation and effort in the control of attention, *Psychol. Rev.*, 82, 116, 1975.

55. Lorist, M.M., Snel, J., Mulder, G., and Kok, A., Aging, caffeine and information processing, *Electroencephalogr. Clin. Neurophysiol.*, 96, 453, 1995.

56. Lorist, M.M., Caffeine and human information processing, in *Nicotine, Caffeine and Social Drinking — Behaviour and Brain Function*, Snel, J. and Lorist, M.M., Eds., Harwood Academic Publishers, Reading, U.K., 1998, 185.

57. Koelega, H., Effects of caffeine, nicotine and alcohol on vigilance performance, in *Nicotine, Caffeine and Social Drinking — Behaviour and Brain Function*, Snel, J. and Lorist, M.M., Eds., Harwood Academic Publishers, Reading, U.K., 1998, 363.

58. Regina, E. G., Smith, G. M., Keiper, C.G., and McKelvey, R. K., Effects of caffeine on alertness in simulated automobile driving, *J. Appl. Psychol.*, 59, 483, 1974.

59. Pons, L., Trenque, T., Bielecki, M., Mopulin, M., and Potier J. C., Attentional effects of caffeine in man: Comparison with drugs acting upon performance, *Psychiat. Res.*, 23, 329, 1988.

60. Roache, J. D. and Griffiths, R. R., Interactions of diazepam and caffeine: behavioral and subjective dose effects in humans, *Pharmacol. Biochem. Behav.*, 26, 801, 1987.

61. Loke, W. H., Effects of caffeine on mood and memory, *Physiol. Behav.*, 44, 367, 1988.
62. Terry, W. S. and Phifer B., Caffeine and memory performance on the AVLT, *J. Clin. Psychol.*, 42, 860, 1986.
63. Casal, D.C. and Leon, A.S., Failure of caffeine to affect substrate utilization during prolonged running, *Med. Sci. Sports Exer.*, 17, 174, 1985
64. Tarnapolsky, M. A., Caffeine and endurance performance, *Sports Med.*, 18, 109, 1994.
65. James, J. E., *Caffeine and Health*, Academic Press, London, 1991, 30.
66. Spriet, L. L., MacLean, D. A., Dyck, D. J., Hultman, E., Cederblad, G., and Graham, T.E., Caffeine ingestion and muscle metabolism during prolonged exercise in humans. *Am. J. Physiol.*, 262, 891, 1992.
67. Tarnapolsky, M. A., Atkinson, S. A., MacDougall, J. D., Sale, D. G., and Sutton, J. R., Physiological responses to caffeine during endurance running in habitual caffeine users, *Med. Sci. Sports Exer.*, 21(4), 418, 1989.
68. Nehlig, A., Daval, J. L., and Debry, G., Caffeine and the central nervous system: mechanisms of action, biomedical, metabolic and psychostimulant effects, *Brain Res. Rev.*, 17, 139, 1992.
69. Lowenthal, D. T. and Kendrick, Z. V., Drug-exercise interactions, *Annu. Rev. Pharmacol. Toxicol.*, 25, 275, 1985.
70. Hill, D. W. and Smith, J.C., Effect of time of day on the relationship between mood state, anaerobic power, and capacity. *Percept. Mot. Skills*, 72, 83, 1991.
71. Anselme, F., Collomp, K., Mercier, B., Ahmaïdi, S., and Prefaut, C., Caffeine increases maximal aerobic power and blood lactate concentration. *Eur. J. Appl. Physiol.*, 65, 188, 1992.
72. Krueger, H., Zülch, J., and Gandorfer, M., Der Einfluss von Kaffein auf die motorische Reaktions- und die visuell-mentale Verarbeitungszeit, *Z. Ernährungswiss.*, 18, 51, 1979.
73. Jacobson, B. H. and Kulling, F. A., Health and ergogenic effects of caffeine, *Br. J. Sports Med.*, 23, 34, 1989.
74. Falk, B., Burstein, R., Ashkenazi, I., Spilberg, O., Alter, J., Zylber-Katz, E., Rubinstein, A., Bashan, N., and Shapiro, Y., The effect of caffeine ingestion on physical performance after prolonged exercise, *Eur. J. Appl. Physiol.*, 59, 168, 1989.
75. Collomp, K., Ahmaïdi, S., Chatard, J. C., Audran, M., and Prefaut, C., Benefits of caffeine ingestion on sprint performance in trained and untrained swimmers. *Eur. J. Appl. Physiol.*, 64, 377, 1992.
76. Graham, T. E. and Spriet, L. L., Performance and metabolic responses to a high caffeine dose during prolonged exercise, *J. Appl. Physiol.*, 71, 2292, 1991.
77. Holding, D. H., Fatigue, in *Stress and Fatigue in Human Performance*, Hockey, G.R.J., Ed., John Wiley & Sons, Chichester, 1982, 145.
78. Klein, R. H. and Salzman, L. F., Paradoxical effects of caffeine, *Percept. Mot. Skills*, 40, 126, 1975.
79. Pollock, V. E., Teasdale, T., Stern, J., and Volavka, J., Effects of caffeine on resting electroencephalogram and response to sine wave modulated light. *Electroencephalogr. Clin. Neurophysiol.*, 51, 470, 1981.

chapter sixteen

The subjective effects of caffeine: bridging the gap between animal and human research

David V. Gauvin and Frank A. Holloway

Contents

I. Introduction 257
II. Analysis 259
III. Synthesis...273
IV. Conclusions..276
References..277

I. Introduction

The drug discrimination task is a psychophysical method of analysis of the subjective effects of drugs. As in all sensory/perceptual systems, these subjective effects are not directly observable events and, as such, are inferred from the behavioral response of the experimental subject. The response choice measure of a drug discrimination task has demonstrated striking similarities across laboratories. The drug discrimination task has been utilized by many researchers to investigate the neuropharmacological mechanisms by which the interoceptive stimulus dimensions of drug administration acquire stimulus control over behavior.

The visual and/or auditory systems have been adopted here as examples of accuracy and specificity of stimulus control. In both of these systems it has been clearly established that the physical or exteroceptive dimensions of the stimuli (i.e., intensity, wavelength, and purity) are not the bases for discriminative accuracy. Rather, visual and auditory discriminations are based primarily on the perceived or psychological dimensions of bright-ness–hue–saturation and loudness–pitch–timbre, respectively.[1-3] The full spectrum of stimulus sensitivity within the visual and auditory systems in both humans and animals has been repeatedly demonstrated to rely on multiple interactive neuronal systems.[4,5] Analogously, we have adopted the view that drug discriminative control, and in particular, caffeine stimulus control is most likely based on the perceived or psychological attributes produced by caffeine and subserved by multiple neuronal systems.

What an animal must do to successfully discriminate between the pres-ence and absence of caffeine or to discriminate between caffeine and another drug state is complex and relies on a number of different hypothetical pro-cessing systems:

1. The subject must learn to attend to specific, relevant internal stimuli and to disregard irrelevant stimuli.[6-8]
2. Since it is impossible to present the subject with two discriminative drug states simultaneously, as in a typical simultaneous "matching" experiment, the subject is presented with sequential stimuli with at least 2 days between the test stimulus and one of the training stimuli. Therefore, the subject must rely on some neuronal "template" or memory of drug action in order to make an accurate discriminative response.[7,8]
3. The subject must rely on previously learned associations between reinforcement contingencies and the location of which specific drug lever will produce food reinforcers. This relies on behavioral, phar-macological, and reinforcement histories.[9]
4. Once the response choice has been made by the experimental subject, it must emit the response operant or class, the distribution of which appears to be governed by a probabilistic matching vs. maximizing response strategy.[10]

With respect to stimulus control by compound or complex exteroceptive stimuli, it repeatedly has been demonstrated that the specific or particular stimulus dimension which controls the behavioral operant is largely beyond experimental control.[8,11-13] Which specific aspect of the multidimensional or multimodal cue (e.g., "form or color," or, "light or tone") becomes the "func-tional" or "effective" stimulus controlling the response can be controlled by a number of possible factors, including: 1) inherent saliency of the dimension (genetic factors); 2) past experience; and/or 3) the behavioral demands of the task. It has been suggested that the functional or effective stimulus which acquires stimulus control may even be idiosyncratic.[8,11,12] While drug dis-crimination tasks have been referred to as "simple," it is generally accepted that drugs are multidimensional stimuli, and as such the specific dimen-sion[s] utilized by an experimental subject to solve the discrimination task may differ across experimental subjects. We have previously suggested that the specific reinforcement schedule used during training of a drug discrimination task may influence which specific dimension(s) of the drug stimulus will capture stimulus control.[10]

This approach may seem to deviate from the majority of researchers who have previously argued that drug discriminative control is primarily based upon a specific neuronal system. However, the perceptual, decision-making, and motor system[s] governing the 1) attention to specific aspects of the drug test stimulus, 2) perceptual processing of relevant stimulus information, and 3) the completion of the discriminative response appear to operate in a fashion similar to all other perceptual systems governing the processing of exteroceptive stimuli and most probably do not rely on a single neuronal system.

The purposes of the present report are threefold: 1) to review the published reports on caffeine stimulus control, limiting our focus to published animal drug discrimination reports; 2) to present data demonstrating the relative similarities between the controlled psychomotor stimulant, cocaine, and caffeine and its metabolic by-products; and 3) to attempt to make sense of the existing animal literature base and to make analogies between these results and the underlying processes which may contribute to the continued high rate of human caffeine consumption.

II. Analysis

When discussing a stimulus or an experience in which it makes sense to ask "How much?" or "How intense?" we have what Stevens and Galanter[14] refer to as a prothetic continuum. Changes from one level of not detecting a stimulus to another level in which detection always occurs is commonly referred to as "quantitative specificity" of the stimulus, and it comes about by adding or subtracting from what is present.[1] Table 16.1 summarizes the data from a number of published reports in which caffeine has been trained as a stimulus in a two-choice (caffeine vs. saline) drug discrimination task in animal subjects. A wide range of doses (10 through 60 mg/kg) have been used as training stimuli and both positive (food) reinforcement and (shock) avoidance paradigms have been utilized. Universally, the drug discrimination task provides intensity functions which clearly demonstrate the quantitative specificity of the prothetic continuum that is both time and dose dependent. In an early study, Modrow and Holloway[18] demonstrated that intraperitoneal injections of caffeine in the rat produced a discriminable cue with a rapid onset (within 5 min) and gradual decline. Importantly, in the same study the plasma caffeine levels on the declining limb of the plasma caffeine curve were not predictive of the caffeine cue. The relationship between plasma caffeine levels and the percentage of total session responses emitted on the caffeine-appropriate lever during caffeine time-effect test sessions suggested that acute tolerance developed to the cueing attributes of caffeine. At 240 min after caffeine was injected, a majority of responses were emitted on the saline-appropriate lever, yet plasma concentrations remained at levels equivalent to plasma levels which engendered predominantly caffeine-appropriate

Table 16.1 Drug Cross-Generalization Profiles to Caffeine Training Stimuli

Caffeine training dose	Schedule of S^R	Test Compound (TC)	Highest % of caf-approp. responding or trials to caf lever	Dose of TC producing max. caf-approp. responding	Generalization? None Partial Complete	Reference
32 mg/kg	FR-30	d-amphetamine	27%	2.0 mg/kg	Partial	Modrow and Holloway, 1981
		methylphenidate	13%	5.0 mg/kg	None	
		nicotine	18%	0.1 mg/kg	None	
		TRH	<1.0%	10 mg/kg	None	
		theophylline	74%	56 mg/kg	Complete (per 70% trng. criteria)	
60 mg/kg	FR-10	d-amphetamine	62%	1.5 mg/kg	Partial	Winter, 1981
		aminophylline	98%	10 mg/kg	Complete	
32 mg/kg	VR-10	theophylline	76%	56 mg/kg	Complete (per 70% trng. criteria)	Carney et al., 1985
		paraxanthine	83%	100 mg/kg	Complete	
		3-methylxanthine	75%	56 mg/kg	Complete	
		theobromine	20%	75 mg/kg	None	
32 mg/kg	VR 5-15	theophylline	80%	100 mg/kg	Complete	Modrow and Holloway, 1985
		d-amphetamine	<20%	1.0 mg/kg	None	
		pentylenetetrazole	<20%	20 mg/kg	None	

32 mg/kg	VR 5-15	l-PIA	<20%	0.05 mg/kg	None	Holloway et al., 1985
		chlordiazepoxide	22%	10 mg/kg	None	
		papaverine	73%	9.0 mg/kg	Complete (per 70% trng. criteria)	
42 mg/kg	VR 5-15	ephedrine	36%	17.8 mg/kg	Partial	Gauvin et al., 1993
		phenylpropanolamine	30%	32 mg/kg	Partial	
		d-amphetamine	26%	1.8 mg/kg	Partial	
		cocaine	50%	17.8 mg/kg	Partial	
10 mg/kg	discrete trial shock-avoidance	theophylline	65%	30 mg/kg	Partial	Holtzman, 1986
		theobromine	30%	300 mg/kg	Partial	
		cocaine	80%	100 mg/kg	Partial	
		methylphenidate	90%	10 mg/kg	Complete	
		St 587	80%	10 mg/kg	Partial	
		d-amphetamine	65%	3.0 mg/kg	Partial	
		ephedrine	85%	100 mg/kg	Partial	
		apomorphine	<10%	0.3 mg/kg	None	
		clonidine	<10%	1.0 mg/kg	None	
		picrotoxin	<10%	1.0 mg/kg	None	
30 mg/kg	discrete trial shock-avoidance	theophylline	80%	175 mg/kg	Partial	Holtzman, 1986
		theobromine	40%	300 mg/kg	Partial	
		cocaine	80%	100 mg/kg	Partial	
		methylphenidate	80%	10 mg/kg	Partial	
		d-amphetamine	25%	1.0 mg/kg	Partial	
		ephedrine	45%	100 mg/kg	Partial	

Table 16.1 (continued) Drug Cross-Generalization Profiles to Caffeine Training Stimuli

Caffeine training dose	Schedule of S^R	Test Compound (TC)	Highest % of caf-approp. responding or trials to caf lever	Dose of TC producing max. caf-approp. responding	Generalization? None Partial Complete	Reference
10 mg/kg	discrete trial shock-avoidance	CGS 15943	100%	3.0 mg/kg	Complete	Mumford and Holtzman, 1991
		theophylline	80%	30 mg/kg	Partial	
		isobutylmethylxanthine	40%	3.0 mg/kg	Partial	
		1,7-dimethylxanthine	95%	56.0 mg/kg	Complete	
		7,(β-chloroethyl) theophylline	80%	10.0 mg/kg	Partial	
		8-chlorotheophylline	<10%	100 mg/kg	None	
		β-hydroxyethyltheophylline	95%	100 mg/kg	Complete	
10 mg/kg	discrete trial shock-avoidance	d-amphetamine	100%	1.0 mg/kg	Complete	Mumford and Holtzman, 1991
		apomorphine	95%	1.0 mg/kg	Complete	
		diethylproprion	100%	10 mg/kg	Complete	
		phendimetrazine	95%	10 mg/kg	Complete	
		methylphenidate	100%	10 mg/kg	Complete	
		cocaine	100%	18 mg/kg	Complete	
		mazindol	85%	3.0 mg/kg	Partial	
		ethylketocyclazocine	<10%	3.0 mg/kg	None	
		fenfluramine	40%	3.0 mg/kg	Partial	
		phencyclidine	<10%	3.0 mg/kg	None	
		pentylenetetrazole	<10%	10 mg/kg	None	

					Mumford and Holtzman, 1991
56 mg/kg	discrete trial shock-avoidance	yohimbine	15%	10 mg/kg	Partial	
		mescaline	20%	30 mg/kg	Partial	
		β-CCE	0%	30 mg/kg	None	
		IBMX	<10%	10 mg/kg	None	
		papaverine	<10%	30 mg/kg	None	
		phentolamine	<10%	10 mg/kg	None	
		CGS 15943	20%	30 mg/kg	Partial	
		theophylline	100%	100 mg/kg	Complete	
		isobutylmethylxanthine	20%	10 mg/kg	Partial	
		1,7-dimethylxanthine	40%	100 mg/kg	Partial	
		7,(β-chloroethyl)theophylline	75%	30 mg/kg	Partial	
		8-chlorotheophylline	30%	300 mg/kg	Partial	
		β-hydroxyethyltheophylline	80%	300 mg/kg	Partial	
		d-amphetamine	15%	1.0 mg/kg	Partial	
		apomorphine	15%	1.0 mg/kg	Partial	
		diethylproprion	<10%	10 mg/kg	None	
		phendimetrazine	20%	18 mg/kg	Partial	
		methylphenidate	<10%	10 mg/kg	None	
		cocaine	40%	30 mg/kg	Partial	
		mazindol	30%	10 mg/kg	Partial	
		ethylketocyclazocine	<10%	1.0 mg/kg	None	
		fenfluramine	<10%	3.0 mg/kg	None	
		phencyclidine	<10%	3.0 mg/kg	None	

Table 16.1 (continued)   Drug Cross-Generalization Profiles to Caffeine Training Stimuli

Caffeine training dose	Schedule of S^R	Test Compound (TC)	Highest % of caf-approp. responding or trials to caf lever	Dose of TC producing max. caf-approp. responding	Generalization? None Partial Complete	Reference
		pentylenetetrazole	<10%	10 mg/kg	None	
		yohimbine	10%	10 mg/kg	None	
		mescaline	20%	30 mg/kg	Partial	
		β-CCE	<10%	30 mg/kg	None	
		IBMX	<10%	10 mg/kg	None	
		papaverine	<10%	30 mg/kg	None	
		phentolamine	<10%	10 mg/kg	None	
		chlordiazepoxide	<10%	10 mg/kg	None	
		desipramine	<10%	30 mg/kg	None	
		FG-7142	<10%	30 mg/kg	None	
		NECA	<10%	0.03 mg/kg	None	
		Ro20-1724	<10%	3.0 mg/kg	None	
10 mg/kg	discrete trial shock-avoidance	methylphenidate	100%	3.0 mg/kg	Complete	Holtzman, 1987
30 mg/kg	discrete trial shock-avoidance	methylphenidate	100%	10 mg/kg	Complete	Holtzman, 1987
20 mg/kg	VI-FR	phenylpropanolamine	40%	32 mg/kg	Partial	Marathasan and Stolerman, 1992
		phenylpropanolamine plus caffeine	80%	5.6 mg/kg plus 5.6 mg/kg	Complete	
		+ amphetamine	86%	0.6 mg/kg	Complete	
		cocaine	70%	4.0 mg/kg	Partial	

From Gauvin, D.V. *Pharmacopsychoecologia*, 1994. With permission.

responding 15 min after injections. These data may suggest that the first self-administered dose of caffeine in humans may be under stimulus control of one specific attribute or behavioral effect of the multidimensional caffeine cue, and all subsequent cups may be under the control of other dimensional aspects or behavioral effects of caffeine. These data also suggest that the subjective effects of a given dose of caffeine may depend on how many other doses have already been self-administered. That is, two cups of coffee self-administered in the afternoon may produce different subjective effects from the effects produced by equivalent doses administered earlier in the morning.

Table 16.1 also lists numerous drugs from diverse pharmacological classes that have been tested for cross-generalization with the caffeine training dose. It is clear from the summarized data that no single neurotransmitter system can explain the complex cross-generalization profile of the multidimensional caffeine cue, which seems to involve at least the adenosine, dopamine, and norepinephrine systems.

Drugs active at putative adenosine receptors and belonging to the family of methylxanthines [theophylline (1,3 dimethylxanthine), paraxanthine (1,7 dimethylxanthine), and aminophylline] have demonstrated complete cross-generalization between a number of caffeine training doses under both schedules of reinforcement. However, the behaviorally-active caffeine metabolite theobromine (3,7 dimethylxanthine) typically has not demonstrated such similarity. More surprisingly, complete cross-generalization with the controlled psychostimulants have been reported from a number of studies, including cocaine, amphetamine, and methylphenidate. These drugs have been linked to reuptake blockade of dopamine, norepinephrine, and serotonin systems but have more often been called "dopamine agonists" (either direct or indirect[40]). Notably however, this cross-generalization profile is not universal.[15,18,20] While Holtzman[21] has concluded that noradrenergic activity underlies the caffeine discriminative stimulus, some noradrenergic pharmacological test probes do not reliably engender caffeine-appropriate responding (i.e., ephedrine, phenylpropanolamine, diethylproprion). Paradoxically, within the same laboratory using the same schedule of reinforcement and identical training doses, the dopamine agonist apomorphine produced <10%[21] and >90%[22] caffeine-appropriate responding. These data highlight an important characteristic germane to both interoceptive and exteroceptive discrimination studies. Mackintosh[6,7] has concluded that an experimental subject will not always solve a discrimination task along the continuum predicted by the experimenter. As discussed above, which specific element of a complex discriminative cue that eventually gains stimulus control over behavior is sometimes idiosyncratic and not always under the control of the experimenter. The Holtzman data discussed above with respect to apomorphine's cross-generalization profile with caffeine is clearly in line with what would be predicted from the exteroceptive literature base.

Interestingly, most tests conducted with generic stimulants, both controlled and over-the-counter, which act through the aminergic neuronal pathways (serotonin, dopamine, and/or norepinephrine) have produced caffeine-appropriate responding in animals trained to discriminate between saline and caffeine. Stimulants acting through other systems (i.e., acetylcholine) have repeatedly failed to produce even partial generalization (for example, nicotine). To summarize, the data in Table 16.1 suggest that the subjective effects of caffeine across a wide range of doses rely on multiple neuronal systems and are relatively similar to the multidimensional stimuli produced by a majority of both controlled and over-the-counter stimulants. This specific aspect of the caffeine stimulus may contribute to its continued use and/or abuse.

In contrast to the prothetic or quantitative continuum of the caffeine stimulus, there is another physical/perceptual continuum.[1] When an experimental subject has a stimulus experience in which the only question that it makes sense to ask is "What kind?," we are dealing with a metathetic continuum.[14] In a reciprocal analysis, Table 16.2 summarizes the drug discrimination literature in which specific doses of drugs from diverse pharmacological classes have been trained as discriminative stimuli and tested for cross-generalization with doses of caffeine to answer the question, "What is caffeine like?" While most studies demonstrated only partial generalization of caffeine to the majority of training drugs, complete generalization was reported between two bupropion training doses (20 and 40 mg/kg) and caffeine test doses (25 and 100 mg/kg, respectively). Bupropion (Welbutrin®) is an atypical antidepressant with stimulant-like properties believed to act through the reuptake blockade of both dopamine and norepinephrine. Other CNS stimulants (cocaine, amphetamine, and cathinone), when trained as discriminative stimuli, failed to produce complete generalization with caffeine and more typically have produced only partial generalization.

One conclusion that can be drawn from the caffeine generalization data is based on the vast literature on the processing of exteroceptive (visual and auditory) stimuli. Data accumulated with regard to exteroceptive stimuli cannot be accounted for by any one single processing or neural model. Most models describing the information processing of multidimensional visual and auditory stimuli assume that the stimulus the experimenter independently manipulates is also operated on independently by the observer.[41] These exteroceptive models also assume that the multidimensional information in the combined or complex stimulus is in no way different from those contained in the separate attributes. Lockhead[42] has proposed that multidimensional exteroceptive stimuli are initially processed by a holistic process, then, and only if the particular task requires it, the individual stimulus attributes or dimensions of the complex stimulus are further analyzed and processed. Relatedly, caffeine acting through a number of neurotransmitter systems, is associated with nonspecific or global stimulant-like interoceptive cues. These global or holistic features of the stimulant cue are similar in nature to a generic class of stimuli such that other CNS stimulants will be perceived to be similar enough in their global features, in a relative sense, to engender caffeine-appropriate responding equivalent to the training dose of caffeine. However, when an animal is trained to a more specific attribute of a CNS stimulant (i.e., cocaine or amphetamine), the global or holistic nature of the caffeine attributes are perceived to be only "similar" to these agents and produce limited or partial similarity or generalization.

One method of analysis of the qualitative or metathetic dimensional attributes of caffeine is to ask the question, "What can block the discriminative caffeine cue?" Table 16.3 lists a number of drugs tested for their ability to block or antagonize the discriminative stimulus properties of the training dose of caffeine in rats. While some drugs had no appreciable effect (pixotyline and spiperone), others have been found to produce complete or partial blockade. Drugs characterized as anxiolytics (chlordiazepoxide, diazepam), CNS depressants (with anxiolytic properties, pentobarbital), and as antihypertensives (with limited use as an anxiolytic, propranolol) have produced partial blockade of the caffeine training cue. The adenosine agonists (CHA, 2-CA, and l-PIA) also have produced partial blockade of the caffeine cue. These agents have potent behavioral effects and produce sedation, sleep-induction, and anxiolysis in rats.[43-45] Interestingly, more selective α-adrenergic-mediated antihypertensive compounds (prazosin and phentolamine), with similar sedative and sleep-induction properties in the rat,[46] produced complete blockade. Yohimbine has been reported to produce anxiety in humans[47] and animals.[48,49] However, yohimbine also has displayed anti-conflict activity in the Geller–Seifter test[50,51] and to produce anxiolytic-like discriminative or subjective effects[52] in rats. Both of these latter results have implicated serotoninergic activity by yohimbine and suggest a unique interaction between behavioral demands, reference stimuli,[9] and the subjective effects produced by yohimbine in the rat. In summary then, Table 16.3 highlights the important finding that all drugs that have demonstrated efficacy in blocking the subjective effects of caffeine (either partially or completely) have also demonstrated some degree of anxiolysis, sedation, and/or sleep induction in the rat.

Recently another approach has been used to examine the pharmacological correlates of the similar qualitative or metathetic dimensions produced by caffeine and cocaine in the rat. Twelve male Sprague–Dawley rats were trained to discriminate between 10 mg/kg cocaine hydrochloride and saline in a standard two-choice food-motivated operant task. Rats responded under a fixed-ratio 10 schedule of food reinforcement during daily 10 min experimental sessions (details of the specific methods used can be found in Gauvin, Peirce, and Holloway).[53] Training was continued until each rat met criteria for stimulus control. Each rat had to emit greater than 90% of the total session responses on the injection-appropriate lever **and** emit less than a total of 20 responses prior to the delivery of the first food pellet. All rats reached training criteria within 50 training sessions. Once stimulus control was achieved by both cocaine and saline injections, test sessions were conducted. Test sessions were identical to training sessions except that, 1) a novel test

Table 16.2 Caffeine Cross-Generalization Profile to Other Training Drugs

Training drug	Training dose	Schedule of S^R	Highest % of training drug-approp. responding	Dose of CAF producing max. drug-approp. responding	Generalization None Partial Complete	Reference
theophylline	56 mg/kg	VR-10	60%	56 mg/kg	Partial	Carney et al., 1985
ephedrine	10 mg/kg	VR 5-15	50%	10 mg/kg	Partial	Gauvin et al., 1993
phenylpropanolamine	17.8 mg/kg		55%	32 mg/kg	Partial	
phenylpropanolamine plus caffeine	10 mg/kg plus 10 mg/kg		25%	17.8 mg/kg	Partial	
ephedrine plus caffeine	5 mg/kg plus 10 mg/kg		58%	17.8 mg/kg	Partial	
phenylpropanolamine plus ephedrine	12 mg/kg plus 6 mg/kg		60%	56 mg/kg	Partial	
caffeine plus ephedrine plus phenylpropanolamine	15 mg/kg plus 3 mg/kg plus 6 mg/kg		30%	56 mg/kg	Partial	
caffeine plus phenylpropanolamine	20 mg/kg plus 20 mg/kg	VI 1-FR10	50%	30 mg/kg	Partial	Mariathasan and Stolerman, 1992
phenylpropanolamine	20 mg/kg	VI 1-FR10	~40%	32 mg/kg	Partial	Mariathasan and Stolerman, 1992
d-amphetamine	0.5 mg/kg	VR 5-15	57%	32 mg/kg	Partial	Holloway et al., 1985
cocaine	10 mg/kg	VR 5-15	63%	56 mg/kg	Partial	Gauvin et al., 1989b

cocaine	10 mg/kg	VR 5-15	58%	32 mg/kg	Partial	Gauvin et al., 1990
cocaine	10 mg/kg	VR 5-15	57%	56 mg/kg	Partial	Harland et al., 1989
cathinone	0.8 mg/kg	FR10	50%	20 mg/kg	Partial	Schechter, 1989
apomorphine	0.16 mg/kg	FR10	33%	30 mg/kg	Partial	Schechter, 1980
d-amphetamine	0.8 mg/kg	VI15s	44%	15 mg/kg	Partial	Schechter, 1977
d-amphetamine	1.0 mg/kg	VI 1-FR10	55%	20 mg/kg	Partial	Kuhn et al., 1974
bupropion	20 mg/kg	FR10	80%	25 mg/kg	Complete	Jones et al., 1980
bupropion	40 mg/kg	FR10	80%	100 mg/kg	Complete	Blitzer and Becker, 1985
yohimbine	3.2 mg/kg	FR10	25%	56 mg/kg	Partial	Browne, 1981
fentanyl	0.04 mg/kg	FR10	0%	20 mg/kg	None	Colpaert et al., 1975
dl-cathinone	2.0 mg/kg	FR10	37.5%	14 mg/kg	Partial	Goudie et al., 1986
midazolam	0.3 mg/kg	FR10	<5%	10 mg/kg	None	Spealman, 1985
chlordiazepoxide vs. saline vs. pentylenetetrazole	5 mg/kg vs. 1 ml/kg vs 15 mg/kg	FR10	0% CDP 6.7% SAL 93.3% PTZ	56 mg/kg	Complete to PTZ	Gauvin and Holloway, 1991

From Gauvin, D.V. *Pharmacopsychoecologia*, 1994. With permission.

Table 16.3   Blockade or Antagonism Tests of Caffeine's Discriminative Stimulus Effects

Reference	Caffeine training dose	Test "blocking" drug	Results
Winter, 1981	60 mg/kg	pizotyline — 5HT antagonist 3 & 10 mg/kg	No effect
		spiperone — DA antagonist 0.05 — 0.2 mg/kg	No effect
Holloway et al., 1985	32 mg/kg	1-phenyl-isopropyl-adenosine (l-pia) adenosine agonist 0.01, 0.05, 0.1 mg/kg	No effect
		chlordiazepoxide — benzodiazepine agonist 1.0, 5.0, 10.0 mg/kg	PARTIAL BLOCKADE of discriminative response without blocking response rate suppression; defined by this lab, as perceptual masking cf Gauvin et al., 1994
Holtzman 1986	10 mg/kg	prazosin — α_1 adrenergic antagonist	COMPLETE BLOCKADE
		yohimbine — α_2 adrenergic antagonist	COMPLETE BLOCKADE
		phentolamine — nonspecific α-adrenergic antagonist	COMPLETE BLOCKADE
		cyclohexyladenosine — adenosine agonist	PARTIAL BLOCKADE
		2-chloroadenosine — adenosine agonist	PARTIAL BLOCKADE
		l-PIA — adenosine agonist	PARTIAL BLOCKADE
		diazepam — benzodiazepine agonist	PARTIAL BLOCKADE
		pentobarbital — CNS depressant	PARTIAL BLOCKADE
		propranolol — β-adrenergic antagonist	PARTIAL BLOCKADE

From Gauvin, D.V. *Pharmacopsychoecologia*, 1994. With permission.

dose or drug was administered prior to the test session, and 2) 10 consecutive responses on either lever produced food. The 10 mg/kg training dose of cocaine engendered a group average of 99.6% (±0.2) cocaine-appropriate responding; saline test injections engendered 0.9% (±0.01) cocaine-appropriate responding. Once these test sessions were completed, cross-generalization tests were conducted with various doses of caffeine and its active metabolites.

Structure activity relationships involving the position of the methyl group distinguish the pharmacological potency of effects demonstrated by caffeine (1,3,7-trimethylxanthine) and its three active metabolites: theophylline (1,3 dimethylxanthine), paraxanthine (1,7 dimethylxanthine), and theobromine (3,7 dimethylxanthine). It has been proposed that methyl substitution on the 1 position is associated with central nervous system stimulation; diuresis is linked to the 3-methyl position; and cardiac stimulation with the 7-methyl position.[54] Therefore, caffeine, with a methyl group at the 1st, 3rd, and 7th positions, shows all three effects; theophylline (methyl groups at the 1st and 3rd positions) has increased CNS stimulation with limited cardiac stimulation; paraxanthine (methyl groups at the 1st and 7th positions) has increased CNS and cardiac stimulation; and theobromine (methyl groups at the 3rd and 7th positions) produces limited CNS stimulation. All but paraxanthine have diuretic properties. From these findings it can be hypothesized that the cross-generalization profile between the cocaine training cue, hypothesized to be based primarily on CNS stimulation, and the methylxanthines would demonstrate a potency relationship of caffeine > theophylline > paraxanthine > theobromine. Further, based on the previous literature cited in Tables 16.1 and 16.2, it was hypothesized that caffeine would produce only partial generalization and theobromine would produce little if any cross generalization with the training dose of cocaine.

Figure 16.1 shows the discriminative stimulus cross-generalization profiles between the four methylxanthines and the training dose of cocaine. Caffeine ($F[6,48] = 2.8$, $p<.03$), theophylline ($F[3,24] = 7.4$, $p<.001$), and paraxanthine ($F[3,24] = 3.2$, $p = .04$) produced significant dose-related increases in the total session responses emitted on the cocaine-appropriate lever. Theobromine did not ($F[4,28] = 2.08$, $p = .11$). The peak in the total session responses emitted on the cocaine-appropriate lever during methylxanthine tests produced a relative potency relationship between them and cocaine as follows: theophylline ($61.6 \pm 14.7\%$) > caffeine (52.3 ± 13.7) > paraxanthine ($47.0 \pm 15.5\%$) > theobromine ($23 \pm 14.8\%$). This relationship was slightly different than the original hypothesis in that theophylline produced a greater degree of cross-generalization than caffeine. These data may suggest that the stimulus attributes associated with the 7th methyl position (cardiac stimulation) may block, mask, or interfere with the perception of the CNS stimulant attributes specifically associated with the methyl 1 position. Theobromine, associated with both diuresis and cardiac stimulation, failed to produce generalization to the cocaine training cue.

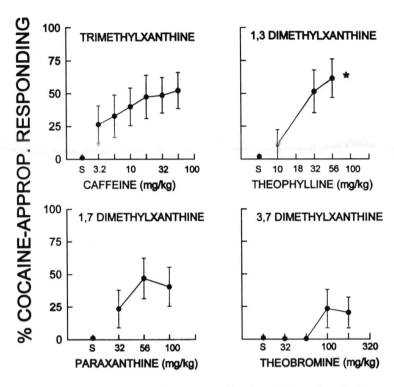

Figure 16.1 Dose–response generalization profiles for caffeine, theophylline, parax-anthine, and theobromine in rats trained to discriminate between 10 mg/kg cocaine and saline in a two-choice operant task under a fixed-ratio 10 schedule of food-maintained lever-press responding. 10 min reinforced test sessions were conducted with each dose of the four drugs only once in each of 10 rats. The group mean percentage of total session responses (±S.E.) emitted on the cocaine-appropriate lever is expressed as a function of test dose. The symbol above the "S" on the abscissa represents the results of saline test sessions. (From Gauvin, D.V. *Pharmacopsychoeco-logia*, 1994. With permission.)

Figure 16.2 shows the rates responding during methylxanthine cross-generalization test sessions. Caffeine (F[6,48] = 14.5, p<.001), theophylline (F[4,32] = 29.8, p<.001), and paraxanthine (F[3,24] = 3.9, p = .02) produced significant dose-related response rate suppression. Theobromine (F[4,28] = 1.06, p = .390 did not produce response rate suppression and at a test dose of 56 mg/kg produced significant rate-increasing effects relative to saline test response rates (p<.05, Duncan's *post hoc* test). The behavioral potency relationship for the rate-altering effects produced by these four methylxanthines was identical to those found for the discriminative response (above).

The conclusions drawn from these new data are similar to the summarized data discussed in Tables 16.1 and 16.2. The relative similarity between the subjective or discriminative stimulus properties of cocaine and caffeine in the rat seem to depend upon global CNS stimulant effects.

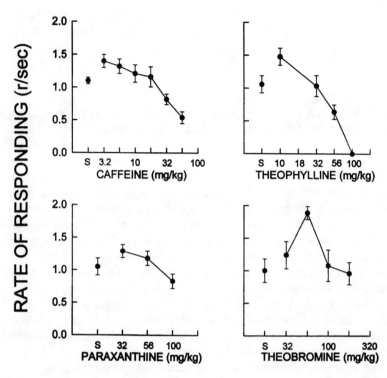

Figure 16.2   Dose-dependent changes in the group mean (±S.E.) rates-of-responding during drug cross-generalization test sessions displayed in Figure 1. Rates-of-responding is expressed in responses per second. The symbol above the "S" on the abscissa represents the results of saline test sessions. (From Gauvin, D.V. *Pharmacopsychoecologia*, 1994. With permission.)

III. Synthesis

The final question is, "What can these animal data tell us about human caffeine consumption?" To begin to answer this, a model is proposed that accounts for the combinations of similarities and differences in the discriminative stimulus effects of caffeine on the basis of the quantitative and qualitative dimensional changes from the normal basal state of the organism. This particular drug model was first proposed by Barry and Krimmer[55] to describe the multidimensionality of interoceptive states produced by drug injections. The model was modified by David Nutt and colleagues[56-58] and extended to the measurement of both the physical domain of the benzodiazepine–betacarboline pharmacological spectrum and the corresponding, albeit hypothetical, psychological subjective continuum perceived by the subject after drug administration. This particular subjective continuum has been used to model the anxiety–relaxation continuum associated with a three-choice chlordiazepoxide–saline–pentylenetetrazole discrimination task.[20,39,59] The utilization of such multidimensional models of stimulus attributes is not new and probably can be traced back to the "bay bridge analogy" for odor research first proposed by John Amoore.[60] A complex multidimensional model of visual stimuli has been detailed by Roger Shepard.[61] Other researchers have proposed similar models for all other perceptual continua.[62-67] With respect to the use of such models to bridge the gap between the subjective or psychological states and the physical pharmacological continuum underlying their production, it is clearly in line with the "sensory model" of drug effects described by Overton.[68,69]

Figure 16.3 details the proposed model which describes the dynamic processes underlying caffeine consumption by humans, which is based on the experimental animal literature discussed above. Three psychological continua are displayed (A, B, and C). The psychological spectra displayed in 3A details a unitary continuum along hypothesized bipolar affective states. One end of the axis is characterized by sleep or unconsciousness and its opponent end by excessive excitation or seizures. These two diametrically opposed psychological states are anchored[9,70] by a neutral centroid region characterized by the range of normal homeostasis. The basic notion that two affective states lie in opponent fashion is not new.[71] Test drugs which cross-generalize to a caffeine training stimulus (Table 16.1) and other training drugs which engender cross-generalization with caffeine test doses (Table 16.2) produce one or more of the subjective effects which lie on the excitation (rightward) continuum. Conversely, drugs which have been found to block caffeine's discriminative stimulus effects in animals (Table 16.3) produce one or more of the subjective effects on the sedation (leftward) continuum.

In the present model, states of lethargy and anxiolysis are located between the normal basal state of the individual and the sleep continua, and states of excitation and anxiety are similarly located along the opponent end. A similar hypothetical psychological continuum is displayed in Figure 16.3B. During the early morning hours soon after waking, or during bouts of somnolence during the day, an individual may find himself or herself positioned along this continuum (dozing icon). Not quite awake, the individual may utilize a cup of coffee for its pharmacologically active ingredient (caffeine) to push him/herself up the psychological continuum to his/her normal basal affective state (standing icon). This is probably the most typical pharmacological use for caffeine consumption[72-76] and can be seen as a form of "functional use" of caffeine and **not** "caffeine abuse," *per se.* During the day as the individual continues to consume caffeine, he/she will push his/her affective state up the psychological continuum. Excessive caffeine consumption will eventually lead to jitteriness (excitation), anxiety, and could eventually lead to seizure activity or overdose. The third continuum (Figure 16.3C) demonstrates what may happen if the normal basal state of the individual has been shifted upward such that the individual's subjective experience is now "anchored"[9,70] at an excited basal state, characterized as a pathological panic or anxiety disorder (left or "excited" icon). The consumption of caffeine at this anchor point would shift the resulting affective

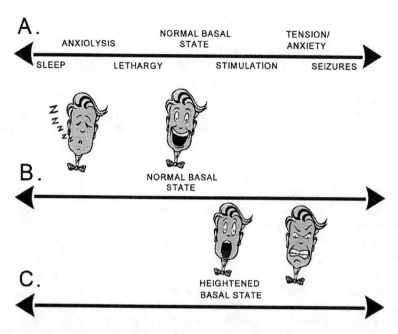

Figure 16.3 Hypothetical affective or subjective continuum which is proposed as a model to explain the results of caffeine drug discrimination tasks and the subjective state changes occurring in human caffeine consumers. (A) The hypothetical continuum is characterized as a single bipolar axis anchored at one end by excessive excitation or seizures and bounded on the opponent end by sleep or unconsciousness. (B) A representation of the state-changes occurring after the "morning cup" of coffee. The interoceptive state of "tiredness" or "lethargy" are attenuated by caffeine by shifting the subjective state up the continuum to the normal homeostatic baseline. (C) A representation of the state-changes occurring after caffeine consumption in a hypothetical individual whose affective basal state has been shifted upward to a more "excited" or "anxious" basal state. Caffeine shifts the internal states further up the continuum resulting in a self-reported shift to "anxiety-like" symptoms. (From Gauvin, D.V. *Pharmacopsychoecologia*, 1994. With permission.)

experience upward such that the same dose of caffeine produces an increased rating in subjective anxiety and could become dysfunctional (right or "anxious" icon). The influence of user status and affective disposition on the psychological response to a given dose of caffeine has been clearly demonstrated in both humans[77-79] and animals.[39] Interestingly, a recent report describes the relationship between caffeine and buspirone, drugs which produce subjective effects along the excitation (rightward) continuum, and a training dose of chlordiazepoxide.[53] The subjective effects of the training dose of chlordiazepoxide, which has been hypothesized to produce and be based on subjective anxiolysis, were blocked or masked by the coadministration of caffeine or buspirone. A result predicted by the bipolar nature of this hypothesized affective continuum.

IV. Conclusions

The review of results from studies examining the subjective effects of caffeine in animals has provided a hypothesized and parallel subjective continuum to that reported by human caffeine consumers. The complex nature of the cross-generalization profile between caffeine and other CNS active drugs suggests that the pharmacologically-induced changes in perceived mood or affect along a "stimulation-sedation" continuum is a common feature of the diverse range of doses and drugs that produce caffeine-appropriate responding. The most pragmatic and intuitive conclusion that can be drawn from the animal literature is that caffeine, acting through a number of neurotransmitter systems, produces a global "stimulant-like" affective state which is subjectively common to the class of CNS stimulants, of which caffeine is an element. Conversely, when animals are trained to more specific stimulants with more limited or focused neuropharmacological mechanisms (i.e., cocaine or amphetamine), caffeine with its global stimulant-like features will produce only partial generalization. These conclusions from the animal literature have led to the suggestion that caffeine in combination with other legal over-the-counter stimulants may function as a replacement-model for the successful treatment of cocaine abuse.[18] Cocaine abusers may have specific interoceptive cues that they utilize to discriminate the quality of cocaine and/or their own basal homeostatic states. Caffeine and/or the other legal stimulants (ephedrine and phenylpropanolamine) may have enough global or holistic features, common to all stimulants, which could abate or diminish the severity of cocaine craving once the stimulant abuser enters treatment and begins to experience cocaine withdrawal.

Holtzman[80] has previously suggested that caffeine is a convenient, relatively safe and useful drug for studying the basic behavioral and cellular mechanisms of drug abuse. Dr. Holtzman's suggestions may be correct, but the most common form of caffeine consumption by humans does not categorize the consumption as "drug abuse," *per se*. Rather, the data reviewed here suggest that human consumption may be viewed as a functional use of a pharmacological agent and may depend upon the subject's affective baseline. The deleterious effects of caffeine consumption may be due to the shift of an individual's interoceptive state along a hypothetical affective continuum. This specific continuum can be characterized as a "sedation–excitation" continuum. Adverse or toxic effects of high dose consumption of caffeine may result from a loss of stimulus control. That is, the affective changes brought about by the first or second cup of coffee in the morning may capture stimulus control over an individual's morning behavioral repertoire. Acute tolerance may develop to these affective state changes during these initial doses of caffeine. As the day and subsequent caffeine consumption progresses, the individual may no longer attend to these same critically relevant interoceptive cues and/or fail to attend to other physiologically relevant stimuli (i.e., motoric fasciculation). Continued consumption may allow for further progression up the "stimulant" dimension to the expression of more deleterious effects of caffeine (jitteriness, anxiety, toxicity, etc.).

While "abuse" models have even been applied to water consumption (i.e., water intoxication[81]) the common form of caffeine use is probably best viewed as a "mood modulator," functionally utilized by the majority of consumers to regulate the circadian variations in affective states. The need for a "state change"[82] at any given time[83] may be the only relevant stimulus that controls the voluntary self-administration of caffeine. Holtzman[80] has pointed out, because of its relative safety, high doses of voluntary consumption, and the development of both acute and chronic tolerance, caffeine may be an ideal drug with which to examine the dynamic processes underlying the use and abuse of other CNS-active drugs.

References

1. Coren, S., Porac, C., and Ward, L. M., *Sensation and Perception*, 2nd ed., Academic Press, Orlando, FL, 1984.
2. Gescheider, G. A., *Psychophysics: Methods, Theory, and Application*, Lawrence Erlbaum Assoc, Hillsdale, NJ, 1985.
3. Matlin, M. W. and Foley, H. J., *Sensation and Perception*, Allyn & Bacon, Needham Heights, MA, 1992.
4. Kuffler, S. W., Nicholls, J. G., and Martin, A. R., *From Neuron to Brain: A Cellular Approach to the Function of the Nervous System*, Sinauer Assoc., Inc., Sunderland, MA, 1984.
5. Kandel, E. R., Schwartz, J. H., and Jessell, T. M., *Principles of Neural Sciences*, 3rd ed., Elsevier Press, New York, 1991.
6. Mackintosh, N. J., *The Psychology of Animal Learning*, Academic Press, London, 1974.
7. Mackintosh, N. J., *Conditioning and Associative Learning*, Oxford University Press, New York, 1985.
8. Gilbert, R. M. and Sutherland, N. S., *Animal Discrimination Learning*, Academic Press, New York, 1969.
9. Gauvin, D. V., Harland, R. D., and Holloway, F. A., Drug discrimination procedures: A method to analyze adaptation level of affective states, *Drug Develop. Res.*, 16, 183, 1989.
10. Holloway, F. A. and Gauvin, D. V., Comments on method and theory in drug discrimination: A potpourri of problems, perplexities, and possibilities, *Drug Develop. Res.*, 16, 195, 1989.
11. Schwartz, B., *Psychology of Learning and Behavior*, W.W. Norton and Company, New York, 1989.
12. Fantino, E. and Logan, C. A., *The Experimental Analysis of Behavior: A Biological Perspective*. W. H. Freeman and Company, San Francisco, 1979.
13. Zentall, T. R., Hogan, D. E., and Edwards, C. A., Cognitive factors in conditional learning in pigeons, in *Animal Cognition*, Roitblat, H. L., Bever, T. G., and Terrace, H. S., Eds., Lawrence Erlbaum Assoc., Hillsdale, NJ, 1984, 389.
14. Steven, S. S. and Galanter, E., Ratio scales and category scales from a dozen perceptual continua, *J. Exp. Psychol.*, 54, 377, 1957.
15. Modrow, H. E. and Holloway, F. A., Caffeine discrimination in the rat, *Pharmacol. Biochem. Behav.*, 14, 683, 1981.
16. Winter, J. C., Caffeine-induced stimulus control, *Pharmacol. Biochem. Behav.*, 15, 157, 1981.

17. Carney, J. M., Holloway, F. A., and Modrow, H. E., Discriminative stimulus properties of methylxanthine and their metabolites in rats, *Life Sci.*, 36, 913, 1985.
18. Modrow, H. E. and Holloway, F. A., Drug discrimination and cross generalization between two methylxanthines, *Pharmacol. Biochem. Behav.*, 23, 425, 1985.
19. Holloway, F. A., Modrow, H. E., and Michaelis, R. C., Methylxanthine discrimination in the rat: Possible benzodiazepine and adenosine mechanisms. *Pharmacol. Biochem. Behav.*, 22, 815, 1985.
20. Gauvin, D. V., Moore, K. R., Youngblood, B. D., and Holloway, F. A., The discriminative stimulus properties of legal, over-the-counter stimulants administered singly and in binary and ternary combinations, *Psychopharmacology*, 110, 309, 1993.
21. Holtzman, S. G., Discriminative stimulus properties of caffeine in the rat: Noradrenergic mediation. *J. Pharmacol. Exp. Therap.*, 239, 706, 1986.
22. Mumford, G. K. and Holtzman, S. G., Qualitative differences in the discriminative stimulus effects of low and high doses of caffeine in the rat, *J. Pharmacol. Exp. Therap.*, 258, 857, 1991.
23. Holtzman, S. G., Discriminative stimulus effects of caffeine: Tolerance and cross-tolerance with methylphenidate, *Life Sci.*, 40, 381, 1987.
24. Mariathasan, E. A. and Stolerman, I. P., Drug discrimination studies in rats with caffeine and phenylpropanolamine administered separately and as mixtures, *Psychopharmacology*, 109, 99, 1992.
25. Holloway, F. A., Michaelis, R. C., and Huerta, P. L., Caffeine-phenylethylamine combinations mimic the amphetamine discriminative cue, *Life Sci.*, 36, 723, 1985.
26. Gauvin, D. V., Harland, R. D., Michaelis, R. C., and Holloway, F. A., Caffeine-phenylethylamine combinations mimic the cocaine discriminative cue, *Life Sci.*, 44, 67, 1989.
27. Gauvin, D. V., Criado, J. R., Moore, K. R., and Holloway, F. A., Potentiation of cocaine's discriminative effects by caffeine: A time-effect analysis, *Pharmacol. Biochem. Behav.*, 36, 195, 1990.
28. Harland, R. D., Gauvin, D. V., Michaelis, R. C., Carney, J. M., Seale, T. W., and Holloway, F. A., Behavioral interaction between cocaine and caffeine: A drug discrimination analysis in rats, *Pharmacol. Biochem. Behav.*, 32, 1017, 1989.
29. Schechter, M. D., Potentiation of cathinone by caffeine and nikethamide, *Pharmacol. Biochem. Behav.*, 33, 299, 1989.
30. Schechter, M. D., Caffeine potentiation of apomorphine discrimination, *Pharmacol. Biochem. Behav.*, 13, 307, 1980.
31. Schechter, M. D., Caffeine potentiation of amphetamine: Implications for hyperkinesis therapy, *Pharmacol. Biochem. Behav.*, 6, 359, 1977.
32. Kuhn, D. M., Appel, J. B., and Greenberg, I., An analysis of some discriminative properties of d-amphetamine, *Psychopharmacologia*, 39, 57, 1974.
33. Jones, C. N., Howard, J. L., and McBennett, S. T., Stimulus properties of antidepressants in the rat, *Psychopharmacology*, 67, 111, 1980.
34. Blitzer, R. D. and Becker, R. E., Characterization of the bupropion cue in the rat: Lack of evidence for a dopaminergic mechanism, *Psychopharmacology*, 85, 173, 1985.
35. Browne, R. G., Anxiolytics antagonize yohimbine's discriminative stimulus properties, *Psychopharmacology*, 74, 245, 1981.

36. Colpaert, F. C., Niemegeers, C. J. E., and Janssen, P. A. J., The narcotic cue: Evidence for the specificity of the stimulus properties of narcotic drugs, *Arch. Intern. Pharmacodyn.Ther.*, 218, 268, 1975.

37. Goudie, A. J., Atkinson, J., and West, C. R., Discriminative properties of the psychostimulant *dl*-cathinone in a two lever operant task, *Neuropharmacology,* 25, 85, 1986.

38. Spealman, R. D., Discriminative stimulus effects of midazolam in squirrel monkeys: Comparison with other drugs and antagonism by Ro 15-1788, *J. Pharmacol. Exp. Therap.*, 235, 456, 1985.

39. Gauvin, D. V. and Holloway, F. A., Cue dimensionality in the three-choice pentylenetetrazole-saline-chlordiazepoxide discrimination task, *Behav. Pharmacol.*, 2, 417, 1991.

40. Cooper, J. R., Bloom, F. E., and Roth, R. H., *The Biochemical Basis of Neuropharmacology.* Oxford University Press, New York, 1991.

41. Lockhead, G. R. and King, M. C., Classifying integral stimuli, *Hum. Percept. Perform.*, 3, 436, 1977.

42. Lockhead, G. R., Processing dimensional stimuli: a note. *Psychol. Rev.*, 79, 410, 1972.

43. Coffin, V. L. and Carney, J. M., Behavioral pharmacology of adenosine analogs, in *Physiology and Pharmacology of Adenosine Derivatives*, Daly, J. W., Kuroda, Y., Phillis, J. W., Shimizu, H., and Ui, M., Eds., Raven Press, New York, 1983, 267.

44. Phillis, J. W. and Wu, P. H., Roles of adenosine and adenine nucleotides in the central nervous system, in *Physiology and Pharmacology of Adenosine Derivatives*, Daly, J. W., Kuroda, Y., Phillis, J. W., Shimizu, H., and Ui, M., Eds., Raven Press, New York, 1983, 219.

45. Chagoya de Sánchez, V., Múñoz, R. H., Suárez, J., Vidrio, S., Yáñez, L., and Múñoz, M. D., Day-night variations of adenosine and its metabolizing enzymes in the brain cortex of the rat — possible physiological significance for the energetic homeostasis and the sleep–wake cycle, *Brain Res.*, 612, 115, 1993.

46. Makela, J. P. and Hilakivi, I. T., Effect of alpha-adrenoreceptor blockade on sleep and wakefulness in the rat, *Pharmacol. Biochem. Behav.*, 24, 613, 1986.

47. Charney, D. S., Heninger, G. R., and Redmond, D. E., Yohimbine induced anxiety and increased noradrenergic function in humans: effects of diazepam and clonidine, *Life Sci.*, 33, 19, 1983.

48. File, S. E. and Pellow, S., Triazolobenzodiazepines antagonize the effects of anxiogenic drugs mediated at three different central nervous system sites, *Neurosci. Lett.*, 61, 115, 1985.

49. Johnston, A. L. and File, S. E., Yohimbine's anxiogenic action: evidence for noradrenergic and dopaminergic sites, *Pharmacol. Biochem. Behav.*, 32, 151, 1989.

50. Baldwin, H. A., Johnston, A. L., and File, S. E., Antagonistic effects of caffeine and yohimbine in animal tests of anxiety, *Eur. J. Pharmacol.*, 159, 211, 1989.

51. Sethy, V. H. and Winter, J. C., Effects of yohimbine and mescaline on punished behavior in the rat, *Psychopharmacology,* 23, 160, 1972.

52. Winter, J. C. and Rabin, R. A., Yohimbine and serotoninergic agonists: stimulus properties and receptor binding, *Drug Develop. Res.*, 16, 327, 1989.

53. Gauvin, D. V., Peirce, J. M., and Holloway, F. A., Perceptual masking of the chlordiazepoxide discriminative cue by both caffeine and buspirone, *Pharmacol. Biochem. Behav.*, 47, 153, 1994.

54. Arnaud, M. J., The pharmacology of caffeine, *Prog. Drug Res.*, 31, 1987, 273.

55. Barry, H. III and Krimmer, E. C., Discriminable stimuli produced by alcohol and other CNS depressants, in *Discriminative Stimulus Properties of Drugs,* Lal, H., Ed., Plenum Press, New York, 1977, 73.

56. Little, H. J., Nutt, D. J., and Taylor, S. C., Kindling and withdrawal changes at the benzodiazepine receptor, *J. Psychopharmacol.,* 1, 35, 1987.

57. Nutt, D. J., Benzodiazepine receptor ligands, *Neurotransmissions, IV,* 1988.

58. Nutt, D. J., Benzodiazepine receptor ligands, *Pharmacol. Therap.,* 47, 233, 1990.

59. Gauvin, D. V. and Holloway, F. A., Cross-generalization between and ecologically-relevant stimulus and a pentylenetetrazole discriminative cue, *Pharmacol. Biochem. Behav,* 39, 521, 1991.

60. Amoore, J. E., *Molecular Basis of Odor.* Charles C. Thomas, Springfield, IL, 1970.

61. Shepard, R., Neural nets for generalization and classification: Comment on Staddon and Reid, *Psychol. Rev.,* 92, 579, 1994.

62. Cain, W. S., The odoriferous environment and the application of olfactory research, in *Handbook of Perception,* Vol. VIA, Carterette, E. C. and Friedman, M. P., Eds., Academic Press, New York, 1978, 277.

63. Gesteland, R. C., The neural code: integrative neural mechanisms, in *Handbook of Perception,* Vol. VIA, Carterette, E. C. and Friedman, M. P., Eds., Academic Press, New York, 1978, 259.

64. Beidler, L. M., Biophysics and chemistry of taste. in *Handbook of Perception,* Vol. VIA, Carterette, E. C. and Friedman, M. P., Eds., Academic Press, New York, 1978, 21.

65. McBurney, D. H., Psychological dimensions and perceptual analyses of taste, in *Handbook of Perception,* Vol. VIA, Carterette, E. C. and Friedman, M. P., Eds., Academic Press, New York, 1978, 125.

66. Wright, R. H., *The Sense of Smell.* CRC Press, Boca Raton, FL, 1982.

67. McBurney, D. H. and Collings, V. B., *Introduction to Sensation and Perception,* Prentice-Hall, Inc., Englewood Cliffs, NJ, 1984.

68. Overton, D. A., State-dependent learning and drug discrimination, in *Handbook of Psychopharmacology,* Vol. 18, Iverson, L. L., Iversen, S. D., and Snyder, S. H., Eds., Plenum Press, New York, 1984, 59.

69. Overton, D. A., Applications and limitations of the drug discrimination method for the study of drug abuse, in *Methods of Assessing the Reinforcing Properties of Abused Drugs,* Bozarth, M. A., Ed., Springer-Verlag, New York, 1987, 291.

70. Helson, H., *Adaptation-Level Theory: An Experimental and Systematic Approach to Behavior,* Harper and Row, New York, 1966.

71. Plutchik, R., *The Emotions:Facts, Theories, and a New Model,* Random House, New York, 1962.

72. Revelle, W., Humphreys, M. S., Simon, L., and Gilliland, K., The interactive effects of personality, time of day, and caffeine; a test of the arousal model, *J. Exp. Psychol. Gen.,* 109, 1, 1980.

73. Leathwood, P. D. and Pollet, P., Diet-induced mood changes in normal populations, *J. Psychiat. Res.,* 17, 147, 1983.

74. Bruce, M., Scott, N., Lader, M., and Marks, V., The psychopharmacological and electrophysiological effects of single doses of caffeine in healthy human subjects, *Br. J. Clin. Pharmacol.,* 22, 81, 1986.

75. Goldstein, I. B., Shapiro, D., Hui, K. K., and Yu, J. L., Blood pressure response to the "second cup of coffee," *Psychosom. Med.,* 52, 337, 1990.

76. Zwyghuizen-Doorenbos, A., Roehrs, T. A., Lipschutz, L., Timms, V., and Roth, T., Effects of caffeine on alertness, *Psychopharmacology,* 100, 36, 1990.

77. James, J. E., The influence of user status and anxious disposition on the hypertensive effects of caffeine, *Intern. J. Psychophysiol.*, 10, 171, 1990.

78. Henry, J. P. and Stephens, P. M., Caffeine as an intensifier of stress-induced hormonal and pathophysiological changes in mice, *Pharmacol. Biochem. Behav.*, 13, 719, 1980.

79. Boulenger, J. P., Uhde, T. W., Wolff, E. A., and Post, R. M., Increased sensitivity to caffeine in patients with panic disorders, *Arch. Gen. Psychiat.*, 41, 1067, 1984.

80. Holtzman, S. G., Caffeine as a model drug of abuse, *Trends Pharmacol. Sci.*, 11, 355, 1990.

81. Edelstein, E. L., A case of water dependence, *Br. J. Addict.*, 68, 367, 1973.

82. Mello, N. K., Stimulus self-administration: some implications for the prediction of drug abuse liability, in *Predicting Dependence Liability of Stimulant and Depressant Drugs*, Thompson, T. and Unna, K., Eds., University Press, Baltimore, 1977, 243.

83. Kaul, B. and Davidow, B., Drug abuse patterns of patients on methadone treatment in New York City, *Am. J. Drug Alcoh. Abuse*, 8, 17, 1981.

chapter seventeen

Overview

B. S. Gupta

Coffee is one of the most widely used beverages throughout the world. Reasons for its widespread use might be its aromatic qualities as well as its assumed stimulating effects. It is not easy to estimate the caffeine intake from the actual intake of coffee because of the wide range of caffeine content generally found in a cup of coffee. Different amounts of ground coffee (Robusta beans contain double the amount of caffeine than the Arabica beans), the use of coffee mixtures with surrogates, and brewing methods (caffeine is lost with heating, hence the duration of contact with hot water is critical) all influence the caffeine content. In a standard American cup of coffee (150 ml) generally the mean caffeine content is 115 mg for drip coffee, 80 mg for percolated coffee, and 65 mg for instant coffee. However, in a standard Canadian cup of coffee (225 ml) the mean caffeine content has been reported as 84 mg for drip, 82 mg for percolated, and 71 mg for instant.

The widespread use of caffeine-containing beverages has led to a vast body of research on the mechanisms underlying the central action of caffeine. Although the effects of caffeine on behavior are complex, it appears that blockade at A_1 through and A_{2A} adenosine receptors are the primary molecular sites of action for caffeine. Other biochemical mechanisms of action of caffeine are: release of intracellular calcium, inhibition of phosphodiesterases, and blockade of regulatory sites of $GABA_A$-receptors. The behavioral stimulation by xanthines perhaps requires blockade of both A_1- and A_{2A}-receptors. It appears that chronic blockade of adenosine receptors by caffeine effects alterations in the central receptors and pathways that are regulated by adenosine through A_1- and A_{2A}-receptors. The currently available evidence does indicate that chronic caffeine ingestion by mice increases the density not only of A_1-adenosine receptors but also of adrenergic, cholinergic, GABAergic and serotoninergic receptors. More research is needed concerning the sites and mechanisms involved in behavioral effects of acute and chronic caffeine.

0-8493-1166-7/99/$0.00+$.50
© 1999 by CRC Press LLC

The role of caffeine as a modulator of neurotransmission also needs to be examined thoroughly on presynaptic adenosine receptors, release and turnover of neurotransmitters, and postsynaptic receptors. This is necessary for understanding more precisely the mechanisms in the brain by which caffeine produces its effect. The effects of caffeine on neurotransmission in brain are quite complex, because caffeine affects a wide range of neurotransmitters, including catecholamines, acetylcholine, serotonin, and amino acids. Moreover, the exact relationship between the neuropharmacological effects of caffeine and its effects on behavior still needs to be ascertained precisely.

The effects of caffeine on cerebral energy metabolism and blood flow, which in fact are closely interrelated, have also been used in experimental research for understanding the behavioral effects of methylxanthines. The methylxanthines, which in general increase cerebral energy metabolism and decrease cerebral blood flow, have been reported to change the regulatory mechanism that couples cerebral blood flow and metabolism levels. This dynamic coupling between cerebral blood flow and energy metabolism, i.e., between delivery and use of oxygen and metabolic substrates, has been reported to be mediated by adenosine. The available evidence also suggests that the adenosine receptor antagonism, which is initiated by caffeine, results in direct changes in dopamine neurotransmission. However, the more recent research shows that serotoninergic involvement in terms of increased activation in the limbic system as well as in the medial prefrontal cortex may also be critically important for the development and expression of conditioned locomotor behavior. The future studies may therefore concentrate on the role of serotoninergic and medial prefrontal cortex neurotransmission in the development of context-specific conditioning of locomotor stimulant responses.

The study of interactions between caffeine and other widely used drugs such as nicotine, benzodiazepines, and alcohol is another fertile area for research. The little information that is currently available does not provide unequivocal and consistent evidence concerning interactional effects of various combinations of these drugs on behavior. Moreover, sometimes, especially when the effects of a substance are biphasic, as has been frequently observed in animal studies with alcohol, it becomes difficult to precisely interpret interactions among substances. Systematic study of these interactions is all the more essential, because caffeine, a most widely used substance, is likely to interact with multiple drugs used by individuals.

Caffeine is a known powerful and consistent arousal agent having physiological, behavioral and emotional consequences. A biobehavioral theory of arousal within the broader context of a dual-interaction theory of physiology and behavior has been proposed. The theory is said to postulate an inverted U-shaped relationship between arousal input and arousal output, additive effects of caffeine and other simultaneous arousal agents, an interaction between habitual use and acute exposure, and individual difference dimensions underlying arousal continua. A substantial body of literature has rendered considerable support for the theory. However, additional research is needed to test the deductions and ramifications underlying this theory.

An important issue that has emerged in recent times is whether caffeine could be considered a drug of abuse or dependence. Caffeine has been proposed to be a model drug of abuse mainly because it possesses some reinforcing and discriminative properties, produces a mild state of physical dependence and leads to the development of tolerance to various sympathomimetic effects, such as increases in heart rate and systolic blood pressure. It has also been proposed that caffeine dependence or withdrawal be included as a diagnosis in DSM-IV and ICD 10. The major criteria, used traditionally, for determining the dependence potential of a drug are: (1) psychoactive effects, (2) reinforcing effects, and (3) withdrawal syndrome and compulsive use.

Studies aimed at examining caffeine's psychoactivity, determined mainly by establishing discriminative stimulus effects employing both two-choice and three-choice discrimination procedures as well as subjective effects, provide only modest evidence of the psychoactive potential of caffeine. Moreover, there is no indication in the literature that pathological psychoactivation takes place with caffeine. Studies done with nonhuman primates and rats do not demonstrate that caffeine functions as a reliable reinforcer. Caffeine's reinforcing effects in humans are also limited to certain conditions. The most frequently reported withdrawal symptoms with caffeine are headache, drowsiness, fatigue, and lethargy. All these symptoms have been found to be of a transient nature. Abstinence leads to neither craving nor an increase in coffee consumption. Moreover, only a minority of regular users have significant problems with caffeine and are unable to abstain. The effects of acute tolerance of caffeine on tasks related to various behavioral domains still need to be explored systematically and precisely. Laboratory studies done with volunteer subjects after coffee abstinence provide little information concerning behavioral sensitivity to chronic or acute tolerance. Data are also needed even for the titration behavior, i.e., increase in the frequency and/or quantity of caffeine intake when the habitual dose is reduced and vice versa. The phenomenon of self-administration, an important aspect of drug tolerance, has also not yet been established for caffeine.

Despite the large number of published articles and research studies, the evidence in regard to caffeine's abuse or dependence potential is not absolutely compelling. The results in general are inconsistent and ambiguous, primarily because changes in behavior are very subtle, transient, and fragmentary. Moreover, these effects are influenced by a complex interplay of factors, both intrinsic and extrinsic to the individual. Thus, the currently available evidence does not indicate that coffee is drunk because it contains caffeine; aroma and taste qualities as well as other constituents might be important elements in the popularity of coffee.

Attempts have also been made to test and verify the deductions obtained from the laboratory studies in the general population by conducting field studies based on various types of surveys. The occurrence of the most frequently reported caffeine abstinence symptoms, i.e., anxiety, depression, and headache, was tested in a 9003-person survey of the general population of Great Britain. The findings provided no evidence of any association between caffeine consumption and depression, caffeine consumption and anxiety, and caffeine avoidance and anxiety in the general population. With respect to headache, drinking 3 to 4 cups of coffee reduced headache, indicating that caffeine in moderate doses may be used as a form of self-medication for headache. The field studies also confirm caffeine's low-dependence potential.

The behavioral effects of caffeine in humans have been examined on a diversity of tasks related primarily to cognitive processing and psychomotor functioning. The tasks used by researchers to investigate the former are the tasks of cognitive ability, information processing, memory, perceptual restructuring, perceptual judgment, vigilance, and divided and sustained attention, while the tasks used for determining psychomotor performance include simple and choice reaction time, simulated automobile driving, tapping, body sway, pursuit rotor, figural after effects, figure tracking, and hand steadiness. Caffeine's effects on affect and mood have also been examined, indicating that caffeine may lead to changes in certain personality characteristics, assumed to be consistent over time, i.e., state variables affecting the trait variables. This stance may be of considerable interest to a person doing research in personality. Also, the drug-effect on behavior may also be influenced by the personality characteristics of the subject. Hence the interaction between caffeine and personality is of interest in two ways: (1) how caffeine changes behavior along the axes defining personality, and (2) how personality determines the way caffeine affects behavior. Another important subjective variable that has recently received attention concerns the expectations or beliefs that individuals have about the effect of caffeine on their behavior. The relevance of this drug-related expectancy in behavioral responses to caffeine has been demonstrated for psychomotor performance. More data are needed for establishing the relevance of this variable, using diversified tasks related to other behavioral domains. Some other factors which affect behavioral responsivity to caffeine are: habituation to caffeine, initial response level, withdrawal symptoms, dosage, diurnal arousal variations, and sex. The effects of caffeine depend upon a complex interplay of these factors. To ignore these factors in research would imply presenting a distorted and equivocal view of the effects of caffeine on human behavior.

Caffeine is believed to enhance alertness and mental efficiency, but the beneficial effects of caffeine may not be uniform for both selective-attention and sustained-attention tasks. The currently available information about caffeine's effects on dual-task performance as well as on tasks employed to investigate sustained attention is limited. More data are needed before any generalized statement concerning caffeine's effects on attention is feasible. Similarly, fatigue is a construct which is not measured directly but is inferred from changes in physiological measures, performance, and mood. In general, the available information points out that caffeine in moderate doses has the potential to alleviate fatigue, although its precise role in producing energizing effects needs to be examined systematically.

Generally, an overnight deprivation procedure is used in laboratory studies to ensure that the behavior measured at the time of testing in the laboratory is not influenced by differences in prior exposure to caffeine. It has been pointed out that the beneficial effects of caffeine on performance might reflect a reversal of the deleterious effects of deprivation rather than a net benefit due to caffeine use. Data on psychomotor performance provide no support for this view; caffeine consumption leads to beneficial effects on performance, whereas caffeine withdrawal produces no impairments. However, more data employing tasks related to other behavioral domains are needed for making conclusive generalizations. Another question that has been raised is that caffeine-effected changes in mood state might induce changes in capacity and quality of information processing. Hence, any beneficial effect of caffeine on cognitive processing might be through induced changes in mood state rather than a direct action on cognitive processing systems. The evidence, though limited, indicates that caffeine's beneficial effects on cognitive processing are independent of changes in mood states. This evidence also needs to be tested employing other diversified behavioral measures.

Although there are basic differences in the behavioral functions tested in animal and human experimentation with caffeine, the collective evidence demonstrates that caffeine acts like a stimulant and its effects are modestly dose dependent. Attempts have been made to find analogies between caffeine-induced state changes detected using discrimination procedures and the perceptual changes by more common exteroceptive stimuli. Of relatively recent origin are the models that relate synthesized experimental findings in animals to the state changes operating in human caffeine consumers. More data from such endeavors are certainly needed so that the gaps between animal and human research are narrowed down even further.

It appears that people generally remain within limits of their individual tolerance for coffee intake. They find absolutely no difficulty in adjusting their coffee intake to their individual tolerance limits. Moreover, the currently available evidence, gathered through rigorous scientific methodology, does not reliably corroborate the accusations leveled against coffee. The positive effects of caffeine as mentioned in the literature, although obtained mostly from laboratory studies where doses in excess of daily-life dosages are generally used, cannot be dismissed outright. More systematic research using doses close to daily-life coffee consumption needs to be carried out. Hence, industry support for the ongoing research efforts as well as the programs being sponsored by various coffee research organizations need not only to be continued but intensified to gain better insight and knowledge of the issues raised in this volume in regard to the consumption, and biological, behavioral, and affective effects of this psychoactive substance consumed so extensively throughout the world.

Index

Page numbers followed by "t" indicate tables; page numbers in italics indicate figures.

A

Acetylcholine, brain, interaction with, 21, 22-25
Addiction, to caffeine, *see* Dependence potential
Adenosine agonists, blockade of caffeine action by, 267
Adenosine analogs, and xanthines, 9-11
Adenosine receptors, *see also* Neurotransmission
  analogs of, and xanthines, 9-11
  caffeine antagonism of, 19-22, 32-33, 93-94, *see also* Neurotransmission
  of central nervous system, 1-12
  chronic caffeine exposure and effect on, 4-9
  in dopaminergic neurotransmission, 49-50, 71
  hippocampal and striatal, 49-50, 71
  methylxanthine action on, 17-25, 265-266
  neurophysiology of, 93-94
Aggression, effect on, 107-108
Alcohol, interaction with, 81-83
  chronic exposure and, 9
  human studies of, 82
  studies of, in humans, 82
Alpha and theta power in EEG, caffeine withdrawal and, 156
Amino acids, brain, interaction with, 21-22
Analogs, adenosine, effects of, 9-11
Anger, effect on, 107-108
Antagonistic activity, of caffeine, *see under* Adenosine receptors
Anxiety, association with, 112-113, 179-180, 184-185, 188, *see also* Mood states
  survey participants/methods/results, 180-182
Anxiety disorders, effect on, 113

Anxiolytic drugs, and caffeine training cue, 267
Anxiolytic effects, of caffeine-benzodiazepine combinations, 78-79
APEC analog, behavioral effect of, 7-8, 10-11
Arousal, caffeine, 88-117
  behavioral effect of, 98
    cognitive, 98-105
    emotional, 105-108
    habitual caffeine ingestion and, 108-110, *see also* Chronic caffeine exposure
  impulsivity trait and, 191-202
  individual differences in, 111-112
  models of, 89-93, *90, 91, 92*
  neurochemical mechanisms of, 93-94
  physiologic effects of, 94-96
  psychopathologic effects of, 112-116
Arousal agents, other, interaction with, 96-98
Ataxia, induced by caffeine and alcohol, 81
Attention, 231-238, *see also* Cognitive performance; Vigilance
  divided, effect on, 234-235
  focused, effect on, 232-234
  selective, effect on, 103, 231-232
  sustained, effect on, 235-237
Avoidance performance, alcohol-caffeine interaction and effect on, 82

B

Behavioral effect(s), of caffeine, 98
  arousal and, 87-116, *see also* Arousal
  dependence and, 137-145, *see also* Dependence potential
  drug interaction and, 75-83, *see also* Alcohol; Benzodiazepines; Nicotine
  drug-related expectancies and, 207-217, *see also* Subjective effects
  emotional, 105-108, *see also* Emotional state
  impulsivity and, 191-202

289

locomotor, 56-71, *see also* Locomotor
    stimulation; Performance efficiency
mood states and, 179-188, *see also*
    Anxiety; Depression; Headache;
    Mood states
neurochemistry of, 17-25, *see also*
    Neurotransmission
neurophysiology of, 17-25, *see also*
    Neurotransmission; Physiologic
    effects
performance and, *see* Cognitive
    performance; Performance
    efficiency
personality traits and, *see* Impulsivity trait
psychopathologic, 87-116, *see also*
    Psychopathologic effects
reward stimulation as, 56-71, *see also*
    Reward behavior
withdrawal and, 151-156, *see also*
    Performance efficiency; Withdrawal
Benzodiazepine receptors
    antagonism of, 19
    interaction with, 22, 80-81
Benzodiazepines
    inhibition of binding of, 3
    interaction with, 78-81, 82-83
        possible explanations of, 80-81
        studies in humans of, 79-80
Biphasic dose-response curve, of caffeine, 4,
    7
Blood pressure, effect on, 95
Brain function, modulation of, 17-25, *see*
    also Central nervous system;
    Cerebral energy metabolism
    neurochemical, 18-22, 93-94
        behavior and, 22-25
    neurotransmission and, 19-22
Bupropion similarity, to effect of caffeine,
    266
Buspirone, relationship between caffeine
    and, 275

C

Caffeinism, 113-115
Calcium mobilization, by caffeine, 2-3, 18,
    94
Catabolism, of caffeine, 94
Catecholamines, brain, interaction with, 20-
    21

Central nervous system, effect on, 266-267
    behavior and, 22-25, *see also* Behavioral
        effect(s); Chronic caffeine exposure
    caffeine arousal and, 93-94, *see also*
        Arousal
    global, 267, 271-273
    neurochemical, 1-12, 19-22, 265-266
    physiologic, 31-43, 94-96
        chronic caffeine exposure and, 108-
            109, *see also* Caffeinism; Chronic
            caffeine exposure
        neurochemical mechanisms of, 18-19,
            49-50, 93-94
Cerebral blood flow, effect on, 38-43
    and cerebral metabolism, 41-42
    general, 38-40
    pathologic conditions of, 42
    vasoconstrictive, 41
    withdrawal, 156
Cerebral energy metabolism, effect on, 33-
    38
    and cerebral blood flow, 41-42
    chronic caffeine exposure and, 37-38
CHA (N6-cyclohexyladenosine) analog,
    behavioral effect of, 7-8, 10-11
Chlordiazepoxide, and caffeine,
    discriminative effects of, 79
Cholinergic systems, chronic effects of
    caffeine on, 8-9, *see also*
    Neurotransmission
Chronic caffeine exposure, effects of, 108-
    110
    acute caffeine exposure added to, 111
    adenosine receptors in, 4-9
    behavioral, 109-110
    cardiovascular, 108-109
    cerebral energy metabolism and, 37-38
    ethanol and, 9
    in mice, 4-9
    nicotine and, 76-77, 114
    in NIH Swiss strain mice, 5-7, 6t
    other studies needed on, 9
    performance efficiency and, 109-110
    physiologic, 108-109
    tolerance and, 70-71
Cocaine
    comparison of locomotor stimulation by
        caffeine to that of, 50-71
    similarity of effect of caffeine to that of,
        267, 271-273

Cognitive performance, effects on, 22-25, 221-224, *see also* Attention; Memory
  arousal and, 98-105, *see also* Arousal
  differentiating effects on mood state and, 224-226
  impulsivity trait and, 191-194, 202
    study methods/results in, 196-202
  reviews of past studies of, 226-227
Complex cognition, effect on, 100-105
Conditioned behavior, effect on, 49-50, 56-71
Conditioned place preference (CPP), 50
Consumption, caffeine, *see* Ingestion, of caffeine
Controlled/compulsive use, of caffeine, 141
Corticosterone, serum, biochemical assay for, 59
CPP/CPA paradigm, 50-52
Critical flicker fusion (CFF) threshold, 235
Cross-generalization profile of caffeine, to other training drugs, 268t-269t
Cueing, caffeine, 265-266

D

Dependence potential, of caffeine, 137-145
  criteria of
    controlled/compulsive use, 141
    drug-reinforced behavior, 141-142
    pleasant/euphoric effects, 142
    psychoactivity, 139-140
    tolerance, 142
  diagnostic classification of, 143-144
  oral vs. direct intake and, 144
  physical dependence and, 143
  primary criteria for, 139-142
  secondary criteria for, 142-143
  subjective variables and, 216
Depressant effect, of adenosine analogs, 9-11
Depression, *see also* Mood states
  association with, 179-180, 182-184, 187-188
    survey participants/methods/results, 180-182
  as effect of caffeine, 152-155
  effect on, 112-113, 115
    benzodiazepine-caffeine interaction and, 78-79

Diagnostic classification, of caffeine dependence, 143-144
Differences, individual, in caffeine arousal, 111-112, 214-216
Differential reinforcement of low-rate (DRL) behavior, *see* DRL performance
Discriminative stimulus effects, of caffeine
  alcohol interaction with, 82
  blockade of, 267, 270t
  chlordiazepoxide interaction with, 79
  dependence potential and, 139-140
DOPAC/DA ratio, limbic, assay for, 58-59
Dopaminergic psychostimulants, differences between caffeine and, 50, 69-71
Dopaminergic system(s), effect on, 71, *see also* Neurotransmission
  adenosine receptor antagonism and, 49-50, 71
  chronic exposure and, 7-8
  mesolimbic, 35-37
  nigrostriatal, 33-35
Dose control, in caffeine studies, 100
Dose-response curve, biphasic, of caffeine, *4, 7*
DRL performance, effect on
  caffeine-alprazolam interaction and, 78
  and fine motor control, 81
Drowsiness, caffeine withdrawal and, 152-155, *see also* Fatigue
Drug abuse, caffeine and study of mechanisms of, 276
Drug cross-generalization profiles, 265, *see also* Training drugs
  to caffeine training stimuli, 260t-264t
Drug dependence, *see also* Dependence potential
  criteria for, 138-139
  definition of, 138
Drug discrimination task, 257
Drug interaction, 75-76
  with caffeine, 75-83, *see also* Alcohol; Benzodiazepines; Nicotine
Drug-reinforced behavior, and caffeine use, 141-142
Drug-related expectancies, 208-209
  of caffeine, 207-217
  role of, *see* Subjective effects
Dual-interaction model, of arousal, 89, *90*
Dual task paradigm, 234-235

Dysthymia, effect on, 115

E

Electrodermal activity (EDA), 95-96
Embedded Figures Task (EFT), 233
Emotional state, effect on, 105-108
Ethanol, *see* Alcohol
Event-related potential (ERP), 243, 216, 243
Expectancies, about drug effects, *see* Drug-
    related expectancies
Exteroceptive stimuli, 257-259
    processing of, 266-267

F

Fatigue
    absence of definition of, 251
    caffeine withdrawal and, 152-155
    effect on, 241-242, 251-252
    experimentally induced, effect on, 244-
        251
    natural, effect on, 242-244
    types of, 242-251
Fine motor control
    benzodiazepine-caffeine interaction and
        effect on, 78
    DRL performance and, effect on, 81
5-HIAA/5-HT ratio, limbic, assay for, 59

G

GABA$_A$-receptors, effect on, 1-9, 94
Gender difference
    in caffeine studies, 101
    in caffeinism, 114

H

Habitual caffeine exposure, *see* Chronic
    caffeine exposure
Headache, *see also* Mood states
    association with, 179-180, 185-187, 188
        survey participants/methods/results,
        180-182
    caffeine withdrawal, 41, 104, 152-153
        postoperative headache and, 155
Health and Lifestyle Survey, 180

*Health Consequences of Smoking: Nicotine
    Addiction* (1988), 138
Heart, effect on, 95
Hippocampal system, effect on, 49-50, 71

I

Impulsivity trait, and behavioral effects of
    caffeine, 191-202
Individual differences, in caffeine arousal,
    111-112, 214-216
Information processing, effect on, 100-101
Ingestion, of caffeine
    effects of acute, 94-108, 111, 223-224, *see
        also* Arousal; Behavioral effect(s)
    physiologic, 31-43, 94-96
    habitual, 108-110, 111, *see also* Chronic
        caffeine exposure
    subjective variables and, 207-208, 216-
        217, 257-277, *see also* Subjective
        effects
Intelligence test scores, effect on, 192-193

L

Liking scales, 140
Limbic system, effect on
    assay for DOPAC/DA ratio in, 58-59
    assay for 5-HIAA/5-HT ratio in, 59
    and dopaminergic neurotransmission, 35-
        37
Locomotor stimulation, effect on, 49-52, 56-
    71, *see also* Performance efficiency
    alcohol-caffeine interaction and, 81
    biochemical assays of, 58-59
    cerebral neurotransmission and, 33-35
    compared to effect of cocaine, 50-71
    conditioning in, 49-50, 56, 58
    place/preference in, 59, 65-69
    reward/aversion in, 49-50, 59-69
    sensitization/tolerance in, 49-50, 56, 59-
        65
    study design/methods in, 52-56

M

Mechanism of action, of caffeine, 18-19, *see
    also* Neurotransmission; Physiologic
    effects

Medial reticular formation (mRF), effect on, 242

Memory, effect on, 100-103, 193-194

Mental fatigue, effect on, 247-249, *see also* Cognitive performance

Mental processing, *see* Cognitive performance

Mesolimbic dopaminergic system, effect on, 35-37

Metabolites, of caffeine, 94, *see also* Cerebral energy metabolism; Neurotransmission

Metathetic continuum, 266

Methylxanthines
  and cerebral metabolism, 31-43, *see also* Cerebral blood flow; Cerebral energy metabolism
  effect of, on neurotransmission, 17-25, 265-266

MHPG, noradrenaline metabolite, 155-156

Micro-sleeps, 247

Modeling
  of caffeine arousal, 89-93, *90, 91, 92*
  of caffeine arousal in humans, 273-276, *275*

Mood states, *see also* Anxiety; Depression; Headache
  association with, 179-188
    study of, 277
  effects on, 105-106, 223
    differentiating effects on cognitive performance and, 224-226
    reviews of past studies of, 226-227
    in withdrawal, 152-155, 176-177

Morphine-Benzedrine Group (MBG) scale, 140

Multiple Sleep Latency Test (MSLT), 235, 243

N

NECA analog, behavioral effect of, 7-8, 9, 11

Neonates, caffeine withdrawal in, 155

Neurotransmission, in CNS, effect on, 1-12, 19-22, 265-266
  behavior and, 22-25
  caffeine arousal and, 49-50, 93-94

Nicotine
  caffeine interaction with, 76-77, 82-83

subjective effects of, 216
and caffeinism, 114

Nigrostriatal dopaminergic system, effect on, 33-35

Noradrenergic cerebral cells, effect on, 35

O

Oddball paradigm, 103

Oral vs. direct intake, of caffeine, 144

P

Paired vs. unpaired, caffeine treatment protocol, 51-52

Performance efficiency, effect on, 98, 161-178, *see also* Cognitive performance; Locomotor stimulation
  chronic caffeine exposure and, 109-110
  personality traits and, 191-192, 194-195, 202
    study methods/results in, 196-202
  study design in, 162-163
  study methods/results in, 163-166
  withdrawal and, 171-174
    acute ingestion after, 166-171, 174-175

Personality traits, and behavioral effects of caffeine, 191-202

Phosphodiesterase inhibition, by caffeine, 3, 18-19

Physical dependence, on caffeine, 143

Physical fatigue, effect on, 249-251

Physiologic effects
  of caffeine, 31-43, 94-96
  of chronic caffeine exposure, 108-109, *see also* Caffeinism
  neurochemical mechanisms of, 18-19, 49-50, 93-94

Physiologic sleepiness, 243

Place/preference conditioning, effect on, 59, 65-69

Pleasant/euphoric effects, of caffeine, 142

Post-traumatic stress disorder (PTSD), effect on, 106, 113, 114, *see also* Stress

Prothetic continuum, 259

Psychoactivity, of caffeine, 139-140, *see also* Arousal; Behavioral effect(s); Reward behavior

Psychomotor performance, *see* Performance

efficiency
Psychomotor task improvement, 175-178,
    see also Performance efficiency
Psychopathologic effects, of caffeine
    arousal, 112-116, *see also* Anxiety;
    Chronic caffeine exposure;
    Depression; Stress
Psychostimulant effects, of caffeine, 35-37,
    see also Arousal; Behavioral
    effict(s), Dependence potential

Q

Quantitative specificity, of stimulus, 259

R

Rapid Information Processing (RIP) task,
    107
Reaction time, and caffeine withdrawal, 156
Receptor sites
    adenosine, 1-12, *see also* Adenosine
    receptors
    benzodiazepine, 19, 22, 80-81
    other than adenosine, 2-3
Reward/aversion conditioning, effect on, 49-
    50, 59-71
Reward behavior, effect on, 49-52, 56-71
    alcohol-caffeine interaction and, 81
    biochemical assays of, 58-59
    cerebral neurotransmission and, 33-35
    compared to effect of cocaine, 50-71
    conditioning in, 49-50, 56, 58
    place/preference in, 59, 65-69
    reward/aversion in, 49-50, 59-69
    sensitization/tolerance in, 49-50, 56
    study design/methods in, 52-56
    study methods in, 52-56

S

Schizophrenia, effect on, 115-116
Seasonal affective disorder, effect on, 115
Selective attention, effect on, 103
Sensitization/tolerance behavior, effect on,
    49-50, 56
Serotonin, limbic, biochemical assay of, 59
Serotoninergic cerebral cells, effect on, 21,
    35

Serum corticosterone, assay for, 59
Skin conductance level (SCL), 95-96
Sleep deprivation, effect on, 244-247
Sleepiness, physiological, 243
Sleep-wake cycle, effect on, 35
Stimulus control, caffeine, in animal studies,
    259-273
Stimulus sensitivity, 257, 259
Stress, effect on, 96-98, 106-107
Striatal system
    adenosine receptors of, effect on, 1-12,
    49-50, 71
    dopaminergic neurotransmission in, effect
    on, 33-35
Stroke, and cerebral effect of caffeine, 42
Subjective effects, of caffeine
    animal and human research on, 257-277
    dependence potential and, 140
    drug-related expectancy and, 207-208,
    216-217
    implications of, 214-217
    placebo control groups and, 216
    research findings on, 210-214
    multiple neuronal systems and, 266
Surgeon General's report (1988), 138
Symptoms, of caffeine withdrawal, 152-155

T

Task performance, *see* Performance
    efficiency
Tolerance, to caffeine, 142
Training drugs, 266
    caffeine cross-generalization profile to
    other, 268t-269t
Training stimuli, caffeine, drug cross-
    generalization profiles to, 260t-264t,
    265

V

Vasoconstrictive effects, of caffeine, 41
Vigilance, effect on, 98-100
Vigilance tasks, 235-237

W

Withdrawal, caffeine
    effect(s) of, 151-156

alpha and theta power in EEG, 156
cerebral blood flow, 156
early studies of, 152
headache as, 41, 104, 152-155
postoperative, 155
in neonates, 155
objective testing of, 155-156
in psychiatric patients, 155
reaction time and, 156
recent studies of, 152-155
performance efficiency in, 161-163, 171-174
effect of acute consumption on, 166-175
study design/methods in, 163-166
symptoms of, 152-155
Work performance, *see* Performance efficiency

X

Xanthines, and adenosine analogs, 9-11

Y

Yerkes-Dodson law, 249